Computational Intelligence Processing in Medical Diagnosis

Studies in Fuzziness and Soft Computing

Editor-in-chief
Prof. Janusz Kacprzyk
Systems Research Institute
Polish Academy of Sciences
ul. Newelska 6
01-447 Warsaw, Poland
E-mail: kacprzyk@ibspan.waw.pl
http://www.springer.de/cgi-bin/search_book.pl?series=2941

Further volumes of this series can be found at our homepage.

Vol. 78. U. Seiffert and L.C. Jain (Eds.)
Self-Organizing Neural Networks, 2002
ISBN 3-7908-1417-2

Vol. 79. A. Osyczka
Evolutionary Algorithms for Single and Multicriteria Design Optimization, 2002
ISBN 3-7908-1418-0

Vol. 80. P. Wong, F. Aminzadeh and M. Nikravesh (Eds.)
Soft Computing for Reservoir Characterization and Modeling, 2002
ISBN 3-7908-1421-0

Vol. 81. V. Dimitrov and V. Korotkich (Eds.)
Fuzzy Logic, 2002
ISBN 3-7908-1425-3

Vol. 82. Ch. Carlsson and R. Fullér
Fuzzy Reasoning in Decision Making and Optimization, 2002
ISBN 3-7908-1428-8

Vol. 83. S. Barro and R. Marín (Eds.)
Fuzzy Logic in Medicine, 2002
ISBN 3-7908-1429-6

Vol. 84. L.C. Jain and J. Kacprzyk (Eds.)
New Learning Paradigms in Soft Computing, 2002
ISBN 3-7908-1436-9

Vol. 85. D. Rutkowska
Neuro-Fuzzy Architectures and Hybrid Learning, 2002
ISBN 3-7908-1438-5

Vol. 86. M.B. Gorzałczany
Computational Intelligence Systems and Applications, 2002
ISBN 3-7908-1439-3

Vol. 87. C. Bertoluzza, M.Á. Gil and D.A. Ralescu (Eds.)
Statistical Modeling, Analysis and Management of Fuzzy Data, 2002
ISBN 3-7908-1440-7

Vol. 88. R.P. Srivastava and T.J. Mock (Eds.)
Belief Functions in Business Decisions, 2002
ISBN 3-7908-1451-2

Vol. 89. B. Bouchon-Meunier, J. Gutiérrez-Ríos, L. Magdalena and R.R. Yager (Eds.)
Technologies for Constructing Intelligent Systems 1, 2002
ISBN 3-7908-1454-7

Vol. 90. B. Bouchon-Meunier, J. Gutiérrez-Ríos, L. Magdalena and R.R. Yager (Eds.)
Technologies for Constructing Intelligent Systems 2, 2002
ISBN 3-7908-1455-5

Vol. 91. J.J. Buckley, E. Eslami and T. Feuring
Fuzzy Mathematics in Economics and Engineering, 2002
ISBN 3-7908-1456-3

Vol. 92. P.P. Angelov
Evolving Rule-Based Models, 2002
ISBN 3-7908-1457-1

Vol. 93. V.V. Cross and T.A. Sudkamp
Similarity and Compatibility in Fuzzy Set Theory, 2002
ISBN 3-7908-1458-X

Vol. 94. M. MacCrimmon and P. Tillers (Eds.)
The Dynamics of Judical Proof, 2002
ISBN 3-7908-1459-8

Vol. 95. T.Y. Lin, Y.Y. Yao and L.A. Zadeh (Eds)
Data Mining, Rough Sets and Granular Computing, 2002
ISBN 3-7908-1461-X

Manfred Schmitt
Horia-Nicolai Teodorescu
Ashlesha Jain · Ajita Jain
Sandyha Jain · Lakhmi C. Jain
Editors

Computational Intelligence Processing in Medical Diagnosis

With 103 Figures
and 49 Tables

Springer-Verlag Berlin Heidelberg GmbH

Professor Dr. Manfred Schmitt
Technical University of Munich
Ismaninger Straße 22
81675 München
Germany
manfred.schmitt@lrz.tum.de

Professor Horia-Nicolai Teodorescu
Romanian Academy
Calea Victoriei 125
Bucharest
Romania
hteodor@etc.tuiasi.ro

Dr. Ashlesha Jain
The Queens Elizabeth Hospital
Woodville Road
Woodville, Adelaide
South Australia 5011

Ajita Jain
Bellevue Residential Care Centre
Bellevue Heights, Adelaide
South Australia 5050

Dr. Sandhya Jain
Julia Farr Services
Fullarton Road, Adelaide
South Australia 5063

Professor Lakhmi C. Jain
University of South Australia
Knowledge-Based Intelligent
Engineering Systems Centre
Mawson Lakes, Adelaide
South Australia 5095
Lakhmi.Jain@unisa.edu.au

ISSN 1434-9922

DOI 10.1007/978-3-7908-1788-1

Cataloging-in-Publication Data applied for
Die Deutsche Bibliothek – CIP-Einheitsaufnahme
Computational intelligence processing in medical diagnosis: with 49 tables / Manfred Schmitt ... (eds.). – Heidelberg; New York: Physica-Verl., 2002
(Studies in fuzziness and soft computing; Vol. 96)

Originally published by Physica-Verlag Heidelberg in 2002
MyCopy version of the original edition 2002

Hardcover Design: Erich Kirchner, Heidelberg
www.springer.com/mycopy

DEDICATION

This book is dedicated to our students.

The proceeds of royalty will be donated to a Charity.

Foreword

Health systems today suffer from runaway costs, inconsistent practice at different sites, lengthy service delays, medical errors, failure to serve remote regions, and the need to use generalists as gatekeepers even though this leads to large diagnostic quality disparities when compared to specialists' decisions. Computerized assistance for medical diagnosis and treatment is a key to solving each of these dilemmas. Medical doctors often indicate that diagnosing disease is the fun part of their job and they aren't interested in having it automated. Like so many other fields, however, the goal is not to eliminate the professional but to extend their reach, and improve overall system performance. Ideally, an array of diagnostic and treatment software programs could act like a force multiplier and help a health system to develop and prosper. Consumers would benefit as well.

Seems like a "no-brainer" as my students would say. After all, automatic diagnostic software is embedded in everyday kitchen appliances, office equipment, and industrial machinery. Even car mechanics use a diagnostic machine to figure out what's wrong with your car. And, airplanes these days come with automatic takeoff and landing software as well as auto-piloting enroute, yet pilots have gotten over the issue of automation helping with the 'fun part' of the job. So why don't doctors already have and use an array of diagnostic programs?

Everyone knows that mechanical systems are far simpler than animate or biological ones to diagnose when trouble arises. The former tend to be deterministic or, at worst, have known probability distributions. The latter are indeterminate and often complex. The complexity arises since there are potentially so many interacting subsystems, co-occurring diseases, and scientific unknowns. Many conditions are not well understood, and for those that appear to be understood, the half-life of knowledge is relatively short. New evidence comes out all the time that replaces the old rules, a fact that makes it equally difficult for either software or human diagnosers to keep up-to-date. In sum, these combined set of issues have lead many people to realize that viable diagnostic and treatment software for the medical field is a "grand challenge" problem.

Given this backdrop, one can't help but be impressed with the extent of intellectual effort by the authors in the current volume. When tackling stubborn problems, one of the primary tactics of systems thinkers is to look beyond your own narrow discipline, to think holistically, and to integrate a larger synthesis. By combining many alternative views of the same problem space – neural nets, fuzzy sets, evolutionary techniques, knowledge based approaches, nonlinear dynamics, and so on – Professor Jain has striven for such a synthesis. He calls this synthesis "computational intelligence," but those of us in the systems field view it as a homerun, a ball knocked into the systems thinking bleachers. Readers should enjoy this collection on that level as well as for the details in each of the many excellent chapters. This book documents an exciting curve in the ongoing stream of efforts to improve health care systems. It should help readers to see larger vistas and new directions worth pursuing.

Barry G. Silverman, PhD
Professor and Director
Ackoff Center for Advancement of Systems Approaches (ACASA)
University of Pennsylvania, Philadelphia, PA 19104-6315
U.S.A.

Preface

Computational intelligence techniques are gaining momentum in medical prognosis and diagnosis. Computational intelligence paradigms include expert systems, artificial neural networks, fuzzy systems, evolutionary computing, data mining and knowledge discovery. There are many real world problems, including medical diagnosis problems, which do not provide the needed information, or the systems under consideration is not well defined. These problems are not easy to solve using conventional computing approaches but computational intelligence may play a major role in these areas.

Increasing numbers of physicians rely on tools and methods that are "computational-intensive," and moreover "knowledge processing-intensive." To cope with the demand of over increasing knowledge, physicians and bio-medical engineers have reached a new realm where computers are not tools, but partners in the medical act. New computational intelligence paradigms are always emerging in medicine. The successful manipulation of these paradigms relies on computational intelligence tools and on the understanding of their basics by the human partner.

In a domain like medicine where knowledge is generated at an exponentially growing pace, the only choice left to practitioners is to correspondingly increase their use of computer-based facilities and to supplement their knowledge and skills by the support offered by computers and computational intelligence. Current high prices of the medical care and increasing complexity of the medicine could encourage a widespread switch to alternative, sometimes new and today strange-looking, possibilities for practicing medicine – from tele-diagnosis to the nursing robot to remote robotic surgery. While these are currently very expensive, the computational intelligence price tag is constantly decreasing.

The topics in this volume have been carefully selected to offer a global overview of the state of the art in the domain of computational intelligence in medicine. All chapters focus on the medical applications, while providing a comprehensive description of technical tools. Essentials on both medical and technical aspects are provided in each chapter to make them easy to read and consistent. Clear and concise explanations for learning paradigms are provided on all topics. Moreover, the contributors of the chapters explain what, when, how, and why information technology solutions are of value. The contributors consists of an international pool of recognized experts in the field. All in all, this is a highly practical book matching the needs of many categories of readers.

The volume is addressed to physicians who use advanced computer-based tools and intelligent devices in their daily practice and research to bio-medical engineers, to computer scientists applying artificial intelligence in medicine or specializing in this field. The volume will also appeal to students in the medical sciences, bio-medical engineering and computer science.

This volume provides medical practitioners with a crop of new tools for their work, whilst also providing the engineer with an array of applications for their knowledge.

We believe that this book will be of great value to researchers, practicing doctors and research students. Most of all, however, this book aims to provide the practicing doctor and scientist with case studies using the most recent approaches.

We are grateful to the authors and reviewers for their valuable contribution. We are indebted to Berend Jan van der Zwaag for his help in the preparation of the manuscript. We also thank the editorial staff of Springer-Verlag for their excellent editorial assistance.

Contents

Chapter 1.
An introduction to computational intelligence in medical diagnosis
H.-N. Teodorescu and L.C. Jain

Chapter 2.
Computational intelligence techniques in medical decision making: the data mining perspective
V. Maojo, J. Sanandres, H. Billhardt, and J. Crespo

Chapter 3.
Internet-based decision support for evidence-based medicine
J. Simpson, J.K.C. Kingston, and N. Molony

Chapter 5.
Case-based reasoning prognosis for temporal courses
R. Schmidt and L. Gierl

Chapter 6.
Pattern recognition in intensive care online monitoring
R. Fried, U. Gather, and M. Imhoff

Chapter 7.
Artificial neural network models for timely assessment of trauma complication risk
R.P. Marble and J.C. Healy

Chapter 8.
Artificial neural networks in medical diagnosis
Y. Fukuoka

Chapter 12.

Septic shock diagnosis by neural networks and rule based systems

R. Brause, F. Hamker, and J. Paetz

Chapter 13.

Monitoring depth of anesthesia

J.W. Huang, X.-S. Zhang, and R.J. Roy

Chapter 14.

Combining evolutionary and fuzzy techniques in medical diagnosis

C.A. Peña-Reyes and M. Sipper

Chapter 15.

Genetic algorithms for feature selection in computer-aided diagnosis

B. Sahiner, H.P. Chan, and N. Petrick

Chapter 1

An Introduction to Computational Intelligence in Medical Diagnosis

H.-N. Teodorescu and L.C. Jain

In this chapter, we advocate the use of Computational Intelligence (CI) in diagnosis, in the context of using artificial intelligence in medicine and in health management. The methodological advantages, economic benefits, main trends and perspectives of using CI in health care and management are discussed.

1 What is Computational Intelligence?

Computational intelligence (CI) is one of several terms coined to name a blurring and continuously evolving domain, essentially limited to artificial intelligence (AI). Because of the swift development of this field, scientists tried to make a difference between the "earlier" methods of AI and the emerging ones, and dubbed the "new AI" "computational intelligence". Currently, CI is used to name both the wider field of AI – with a flavor of "new methods" – and a specific group of AI techniques that recently emerged. Under the restricted interpretation, CI chiefly means the group of neural networks, fuzzy sets, and genetic algorithms. Both meanings remain unsatisfactorily defined, and scientists and users take a pragmatic approach by leaving the evolution of the domain to crystallize the concepts, while focusing on filling the concepts with useful means. In this chapter, we assume the restricted meaning with an enlargement to accommodate most recent trends, like data mining and internet-based developments.

CI sweeps over a large number of heterogeneous paradigms, methods, and techniques (see Figure 1), aimed to imbed into man-made systems, the capabilities of humans and to endow these systems with abilities that are human-like.

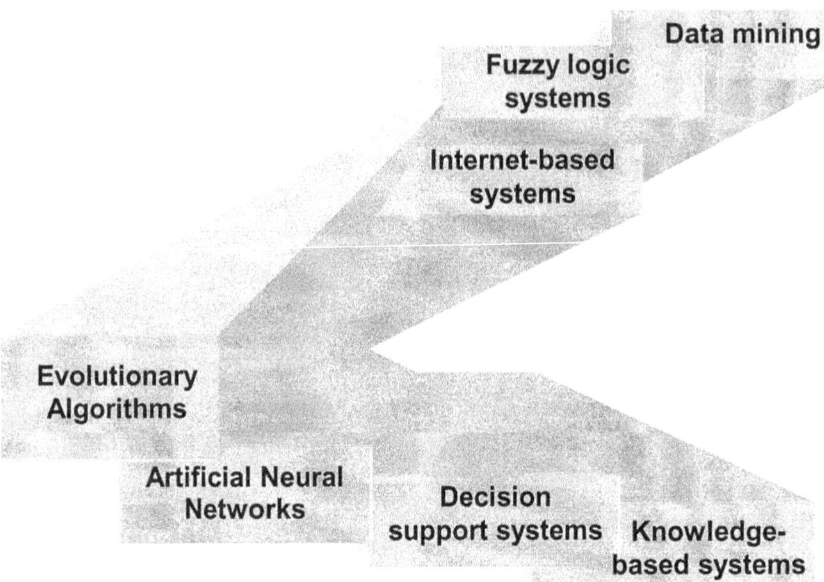

Figure 1. The heterogeneous structure of computational intelligence (only a few paradigms are illustrated).

These abilities include data processing and structuring, pattern recognition, knowledge representation, knowledge processing, learning, knowledge aggregation, knowledge discovery, reasoning, decision making, predictive actions, adaptation and evolution, behavior patterns generation, expert advising, planning, and proactive reasoning. The methods to bring to life these abilities include several competing and often overlapping paradigms, among others neural network-based systems, knowledge-based systems – including expert systems – fuzzy logic and its derivatives, nonlinear dynamics, and evolutionary algorithms. The borders of these domains are frequently "fractal" knowledge borders, blurred and they rather connect that distinguish the corresponding domains.

Sometimes, the methods in the CI are listed as "alternative approaches" to emphasize their "deviation" from "regular" AI methods. Under alternative approaches, most quoted are fuzzy systems, genetic algorithms, neural networks, probabilistic systems, and various combinations of them (hybrid systems). In some applications indeed, one methods can successfully replace another methods. For example, neural networks may replace traditional statistical tools in determining the most relevant features in data. Genetic algorithms can successfully replace and surpass statistical or deterministic (gradient-based) methods in systems adaptation, while fuzzy systems and neural networks are mathematically proved to be able to perform the same tasks in many circumstances.

For an extended history of the use of fuzzy logic and neuro-fuzzy systems in medicine, see [1]. For an excellent yet brief history of the Artificial Intelligence domain, see [2]. For the state of the art of fuzzy logic and neuro-fuzzy systems applications in medicine, see [1], [3]-[5]. A good presentation of early developments in neural networks, before 1985, is presented in [6]. Detailed presentation of the field of computational intelligence and of several applications are presented in [7]-[10].

2 Why CI in Medicine and Especially in Medical Diagnosis?

The need for CI in medicine in general and specifically in diagnosis is due to several factors:

- The increase in complexity of collected data
- The tremendous increase of domain-specific knowledge and, as a consequence, of the number of diseases that can be diagnosed
- The exponential increase in interdisciplinary knowledge
- The need to unceasingly increase the efficiency of the medical act to cope with the economic pressure, the increasing number of the human population, and the steadily increasing age of the population.
- The peculiarities of medical sciences that extensively use uncertain data and knowledge and reasoning methods that are fitted to such data.

The prevalence of knowledge on the data in medicine is put forward by the aims of medicine, which subsequently become aims of using CI in medical practice. Indeed, the aims of CI in medicine complement the objectives of the medical act:

- Increase age expectancy
- Improve life quality

- Improve efficiency of the health care
- Increase the economic and social capabilities of the individual and population as a whole

Finally, the objective of CI in medicine is health-equality (equal chances) in the society. Actually, applying CI in medicine represents a response, expressed by a set of computer-based approaches, to human health, economic and social needs.

Figure 2 shows the number of papers published between January 1995 and April 2001 in various fields of medicine, and related to various CI techniques, are shown. The search on this database of papers quoting the "computational intelligence" term provides only 166 items with the publication date from January 1990 to May 2001, almost all quotations during the last few years. This shows that the term is not yet widely used in the medical field.

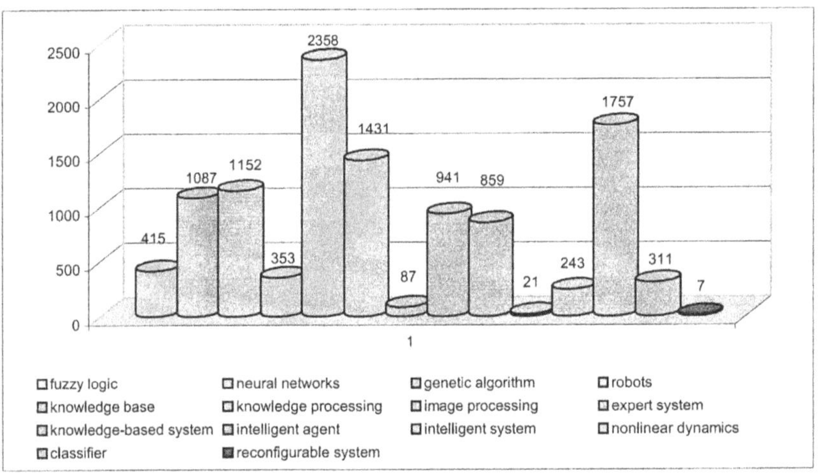

Figure 2. Number of papers in the medical domain reporting research based on CI. According to MEDLINE© database [11].

3 CI in Medical Diagnosis

From the point of view of data and knowledge manipulation, diagnosis includes a set of steps from data collection to inference that translates into the actual diagnosis. The flow of operations is sketched in Figure

3, for a typical situation, with a few typical additional steps, like new knowledge extraction, decision making, and cure strategy.

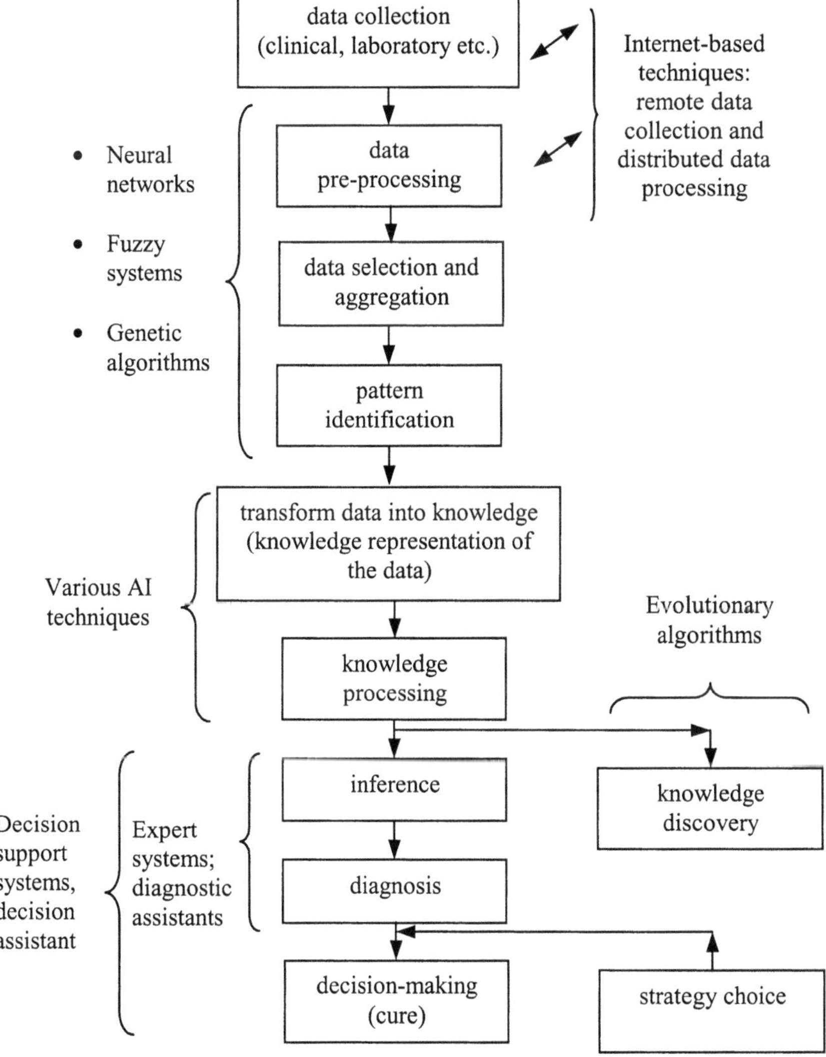

Figure 3. Main data and knowledge processing operations performed in relation to diagnosis.

In Figure 3, several established CI techniques are associated to the steps of diagnosis where they have proved to be most relevant. However, there are numerous researchers attempting the use of each CI technique at every stage of the diagnosis. Notice that for brevity, several data collection and processing methods in diagnosis have not been included in this figure.

4 Data Mining and Knowledge Discovery

Data mining aims at finding relevant information in a mass of data, while knowledge discovery aims at producing new knowledge from data. Both techniques have recently seen a tremendous development, aimed to cope with huge amounts of data with low or inadequate information content, from the point of view of a certain task. A large number of CI techniques have been applied for data mining and knowledge discovery, including evolutionary algorithms, neural networks, and fuzzy logic.

5 Qualitative Reasoning Methods

Qualitative reasoning represents a natural way in human expert reasoning. This paradigm eliminates numbers, operating with qualitative attributes to derive consequences of practical use. Modeling qualitative reasoning has recently been one of the "exotic" tracks pursued in computer science and sound applications have already been demonstrated. Progresses in this domain and combining qualitative reasoning with various other CI approaches may potentially have implications ranging from abandoning the traditional limits between "numerical" and "qualitative" sciences, to blurring the crisp limits between computational intelligence and human intelligence. In medicine, it may open the way to improved cooperation between the human expert and the machine.

6 Issues Related to CI Management in Medicine

CI may contribute in many ways to improve the efficiency of health care. Among others, CI contributes to establishing new metrics in evaluation of the medical performance and its output. On the other hand, no doubt, CI requires well-established benchmarks to test its appropriateness and matching to the goal, and appropriate management of the CI resources, including important computational and interconnection network resources and leading edge software resources. Also, CI requires appropriately trained personnel in leading fields of computer science. Managing the CI for medicine includes reliable estimation of the CI cost and the related services costs, and estimation of the benefits of using CI. Different strategies are needed, depending on the dimension of the health care facility. Today, CI-based medicine looks most effective for large hospital facilities and governmental or large professional associations. However, there are many niche applications where expert systems, decision support systems, and Internet-based data processing that are suitable for smaller facilities and for small groups of practitioners too. Linking goal-setting for the CI tools, process development, and implementing CI tools in the health facility operation are the main challenges in the management of CI interdisciplinary projects.

For a general presentation of the applications of artificial intelligence in medicine and of related problems, see [12].

7 The Prospects of CI in Medicine

The health state includes many facets: individual, group levels, regional, national, sub-continental and continental, and humanity-level health state (see Figure 4). At every level, health problems are perceived differently; moreover they may actually differ in several respects. Optimization of the health state at all levels is probably impossible, while no attempt to characterize at such minute level the health state does exist today. This situation may dramatically change with the advent of genomics and its application to medical sciences. Computational intelligence is *the* tool that helped mapping and information extracting in the genome project. It will become even more important in the field during the search for connections between genome peculiarities and the individual and populational health.

How does computational intelligence fit this scheme? To respond to that surge in demand of CI for medical applications, there are still huge problems to be solved.

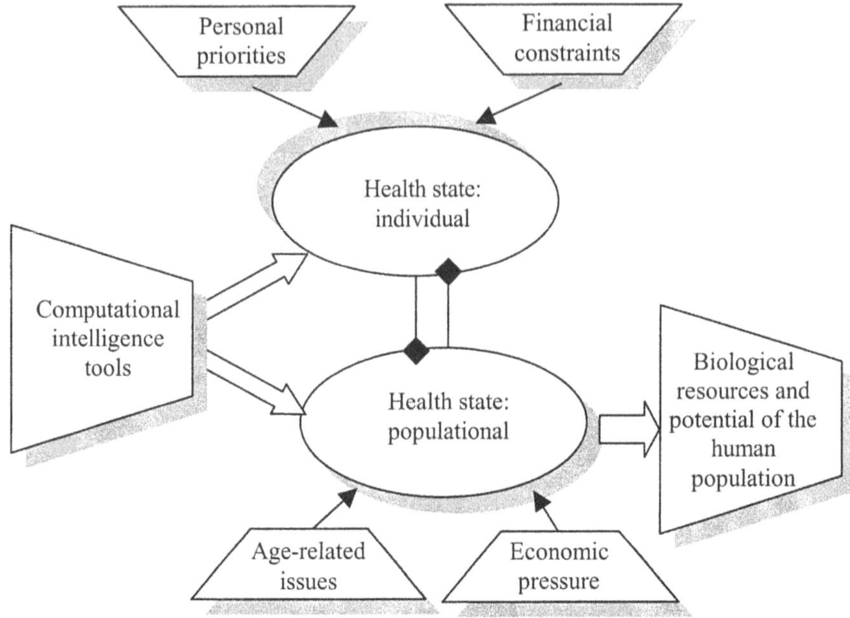

Figure 4. Computational intelligence in the context of health management.

Today, we can talk about computational intelligence-enhanced medicine. In the near future, *computational medicine* will represent a strong branch of medicine and soon, the whole medicine may become computational. Among the medical routines that will first benefit of this evolution, those related to diagnosis, monitoring and rehabilitation have the potential of becoming the heaviest in CI content.

Society has been long driven by knowledge, while data where scare, because of lack of appropriate measuring tools. This explains while knowledge has been generated, in the form of philosophy and religion, long before physics develop as a science. On the other hand, empirical knowledge has been accumulated in the craftsmen' professions which have been seen rather "arts" than "technologies," because of the disparity between empirical knowledge accumulation and experimental and theoretical knowledge. That was the state of the art for as long as the whole Antiquity and Middle Age. The trend has been reversed in the Renaissance and by the middle of the 20th century, huge amounts of data existed and have become difficult to process, especially in "real-time." This has much contributed to the advent of the computer –

which, in various forms, has been invented several times, but with no economic use.

Computers have been useful in data processing, but until recently, they have had only limited power in generating knowledge. If this trend continues, there will be soon a knowledge shortage and data may become useless. The trend is particularly clear in medical sciences, where nowadays equipment is able to collect huge amounts of data on patients. This data may remain of no use if they can not be immediately processed to derive sound knowledge about the patient's state and use knowledge to instantly act in dramatic circumstances like heart attack or epileptic crises. The computers need to be intelligent enough to process data and derive useful knowledge, beyond collecting it and keeping it ordered in databases. High complexity software applications for the medical field can not be created without the correspondingly increased intelligence of the computer itself. Today, knowledge extraction is mostly human expert driven. Autonomous knowledge processing and discovery will be the next step.

A brief glance to the main applications of the computers (Figure 5) shows that the main tasks of the computers evolved from basic numerical computations to data processing, control and communications to data structuring and organizing, while today the main tasks are represented by knowledge extraction, processing and data mining. Tasks foreseeable in the years to come will relay in this last category, transforming the society from a data-driven society to a knowledge-driven one.

The difficulty and cost of knowledge management is the key barrier in the development of the future knowledge-driven medicine, but the foreseeable benefits will undoubtedly drive the development toward this goal.

Acknowledgments

Professor Teodorescu acknowledges the support of a grant from the Romanian Academy and of a grant from the Swiss National Funds, respectively. Part of this chapter summarizes the core report written by him for the *Romanian National Strategic Plan for Health Care and*

Management of the Biological Resources of the Population in the Knowledge Society, commended by the Romanian Academy for the Romanian Government.

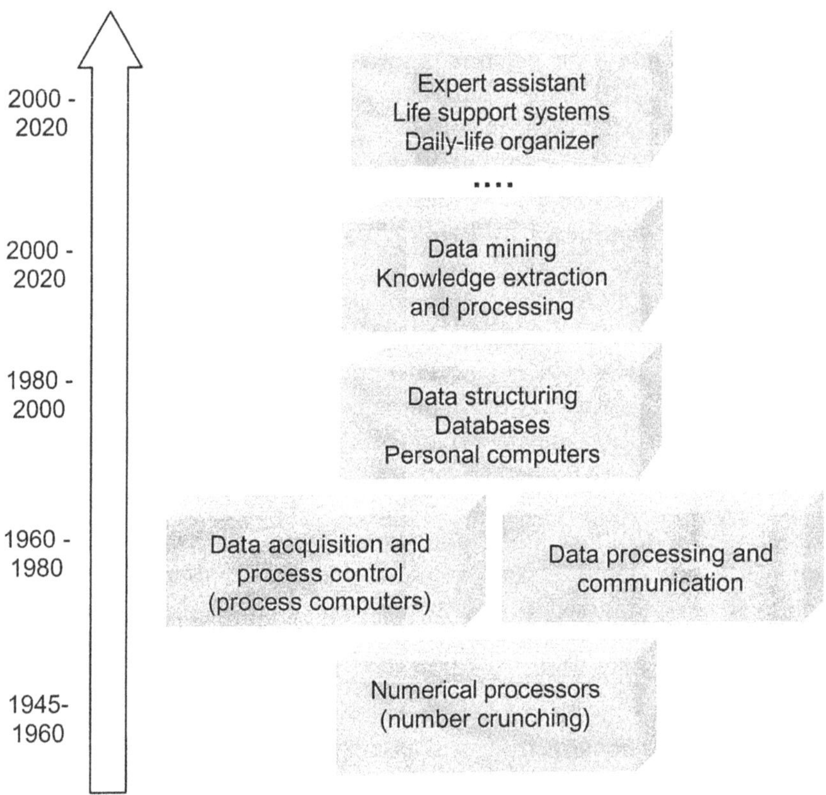

Figure 5. Evolution of computer applications.

References

[1] Teodorescu, H.N., Kandel, A., and Jain, L.C. (Eds.) (1998), *Fuzzy and Neuro-Fuzzy Systems in Medicine*, CRC Press, Boca Raton, Florida, USA.

[2] Durkin, J. (1994), *Expert Systems. Design and Development*, Macmillan Publishing Co., New York, USA.

[3] Teodorescu, H.N., Kandel, A., and Jain, L.C. (Eds.) (1999), *Soft-Computing in Human-Related Sciences*, CRC Press, Boca Raton, Florida, USA.

[4] Steinmann, F. (Guest Ed.) (2001), Special Issue on Fuzzy Theory in Medicine, *Artificial Intelligence in Medicine*, vol. 21, no. 1-3, January-March.

[5] Teodorescu, H.N. and Jain, L.C. (Eds.) (2000), *Intelligent Technologies in Rehabilitation*, CRC Press, Boca Raton, Florida, USA.

[6] Hecht-Nielsen, R. (1988), "Neurocomputing: picking the human brain," *IEEE Spectrum*, vol. 25, no. 3, March, pp. 36-41.

[7] Teodorescu, H.N., Mlynek, D., Kandel, A., and Zimmermann, H.J. (Eds.) (2000), *Intelligent Systems and Interfaces*, Kluwer Academic Press, Boston, USA.

[8] Pedrycz, W. (Ed.) (1997), *Computational Intelligence: an Introduction*, CRC Press; Boca Raton, Florida, USA.

[9] Mackworth, A.K., Goebel, R.G., Poole, D.I. (1998), *Computational Intelligence: a Logical Approach*, Oxford University Press, Oxford, UK.

[10] Teodorescu, H.N., Jain, L.C., and Kandel, A. (Eds.) (2001), *Hardware Implementation of Intelligent Systems*, Springer Verlag, Heidelberg, Germany.

[11] MedLine database, http://www.ncbi.nlm.nih.gov/PubMed/ .

[12] Teodorescu, H.N. and Kandel, A., "Artificial intelligence in medicine," in *Encyclopedia of Information Systems*, Academic Press. (In press.)

Chapter 2

Computational Intelligence Techniques in Medical Decision Making: the Data Mining Perspective

V. Maojo, J. Sanandres, H. Billhardt, and J. Crespo

In this chapter, we give an overview of computational approaches to medical decision making, with particular emphasis on data mining methods. Medicine has been one of the most challenging application areas for Artificial Intelligence since the 1970s. The first generation of expert systems was an academic success. However, these systems failed to have a clinical impact. Researchers realized that medicine, and particularly patient care, is a complex domain, where requirements are different from other areas. A host of approaches were later adopted, including extracting objective information and knowledge from institutional and clinical databases using data mining. We give an overview of data mining methods, including neural networks, fuzzy sets and other machine learning approaches. Several examples of practical medical applications are presented. Finally, we discuss several limitations that must be overcome to effectively apply data mining methods and results to patient care.

1 Background – Artificial Intelligence in Medicine

In the years prior to the introduction of Artificial Intelligence (AI) in medicine, there were different approaches for creating computerized medical decision-making applications [1], [2], such as: (1) Database analysis; (2) Mathematical modeling; (3) Clinical algorithms; (4) Clinical decision analysis. None of them had a real impact in clinical medicine.

By the end of the 60s, cognitive scientists had closely examined the complexity of medical reasoning. This process was classed as hypothetico-deductive [3], [4]. Physicians collect and examine patient data and formulate one or more hypotheses. They study different symptoms, signs and diagnostic tests to refine and test all the hypotheses. Finally, a definitive diagnosis is reached by comparison with other diseases. Once researchers had discovered this process, it could be simulated using computers.

AI researchers had begun to develop computer programs to capture the intelligent qualities of human thought as early as in 1956. The symbolic approach taken by some researchers led to the development of expert systems. The pioneer system was MYCIN [5]. Later, other systems such as INTERNIST [6], Present Illness Program [7], and CASNET [8], were developed at other universities.

MYCIN was designed for the diagnosis and treatment of infectious diseases (mainly meningitis). Its knowledge base contained around 450 IF THEN rules, elicited from experts in the field. It had an innovative inference engine, which was later used separately to build other expert systems. MYCIN's developers also created a new ad hoc method for managing the probabilities associated to rules.

Nevertheless, MYCIN was never used in clinical practice. Although it was a great academic success, it was not designed to address a real medical problem. Some of its developers soon realized that clinicians refused to use it for several reasons, such as: (1) the interfaces were not user friendly and systems were too difficult to use; (2) physician interaction was too slow and long; (3) data were entered by the physicians themselves; (4) it was impossible to achieve a 100% accurate diagnosis; and (5) knowledge was not properly validated and it did not include "deep" concepts related to the pathophysiological questions of the human body and diseases.

Other expert systems developers in the medical area reaped a similar result. Paradoxically, the models and tools developed for these medical applications were successfully applied in other fields outside medicine.

Traditional consultation–based systems were designed like "oracles," covering broad areas and created for academic, not clinical, purposes.

The first-generation systems were based on heuristic knowledge and, therefore, could not capture the complexity of medical reasoning. Knowledge acquisition from experts, carried out by knowledge engineers, was a difficult task, since experts were unable to verbally express compiled and automated knowledge.

A new generation of systems included new reasoning models. For instance, ABEL [9], a system created for advising physicians on acid-base disorders, contained knowledge about the causal processes of a disease. ROUNDSMAN contained a library of knowledge about medical literature. The system looks for an article related to a specific patient, selecting clinical recommendations and evaluating alternative choices [10]. Another system, KARDIO, employed a different approach that can be related to the modern concept of data mining [11].

These second-generation expert systems were conceived as critiquing programs, which could analyze and discuss physicians' decisions. These new programs were more efficient than first-generation applications, but they were not successfully deployed in clinical settings either.

Within AI, several directions received significant attention from the scientific community. Some researchers suggested that hybrid systems, combining several AI techniques could be more useful in medicine. An example is PERFEX, a rule-based expert system for interpreting SPECT images of the heart. Its developers combined neural networks and rules to improve the diagnostic and prognostic capabilities of the original decision support system [12].

There are currently many intelligent programs or modules integrated into other larger clinical systems. They are built to assist in small, specialized tasks, providing information and knowledge. An approach launched in the early 90s was to create standards, like the Arden Syntax for building Medical Logic Modules. Using this PASCAL-like syntax, small decision support systems could be specified, integrated into larger information systems and exchanged by research and clinical institutions. It had limited success, due to the development of new formalisms and methods.

As regards representation mechanisms, ontologies have an increasing interest within the medical informatics community. From a software engineering perspective, a new approach appeared with the development of software components and intelligent agents. They can exchange messages and collaborate across communication infrastructures (e.g., over the Internet using CORBA and JAVA), improving systems development and reuse.

Evidence-based medicine has provided clinicians with objective knowledge as a scientific foundation for patient care. Practice guidelines, protocols or appropriateness criteria are based on a review of the scientific literature –especially considering randomized clinical trials- and expert opinions, gathered using a panel methodology, such as RAND's or Delphi. New standards for representing or tools for visualizing practice guidelines [13],[14], were created to disseminate guidelines over the Internet and link the tools to medical records for patient care.

Another research approach is related to the automated process of extracting information and knowledge from medical databases. This promising area, labeled as data mining, and its applications in medicine, are explained in more detail below.

2 Data Mining

Data mining is a broad and generic concept, with different taxonomies and classifications. Many methods are included under the generic term of data mining, which can be carried out using different commercial and public tools. We will consider data mining as a step in the process of knowledge discovery in databases, which is described below.

2.1 Knowledge Discovery in Databases

Knowledge discovery in databases (KDD) is the process of exploiting the possibilities of extracting knowledge that is implicitly contained in a collection of data: a database. A commonly accepted definition for KDD is: "the nontrivial process of identifying valid, novel, potentially useful and ultimately understandable patterns in data" [15].

Data mining and KDD are sometimes confused. However, KDD is a process composed of several steps (Figure 1), whereas data mining is the step in which patterns or models are extracted from data using some automated technique.

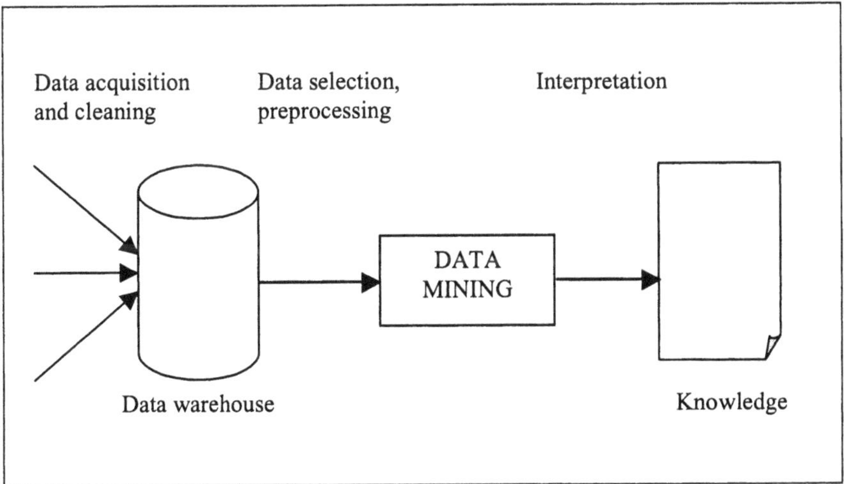

Figure 1. Data mining in the context of knowledge discovery in databases.

Data acquisition and cleaning refers to the process of gathering and combining data from multiple data sources and storing them in a data warehouse (DW). Inconsistent and irrelevant data can be removed in this phase. Most prediction methods require data to be in a standard form that can be processed later. Data selection and pre-processing is the step where appropriate data are selected for analysis and transformed to representations for the data mining process. The data mining process itself tries to identify and extract new patterns from the selected data.

Since the process of data mining depends on the quality and quantity of the available data, the selection and pre-processing step is essential for assuring results reliability. In the interpretation phase, these patterns are analyzed and the useful, understandable and novel patterns are selected. These patterns become knowledge and are either presented to the user or used otherwise.

Data mining depends on the quality and quantity of the available data. Thus, a DW is essential to facilitate the data mining process. The DW is an integrated database created to enter, store and search data coming from local or remote sources within a whole institution. A DW can be centralized or distributed, having a logical and uniform view. It is organized by themes, and is consistent and time-dependent. If a DW is designed and maintained correctly, the data-mining step will be easier.

Although the concept of a DW is different to an integrated electronic patient record (EPR), some of the generic goals, methods and tools for maintaining data and extracting information are similar. Data mining methods can be applied to extract knowledge from institutional DWs (e.g., to identify patterns in increasing costs) or in electronic patient records (e.g., to find prognostic rules).

Once data are available and prepared for analysis, many different methods can be used, as shown below.

2.2 Methods

Han and Kamber [16], classify data mining from different points of view: (1) by the types of databases to be mined, (2) the kinds of knowledge to be discovered, (3) the techniques to be used, and (4) the kinds of applications. Data mining is often classed by functionality into predictive and informative data mining.

In this section, we concentrate on the techniques and methods applied in data mining. More precisely, we discuss Artificial Intelligence techniques used for classification, clustering and prediction.

There are several classifications of data mining methods in the literature [17]-[19]. These methods are classed according to several criteria, such as: (1) the kind of patterns discovered: predictive or informative, (2) the representation language: symbolic or subsymbolic, or (3) the data mining goal: classification or regression.

We divide the methods into three families, each with several subfamilies: (i) statistics and pattern recognition, (ii) machine learning (ML), and iii) artificial neural networks (ANNs). However, there is certainly no clear cut between these approaches. Many statistical or pattern recognition approaches can be considered as machine learning techniques and vice versa. Also neural networks can be classed as belonging to ML.

Statistical methods have been used for a long time, whereas the use of ML and ANN applications has increased in the last few years. Solutions offered by all methods can be compared considering performance (classification and prediction accuracy, sensitivity, specificity or learning speed), as well as by the comprehensibility and significance of the extracted knowledge [20], [21].

Comprehensibility is an important issue in data mining. If the user can easily understand the results provided by the system, the knowledge is much easier to validate and use and the system is more likely to be accepted. Also, techniques that provide comprehensible results can be used for tasks such as characterization, discrimination and association analysis.

In domains like medicine, accuracy and comprehensibility are the most important parameters. Results presented in comparative studies [19], suggest that accuracy is similar for the three families, but comprehensibility is better for ML methods. Weiss and Indurkhya also point out that decision tree induction is the choice when easily explained solutions are a goal, even though results are slightly weaker than those provided by other methods.

Methods from the three families have been applied in several real and artificial domains [18], [22], and their performance has been compared in several studies [19], [23], [24]. There is apparently no generally best method [25], because the performance of a method is strongly related to the application domain.

The general paradigm for inductive inference [26], can be used to explain the goal of a DM algorithm. This is to find one hypothesis from all the possible hypotheses, i.e., a description that is consistent with the set of facts under analysis using some background knowledge. For classification and prediction, this is done by supervised learning algorithms that take a set of already classified training examples and try to find a classifier of some sort (e.g., a decision tree, induction rules, etc.) that is consistent with these examples and is highly likely to classify or predict new cases correctly.

The emphasis on the classification of new cases is important, since we obviously want to generalize the information extracted from the training cases so that it can be applied to classify or predict new cases correctly. This is the inductive part of learning.

For clustering, the goal of the data-mining algorithm is to organize the data into classes that are previously unknown. In this case, there is no set of training examples with known class labels. Therefore, clustering is also called unsupervised learning. A clustering algorithm tries to find some kind of relationships or similarities among the data instances and groups similar cases together. The results are usually informative and can provide novel, potentially useful information about the characteristics of particular groups of data instances or cases.

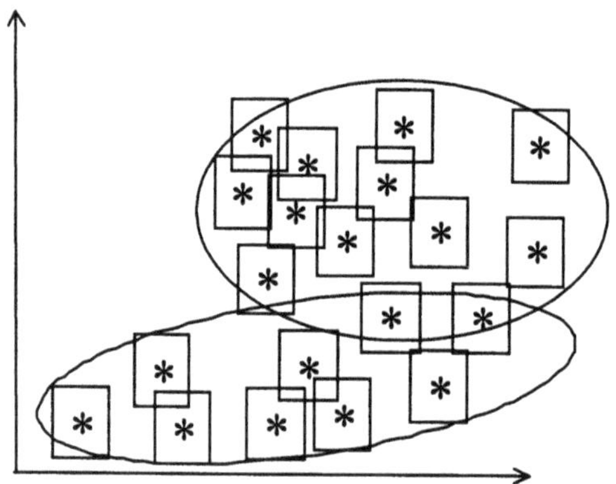

Figure 2. Representation of clustering of data into 2 different clusters.

Below, we give a brief description of data mining techniques. All these methods consider the database records selected for data mining as examples, cases or instances. The examples are described by sets of features or attributes (i.e., the fields in the database). Thus, a particular example is an instantiation of these features. Examples can be either labeled with the class they belong to in supervised classification or unlabelled, in clustering. Many techniques consider the examples as vectors in an n-dimensional hyperspace, where each dimension corresponds to one feature.

2.2.1 Statistics and Pattern Recognition

Statistical methods rely on there being an explicit underlying probability model. This provides a measure of the membership of an example to a class. The application of these methods requires human intervention to define models and hypotheses. Classical methods are based on Fisher's work on linear discrimination or linear regression. The main methods are:

- *Linear discriminants*. A hyperplane in the space defined by the features is searched to separate the classes minimizing a quadratic cost function.
- *Quadratic discriminants*. Similar to linear discriminants, the case classification regions are separated by quadratic surfaces.
- *Logistic discriminants*. They use quadratic surfaces that maximize a conditional likelihood.

These methods are supervised and can be used to classify cases or to predict some unknown data values.

Modern statistical methods use more flexible models to provide an estimate of the joint distribution of the features in each class. They are independent of the density functions of the classes of the problem. This characteristic means that they can be used in cases where the classical methods are not applicable because the distributions involved are unknown. They are also known as non-parametric methods.

Kernel Density Estimation

Kernel density estimation is based on the estimation of the density functions by means of non-parametric procedures. Once these functions have been obtained, classical Bayesian procedures can be used to classify new cases, minimizing the risk or minimizing the error. The density functions are estimated using a specific kernel function for each type of variable: continuous, binary or nominal. An important element is the "smoothing parameter" λ, which determines the scope of the function searched. Higher values of λ produce "smoother" functions, whereas low values have the opposite effect.

K-nearest Neighbor (K-NN)

K-NN classifies cases with respect to their neighbors in the hyperspace. The training data are not used to create a classification model. The estimation of the density functions is based on the number of examples of each class in a hypersphere centered on the example that is to be classified. The example is classified within the class that has the highest number of representatives in the studied space. In the case of k=1, we choose the class of the nearest example. This approach calls for a careful study of the distance measure to be used and of the standardization processes when heterogeneous variables are considered. The search of the k nearest neighbors must be optimized, especially when the number of examples is high and k>1.

Naive Bayes
Naive Bayes is based on the assumption of independence among the features used to describe the examples. The conditional probabilities of an example belonging to a certain class, given a particular value of a feature, can then be used directly. These probabilities are easily calculable. Another advantage of this method is that once the probabilities have been estimated, not all the examples have to be retained in memory. The application of this principle to causal networks results in Bayesian networks. They are directed acyclic graphs of causality, where the inferences are made using the conditional probabilities of the variables [27]. Implementations of the automatic generation of causal models are BIFROST [28], and Bayesian Knowledge Discovery [29].

Alternative Conditional Expectation (ACE) [30]
The objective of ACE is to find the non-linear transformations of the variables that produce an acyclic model that best matches the distribution under study. Transformations are searched according to a simple iterative schema.

Multivariable Adaptive Regression Spline (MARS) [31]
MARS is based on "spline" functions for function adjustment. MARS implements a selection strategy that alternates a phase where basis functions are added to the model with a phase where pruning eliminates terms. Splines of degree three or less are usually employed. The calculation and choice of the number of control knots is automated to make the adjustment process more flexible. The best model is determined using a cross validation criterion.

2.2.2 Machine Learning

ML is an AI research area that studies computational methods for improving performance by mechanizing the acquisition of knowledge from experience [18]. ML algorithms enable the induction of a symbolic model, decision tree or set of rules, of preferably low complexity, but high transparency and accuracy [21]. The most relevant field of ML in DM is Inductive Learning (IL). IL is a process of acquiring knowledge by drawing inductive inferences from teacher or environment-provided facts [26].

Decision Trees
Decision trees are an easily understandable representation of classifiers. Their nodes represent tests of particular attributes and their leaves correspond to class labels. Examples are classified by passing from the top of the tree to a leaf. The branch whose test is passed by the example is selected in each node. The example is classified in the class that is marked at the leaf. Decision trees can be generated or learnt using algorithms, such as Quinlan's ID3 and C4.5 [32], or a recursive-partitioning method, as in CART [33].

Recursive-partitioning algorithms build a tree by recursively dividing a population (set of training examples) into smaller subsets. Using this method, it is possible to identify the best predictor of a disease in a patient population. The population is divided into two groups, with and without the predictor. The best predictor is obtained using a measure of difference (usually entropy). The process is repeated for all predictors, creating a decision tree with associated probabilities.

ID3 and its successors are very popular methods, although they have several problems that remain to be solved, such as: example selection for the training sets, pruning methods, generalization to independent test sets and conversion to classification rules.

Rule Induction
Decision rules are another easily understandable form of classifier representation. They are quite similar to decision trees, but model a classifier as a set of IF THEN rules. Each rule has a set of conditions on the features and returns a class label. An example is classified in the

class whose conditions are met. There are different strategies for dealing with competing rules (e.g., assigning probabilities).

AQ15 [34], [35], is a method for learning strong rules using constructive induction. It looks for rules with few errors that create new attributes. Moreover, AQ15 has an incremental learning facility for using an initial hypothesis about the domain being modeled.

CN2 [36], [37], is a variant of AQ that combines the best features of AQ and ID3 to solve AQ's problems for dealing with noisy data. Another drawback of AQ solved by CN2 is order dependency in the training examples. CN2 produces probabilistic rules.

Inductive Logic Programming (ILP)
These techniques are used in abstract computational problems, where learning of recursive rules might be needed to address specific problems (e.g., problems that can generate infinite data). ILP processes the induction of concepts with logic programs usually containing Horn clauses. Most methods employ PROLOG. This language has a higher expressive power than a description based merely on value–attribute formalisms, since it allows recursive definitions of predicates. Recursive definitions are a means of defining predicates in a clear and generalized way. However, the descriptive power may have negative effects when a description of a predicate is too specific or too detailed. In this case, the predictive power, that is, the generalization of the description to new, unseen cases, is lost.

ILP algorithms start with a set of positive examples E^+, a set of negative examples E^-, and a logic program B that contains the existing knowledge on the domain. The aim of the algorithms is to find a program P that is complete and consistent with the knowledge on the domain and the given examples. Both bottom-up or top-down approaches can be used to search the solution. A bottom-up approach is employed in systems like GOLEM [38]. An implementation of the top-down approach is FOIL [39].

Case-Based Reasoning (CBR) and Instance-Based Learning (IBL)
A new trend appeared in AI at the end of the 80s. It considered that the key question in reasoning may be neither a process of heuristic search (as in expert systems) nor a pattern recognition task (as in neural

networks). This appears to be particularly applicable from the viewpoint of medicine. A medical specialist gains expertise through the experience of dealing with many cases over the years. Thus, a cognitive approach should consider storing cases -patients- and making inferences on this basis.

Case-based or instance-based methods take advantage of previous cases that have been solved by storing known cases or examples in a case base. When a problem is presented to the system, it looks for previous cases that are similar to the new one, analyses their solutions establishing differences and creates a solution adapted to the new case. CBR stores training instances in memory, and its performance relies on the indexing scheme and the similarity metric used. New cases may be added to the library of cases if they provide new and useful information. Thus, the case base grows over time [40].

One of the basic problems of these systems is to find an appropriate similarity measure for comparing the examples. Another problem is that the accumulated knowledge cannot usually be understood directly by humans. This is because a case base containing a lot of examples is not suitable for describing the principle characteristics of classes or categories. Also, it can be quite difficult to find appropriate mechanisms that adapt the solutions of known cases to the new cases, if there is no perfect match and differences appear.

From the point of view of medicine, identical cases are seldom found, except in common diseases where physicians do not need computer advice. Physicians seem to use case comparison as an additional strategy to heuristic search, pattern recognition or causal reasoning, rather than as a standalone technique [41]. An advantage of this kind of systems is that knowledge acquisition is incremental, since new cases can be added to the systems' knowledge.

Examples of CBR and IBL systems are PROTOS [42], CASEY [40] and IBL [43].

Genetic Algorithms [44]

The process of learning can be understood as a process of searching all possible hypotheses for the hypothesis that best matches the given examples. This means that learning can be considered as an optimization problem, where the search space is usually quite complex. Genetic Algorithms can explore complex search spaces efficiently and, therefore, they are suited for solving this kind of optimization problems.

The origin of GAs is the study of cellular automata [44], [45]. GAs simulate the mechanism of natural evolution. GAs explore search spaces in an adaptive manner, generating and testing new possible solutions. A GA starts with a population of individuals, where each individual represents a possible solution to the problem at hand. This population evolves in successive generations by applying genetic operators like reproduction, crossover and mutation.

In each generation, the individuals are evaluated using a fitness function to determine their suitability to the problem. Then, a set of individuals is selected for reproduction and the new generation is obtained by applying the genetic operators. Better individuals have a higher probability of being selected for reproduction and, thus, bad individuals will "die out." The main difficulties of GAs are, on the one hand, to find a codification of the hypothesis in the form of chromosomes - strings of bits are usually used- and, on the other, to define the fitness function.

Examples of systems that implement genetic algorithms are DARWIN [46], and Rosetta [47].

Fuzzy Sets [48], and Rough Sets [49]

Fuzzy sets and rough sets provide representation formalisms that can be used to model uncertainty relatively simply. Therefore, they are especially suitable in domains with noisy or inconsistent data or in domains where other solutions would be excessively complex. This is also the reason why fuzzy sets and rough sets are often employed as representation formalisms in other types of reasoning systems.

Fuzzy sets are based on the idea that the human mind does not reason using strict binary or multivariant logic. Instead, we use rather vague descriptions of measures, features and attributes (e.g. "small," "high," "nice," etc.). Fuzzy set theory models this type of uncertainty using set membership functions in the interval 0 to 1. Thus, the truth value of predicates can be described by a number between 0 and 1 and not just by true or false. For example, in traditional logic, the fact that blood pressure is high is either true or false. Humans, however, use the word high with some uncertainty. There is usually no clear cut between high and "not high" blood pressure. In fuzzy logic, the truth value for the predicate blood pressure is high will be a number between 0 and 1, where 1 means that the blood pressure is really high and 0 that it is certainly not. A number between 0 and 1 indicates how certain we are of blood pressure being high.

Rough sets can be considered as a special type of fuzzy sets, where the truth values of predicates are trivalued with the values true, false and perhaps. A system that implements rough sets is ROSETTA [47].

2.2.3 Artificial Neural Networks (ANNs)

Although ANNs are usually included within the broad concept of "machine learning," we will consider them separately, because of to their importance.

ANNs are based on models of the brain. They represent knowledge as a network of units, or neurons, distributed in one or several layers, which transmit the activation values from the input units to the output units. There are rules to define the weighting of the transmission, the activation of the units and the connection pattern of the network. Different settings of these configuration rules produce different types of ANNs. Learning can be supervised (for classification and prediction) or unsupervised (for clustering).

Artificial neural networks have been used, with relative success, in applications such as diagnosis in medical images (e.g., tumors in mammographies), identification of certain features in electrocardiogram signals, electroencephalograms or monitoring systems in Intensive Care Units [50]. They can be used, too, to output predictive rules in large databases such as electronic patient records.

2.2.3.1 Supervised Artificial Neural Networks

Perceptron and Multi-layer Perceptron (MLP)

Rosenblatt proposed the Perceptron Learning Rule in the early 1960s [51], to learn adequate weights in learning problems. The perceptron produces a segmentation of the hyperspace defined by the attributes, but can only learn linearly separable functions. Minsky and Papert [52], showed these limitations with the "exclusive OR" problem. This problem can be solved, however, by multilayer perceptrons, also called Backpropagation Networks. They can learn arbitrary classification functions. A MLP has hidden layers of neurons that neither receive external input nor give any external output. In 1985, a learning rule, known as the Generalized Delta Rule [53], was created for the MLP. Using this rule, different architectures of connections can be trained.

Radial Basis Function Networks (RBFN) [54]
A RBFN is similar to a MLP with one hidden layer. RBFNs can also learn arbitrary mappings. The basic difference is the hidden layer. RBFN hidden layer neurons implement radial basis functions. They have a center for the input at which their output value is maximal. The output value falls as the input value moves away from the center. The functions in the output layer are simple linear transformations that can be optimized using linear techniques. This has the advantage that they are fast and overcome problems like local minima. Another advantage with respect to the MLP is that RBFNs can model any non-linear function with a single hidden layer.

Cascade Correlation Learning Architecture [55]
The above types of neural networks are based on fixed architectures and learning involves adjusting the parameters of the connections. There are also techniques that allow the network structure to be modified during learning. One example is Fahlman and Lebiere's cascade correlation learning architecture. The process starts with no hidden layers and adds new hidden units in an iterative process. In each phase, the network is trained and the residual error is measured. If this error is below a certain threshold the algorithm stops. Otherwise, a pool of unrelated candidate units are trained in parallel. The unit, whose output has the highest correlation with the residual error to be eliminated, is selected for addition. This technique reduces learning time and determines the size and structure of the network automatically.

Random Access Memory Nets (RAMnets) [56], [57]
RAM-nets are based on a representation of input data as strings of bits. The input pattern is mapped to k n-tuples, where each n-tuple represents n bit positions in the input pattern. So, the network consists

of k RAM-nodes, each with an n-tuple input. For each node, 2^n memory locations are allocated such that a binary signal on the input will address exactly one memory location for this node. Before training starts, all $k2^n$ memory locations are set to 0. Training is carried out with positive examples only. Each training example is presented to the network in the form of k n-tuples, and all memory locations accessed in all k nodes are set to 1.

When a new instance is presented, each RAM-node responds with the value of the accessed memory location. That is, it responds with a 1 only for the patterns seen in the training set. The sum of the responses is returned as the response of the network. Thus, if a training sample is presented after learning, the response will be k.

In this scenario, the RAM-net is taught to classify possible cases into cases that belong to the class and cases that do not. If more categories are required, one RAM-net is learned for each category.

The basic advantage of RAM-nets over other artificial neural networks is that learning performance is very fast, and the results are comparable to other techniques.

2.2.3.2 Unsupervised Artificial Neural Networks

Unlike supervised ANNs, unsupervised ANNs explore the network structure without previous knowledge of the classes and can discover hidden features of the data under study. On the other hand, their computational costs are sometimes higher, and an appropriate selection of the initial parameters might be difficult.

K-means Clustering [58]
This technique divides the set of examples into k clusters by minimizing some distance measure between the examples in each cluster and such that the inner-cluster distances are globally minimized. The user must provide the number of clusters. The algorithm starts by selecting k training examples to determine the centers of the initial clusters. These centers are called code vectors. An iterative process moves the cluster centers around the hyperspace until a good model of the probability distribution function of the input space is obtained. In each iteration, the distances between each example and the current cluster centroids are calculated. The example is assigned to the closest

cluster. Then, the centroid of each cluster is recalculated as the mean of all examples of that cluster, and the process is repeated. The iteration stops at some termination criterion, usually when the changes in the clusters are below a given threshold.

The Gaussian Mixture Model is a generalization of this model, where a variance is assigned to each center.

Kohonen Nets [59]
Kohonen nets provide a way to learn topographic mappings from the input space usually to a one- or two-dimensional cluster space. In Kohonen nets, there are one- or two-dimensional layers of neuron clusters. Each neuron is connected only to its nearest neighbors. Furthermore, each unit is connected to all input nodes. As is the case with k-means, each cluster node has an associated weight vector (code vector), where each weight corresponds to one of the input features. The iterative learning process involves presenting the training examples and calculating some measure of distance between the input vectors and the code vectors. For each example, the node whose code vector is closest to a presented input vector is identified. Then, the weights of this node and of the nodes in its neighborhood are modified to reflect the input vector more closely. The learning rate (e.g., the amount of modification) and the size of the neighborhood decrease over time. After learning, similar input patterns will activate the same cluster nodes or nodes in the same region, and the network represents a topographic map of the feature space.

2.3 Data Mining Tools

Numerous public domain and commercial software products for data mining are now available to researchers and medical practitioners. For instance, products such as Intelligent Miner© from IBM, MineSet © from Silicon Graphics, Clementine© from SPSS, various programs from SAS, and other off-the-shelf software products are used by an increasing number of people and organizations.

Some of these commercial tools combine features that are designed to mine data sets, as well as text collections and Web documents. In medicine, where there are more than 20,000 medical Web sites, as well as large text databases, such as MEDLINE, more tools are needed to extract knowledge from these sources.

For additional information on data mining tools, see [60].

3 Applications in Medicine

The main application of data mining is in economics and enterprise analysis, where managers need to analyze an institution's performance and customer behavior. Since there is a huge amount of raw data, organizations look to computing methods to mine their databases and extract meaningful information.

The health sector has similar needs. The amount of available clinical data (e.g., from medical records, clinical trials or research studies) has increased sharply in the last few decades. Thus, new methods are needed to collect, store, search, retrieve and analyze data, locally or over the WWW [61]. Furthermore, physicians spend a large percentage of their time working on information management tasks. Thus, it is only logical that new data mining approaches focus on developing methods to extract information and knowledge from raw medical data.

During the 80s, numerous studies were published in many journals, including the New England Journal of Medicine, JAMA, Annals of Internal Medicine and the British Medical Journal, about clinical prediction rules extracted from large clinical data sets. These reports presented research based mainly on statistics and CART [33], a program based on a recursive-partitioning algorithm.

The variety of techniques and criteria made it difficult to create a standard methodology, although some were proposed [62], Some of these methodologies suggested using classical and AI techniques, such as neural networks and induction. Most of the studies were related to triage applications for tasks such as hospital admissions or heart disease diagnosis. For instance, research at the Brigham and Women's Hospital, Boston, USA, to create decision trees using recursive-partitioning methods for use in heart disease diagnosis has been ongoing for over 20 years. This technique was applied to more than 5000 patients with myocardial infarction [63].

An example of pioneering work in machine learning for medical applications was the KARDIO system [11]. This program, for cardiological diagnosis and treatment, used an inductive algorithm to extract rules from large clinical databases. The approach included objective qualitative and causal knowledge extracted automatically by computers. It aimed to eliminate subjective biases in the knowledge base contents. Hundreds of medical data mining applications were designed later, mainly for academic purposes.

By way of example, a recent medical informatics conference, organized by the American Medical Informatics Association (AMIA) included eight papers labeled as "data mining." Topics included, for instance, text mining in genomics, classification of x-ray reports using machine learning, Bayesian networks to analyze laboratory data, querying medical records with SQL for genetic research and in hospital DWs [64].

Although there are numerous examples of research, not many are based on scientifically solid studies. It is quite common to find errors in the study design from an epidemiological point of view. These pitfalls are discussed later. We will detail a few selected research projects, concerning the analysis of clinical data sets. These projects have the added value of having used a large amount of clinical cases, as compared to most data mining projects, where sample sizes are generally much smaller.

Cooper and his colleagues carried out a research study at the University of Pittsburgh and Carnegie-Mellon University to extract clinical predictors for pneumonia mortality [65]. They applied eight statistical and machine learning methods to a training set of 9847 cases and tested those methods with a set of 4352 cases. They used BDMP©, commercial and their own machine learning methods, such as logistic regression, decision trees, Bayesian networks, rule-based models, neural networks and k-nearest neighbors. The results of the different techniques were quite similar. Hybrid methods -the combination of some of the above methods- were proposed.

A study carried out at MIT and NEMC in Boston, USA, for classification in cardiac ischemia [20], was based on a clinical database of 5773 patients. They used logistic regression to extract significant clinical predictors. Results were compared to decision trees generated by C4.5. Both methods performed slightly worse than physicians. A combination with neural networks was proposed to improve system performance.

Several US hospitals carried out a study for predicting mortality after coronary artery bypass surgery [66]. They used artificial neural networks to predict clinical outcomes. The neural nets were trained with 4782 patients and tested with two sets of 5309 and 5517 patients, respectively. Comparison with logistic regression showed similar results. They refuted previous research studies claiming that neural networks performed better than statistics. They suggested that the experiments conducted in the above studies failed to detect some errors.

As explained above, a common characteristic of these three studies is that they used large clinical databases. It seems that large sample sizes did not improve the research results. In section 4, we will show that there are other factors that might be important for a data mining project to be successful.

A research project carried out by the authors [67], aimed to extract clinical prediction rules for prognosis of patients with rheumatoid arthritis. 1000 paper-based medical records, from a hospital in Madrid were analyzed. Since most of them had important errors, missing data, or the current status of the patients could not be evaluated, only 374 cases were selected. Data from these patients, acquired over 20 years, were transferred to an electronic support using a standard database management system.

We created a new method of constructive induction of machine learning. We studied the relationship among attributes evaluated at the beginning of the disease and attributes that measure the long-term status in patients. We combined various ML and clustering algorithms an we applied them to the patient database. An extended description of the methods we applied has been reported elsewhere [68].

Within this retrospective cohort epidemiological study, medical experts chose 21 predictive variables among data. These variables were evaluated in the first year. The 7 present numerical features were used for constructive induction. A panel of rheumatologists selected 2 outcomes, death and health status, based on widely used rheumatological questionnaires.

Using some software programs that we developed as well as public domain tools, we were able to obtain several prediction rules. These prognostic rules relate clinical variables and outcomes (e.g., death). Thus, they could assist physicians in medical practice, by identifying clinical predictors and suggesting the best treatment options for each patient.

Our research can be compared to many other projects that have been carried out during the last 15 years. In fact, we made a strict patient selection, since only 1/3 of the original cases were selected for the study. We tested many different algorithms and tools, and the results were analyzed by rheumatology experts.

Nevertheless, we realized that there are many other factors that should have been considered from the beginning of the study. We explain below some of these drawbacks, that restrict the clinical application of many data mining projects in medicine.

4 Limitations of Data Mining in Medicine

Since MYCIN, many research reports claimed that AI systems outperformed physicians in diagnostic accuracy [69]. Machine learning reports have shown a similar trend. It seems that some requirements and biases were not considered correctly in many experiments [66], [70]. Thus, most data mining systems are still not used routinely in clinical practice. Below, we will detail some of the limitations that have prevented data mining from having a clinical impact.

At the sessions of the 1st International Symposium on Medical Data Analysis, ISMDA 2000, held in Frankfurt, there were some heated debates about the limitations of data mining methods in medicine. Most of the participants in the sessions were mathematicians, statisticians and computer scientists, while only a few were physicians. A significant statement of one of the participants was: “I think that my research was correct, results were excellent, the software is user friendly, doctors like it, but they do not use our program in patient care. Why?”

Based on our own research results, we suggest that there are many issues that are not usually considered properly when applying computing methods to medical data analysis. The methodologies used by health professionals and data analysts are different. Educational approaches on both sides are still divergent. Physicians are trained to reason "medically" [3], whereas computer scientists, engineers and mathematicians learn a completely different way of thinking. Differences grow even more later on, as research designs in both areas are usually diverse.

From our experience of working at a school of computer science, where there is little emphasis on a medical informatics curriculum and no clinical teaching at all, it is common to see the difficulties faced by computer science students. They do not understand many of the clinical issues involved in medical informatics project. For example, they find it difficult to comprehend why epidemiological factors have to be considered in applied AI research.

As stated above, other professionals [66], have suggested that many publications of data mining results were too optimistic. A book on medical data mining [71], also points about the difficulties that researchers face when they work on clinical applications. We outline a set of factors that lead to the current limitations of data mining in medical applications. These factors should be considered in medical data mining:

- One of the most important pitfalls in data analysis, and particularly data mining, is a correct study design. Randomized clinical trials are the most objective and reliable method of collecting data in medical research. However, they are difficult, costly and take a long time to complete. Thus, it is quite common for researchers to use small clinical data sets or databases in the public domain. They are useful for "toy" projects or academic exercises, but not for real medical applications.
- Questions such as patient eligibility, biases, gold standards, sample sizes or the selection of variables and predictors must be correctly considered. It is quite difficult to find an adequate sample, in terms of data quality and quantity, for performing a correct analysis.
- Data pre-processing is important since noise and missing data are quite common in medical databases.

- Some methods, such as neural networks, behave like black boxes and give no explanation of the reasoning used. Their acceptance is, therefore, limited.
- Lack of comparison with traditional statistical methods, which are more familiar to most physicians.
- Need for a formal validation of results, considering both clinical and software results.

5 Conclusions

The patient is at the center of clinical practice. Several challenges concerning patient care will arise in the coming years, related to many issues and areas, such as quality of care, cost control and patient satisfaction. New findings from the Genome Project, delivery of health care at home for the elderly and isolated or handicapped people, remote control of surgery or new intelligent programs will be available to improve the quality of life through better health care.

More explicit and objective knowledge is needed for patient management, diagnosis and prognosis. This knowledge can be obtained from the enormous amount of data that is daily generated in a host of clinical settings and in multiple formats. Reliable data can be obtained from clinical trials at medical sites linked over the Internet. Information from genomic and clinical databases can be accessed and pooled from remote heterogeneous sources over the WWW using virtual repositories and communication. Knowledge can be extracted from large clinical and institutional databases using data mining tools and methods. New methodologies are needed to improve the use of data, information, and knowledge concerning medical care by health professionals.

In this chapter, we considered the aspects that are important in medical data mining. The failures of past research in medical informatics and AI in medicine must be avoided and reconsidered. Professionals can select different commercial or public domain methods and tools to carry out their data mining projects, but the research must be designed carefully. A study design in medical research must consider many epidemiological questions right from the start. Computer scientists, engineers or statisticians are often unfamiliar with the kind of constraints and requirements of clinical research. Medical informaticians can fill this gap, bringing together people and methodologies from both sides.

Applications of data mining results in real clinical settings are still few and far between. The limitations stated above have prevented most of the research carried out in the area from being effectively applied for patient care. Medical data analysis, and particularly, data mining research, needs methodological approaches combining epidemiological considerations and traditional statistical or computing methods. If the pitfalls and limitations of past medical data-mining methods are overcome, the end users, health practitioners, will accept the results obtained for use in their clinical routine.

References

[1] Shortliffe, E.H., Buchanan, B., and Feigenbaum, E. (1979), "Knowledge engineering for medical decision making: a review of computer-based clinical decision aids," *Proceedings of the IEEE* 67, pp. 1207-1224.

[2] Szolovits, P. (Ed.) (1982), *Artificial Intelligence in Medicine*, Westview Press, Boulder, CO.

[3] Kassirer, J. and Kopelman, R. (1991), *Learning Clinical Reasoning*, Williams and Wilkins, Baltimore, USA.

[4] Shortliffe, E.H. and Perreault, L. (2001), *Medical Informatics: Computer Applications in Health Care and Biomedicine*, 2nd edition, Springer Verlag, New York, USA.

[5] Buchanan, B. and Shortliffe, E.H. (1984), *Rule-Based Expert Systems: the Mycin Experiments of the Stanford Heuristic Programming Project*, Addison-Wesley, New York, USA.

[6] Miller, R.A., Pople, H.E., and Myers, J.D. (1982), "Internist-1, an experimental computer-based diagnostic consultant for general internal medicine," *New England Journal of Medicine*, vol. 307, pp. 468-476.

[7] Szolovits, P. and Pauker, S.G. (1978), "Categorical and probabilistic reasoning in medical diagnosis," *Artificial Intelligence*, vol. 11, pp. 115-144.

[8] Weiss, S., Kulikowski, C., and Safir, A. (1978), "Glaucoma consultation by computer," *Computers in Biology and Medicine*, vol. 8, no. 1, p. 25.

[9] Patil, R. and Senyk, O. (1988), "Compiling causal knowledge for diagnostic reasoning," in: Miller, P. (Ed.), *Selected Topics in Medical Artificial Intelligence*, Springer Verlag, New York, USA.

[10] Rennels, G., Shortliffe, E., Stockdale, F., and Miller, P. (1986), "Computational model of reasoning from the clinical literature," *Proceedings of the 10th Annual Symposium on Computer Applications in Medical Care*, Washington, DC, pp. 373-380.

[11] Bratko, I., Mozetic, I., and Lavrac, N. (1989), *KARDIO: a Study in Deep and Qualitative Knowledge for Expert Systems*, The MIT Press, Boston, MA.

[12] Pazos, A., Maojo, V., and Ezquerra, N. (1992), "Neural networks in nuclear medicine: a new approach to prognosis in myocardial infarction," *Proceedings of the Seventh World Congress in Medical Informatics (Medinfo 92)*, North Holland, Geneva, Switzerland.

[13] Gordon, C. and Christensen, J.P. (Eds.). (1995), *Health Telematics for Clinical Guidelines and Protocols*, IOS Press, Amsterdam, Netherlands.

[14] Herrero, C., Maojo, V., Crespo, J., Sanandrés, J., and Lazaro, P. (1996), "A specification language for clinical practice guidelines," *Proceedings of the IEEE Conference in Medicine and Biology*, Amsterdam.

[15] Fayyad, U., Piatetsky-Shapiro, G., and Smyth P. (1996), "From data mining to knowledge discovery: an overview," in: Fayyad, U., Piatetsky-Shapiro, G., Smyth, P., and Uthurusamy, R. (Eds.),

Advances in Knowledge Discovery and Data Mining, AAAI Press/ The MIT Press, Menlo Park, California, pp. 1-34.

[16] Han, J. and Kamber, M. (2000), *Data Mining: Concepts and Techniques*, Morgan Kaufmann Publishers.

[17] Goebel, M. and Gruenwald, L. (1999), "A survey of data mining and knowledge discovery software tools," *SIGKDD Explorations*, ACM SIGKDD.

[18] Langley, P. and Simon, H.A. (1995), "Applications of machine learning and rule induction," *Communications of the ACM*, vol. 38, no. 11, pp. 55-64.

[19] Weiss, S. and Indurkhya, N. (1998) *Predictive Data Mining. A Practical Guide*, Morgan Kaufmann, San Francisco, CA.

[20] Long, W.J., Griffith, J.L., Selker, H.P., and D'Agostino R.B. (1993), "A comparison of logistic regression to decision-tree induction in a medical domain," *Computers and Biomedical Research*, vol. 26, pp. 74-97.

[21] Lavrac, N., Keravnou, E., and Zupan, B. (1997), "Intelligent data analysis in medicine and pharmacology: an overview," *Intelligent Data Analysis in Medicine and Pharmacology*, Kluwer, pp. 1-13.

[22] Bratko, I. and Muggleton, S. (1995), "Applications of inductive logic programming," *Communications of the ACM*, vol. 38, no. 11, pp. 65-70.

[23] Lim, T.S., Loh, W.Y., and Shih Y.S. (1997), "An empirical comparison of decision trees and other classification methods," Technical Report 979, Department of Statistics, University of Wisconsin-Madison, Madison, WI.

[24] Dreiseitl, S., Ohno-Machado, L., Kittler, H., Vinterbo, S., Billhardt, H., and Binder, M. (2001), "A comparison of machine learning methods for the diagnosis of pigmented skin lesions," *Journal of Biomedical Informatics*, vol. 34, no. 1, pp. 28-36.

[25] Brodley, C.E. (1995), "Recursive automatic bias selection for classifier construction," *Machine Learning*, vol. 20, pp. 63-94.

[26] Michalski, R.S. (1983), "A theory and methodology of inductive learning," *Artificial Intelligence*, vol. 20, pp. 111-161.

[27] Pearl, J. (1988), *Probabilistic Reasoning in Intelligent Systems: Networks of Plausible Inference*, Morgan Kaufmann, San Mateo, CA.

[28] Højsgaard, S. and Thiesson, B. (1995), "BIFROST – block recursive models induced from relevant knowledge, observations and statistical techniques," *Computational Statistics and Data Analysis*, vol. 19, pp. 155-175.

[29] Ramoni, M., and Sebastiani, P. (1997), "Learning Bayesian networks from incomplete databases," *Proceedings of the Thirteen Conference on Uncertainty in Artificial Intelligence*, Morgan Kaufman, San Mateo, California, USA, pp. 401-408.

[30] Breiman, L. and Friedman, J. (1985), "Estimating optimal transformations for multiple regression and correlation (with discussion)," *Journal of the American Statistical Association*, vol. 80, pp. 580-619.

[31] Friedman, J.H. (1991), "Multivariable adaptive regression splines (with discussion)," *The Annals of Statistics*, vol. 19, pp. 1-141.

[32] Quinlan, J.R. (1992), *C4.5: Programs for Machine Learning*, Morgan Kaufmann, San Mateo, CA.

[33] Breiman, L., Friedman, J., Olshen, R., and Stone, C. (1984), *Classification and Regression Trees*, Wadsworth International Group.

[34] Michalski, R.S., Mozetic, I., Hong, J., and Lavrac, N. (1986), "The AQ15 inductive learning system: an overview and experiments," *Proceedings of IMAL 1986*, Orsay, Université de Paris-Sud.

[35] Michalski, R.S., Mozetic, I., Hong, J., and Lavrac., N. (1986), "The multi-purpose incremental learning system AQ15 and its

testing application to three medical domains," *Proceedings of the 5th national conference on Artificial Intelligence (AAAI-86)*, Philadelphia, pp. 1041-1045.

[36] Clark, P. and Boswell. R. (1991), "Rule induction with CN2: some recent improvements," Kodratoff, Y. (Ed.), *Proceedings of the European Working Session on Learning (EWSL-91)*, Porto, Portugal, Springer Verlag, pp. 151-163.

[37] Clark, P. and Niblett, T. (1988), "The CN2 induction algorithm," *Machine Learning*, vol. 3, pp. 261.

[38] Muggleton, S. and Feng, C. (1990), "Efficient induction of logic programs," *Proceedings of the 1st Conference on Algorithmic Learning Theory*, Ohmsma, Tokyo, Japan, pp. 368-381.

[39] Quinlan, J.R. (1990), "Learning logical definitions from relations," *Machine Learning*, vol. 5, pp. 239-266.

[40] Koton, P.A. (1988), *Using Experience in Learning and Problem Solving*, Ph.D. Thesis, Massachusetts Institute of Technology, Laboratory of Computer Science, MIT/LCS-TR441.

[41] Evans, D. and Patel, V. (Eds.) (1989), *Cognitive Science in Medicine. Biomedical Modeling*, MIT Press, Boston, MA.

[42] Bareiss, R. (1989), *Exemplar-Based Knowledge Acquisition: a Unified Approach to Concept Representation, Classification and Learning*, Academic Press.

[43] Aha, D.W., Kibler, D., and Albert, M.K. (1991), "Instance-based learning algorithms," *Machine Learning*, vol. 6, pp. 37-66.

[44] Holland, J. (1975), *Adaptation in Natural and Artificial Systems*, University of Michigan Press, Ann Arbor, USA.

[45] Holland, J.H., Holyoak K.J., Nisbett, R.E., and Thagard, P.R. (1987), *Induction: Processes of Inference, Learning, and Discovery*, MIT Press, Cambridge, MA.

[46] Bourgoin, M. and Smith, S. (1995), "Big data – better returns, leveraging your hidden data assets to improve ROI," in: Freedman et al. (Eds.), *Artificial Intelligence in the Capital Markets*, Probus Publishing company.

[47] Øhrn, A., Komorowski, J., Skowron, A., and Synak, P. (1998), "The design and implementation of a knowledge discovery toolkit based on rough sets: the ROSETTA system," in: Polkowski, L. and Skowron, A. (Eds.), *Rough Sets in Knowledge Discovery. Studies in Fuzziness and Soft Computing*, Physica-Verlag, vol. 18, pp. 376-399.

[48] Zadeh, L.A. (1978), "Fuzzy sets as a basis for a theory of possibility," *Fuzzy Sets and Systems*, vol. 1, pp. 3-28.

[49] Pawlak, Z. (1991), *Rough Sets: Theoretical Aspects of Reasoning about Data*, Kluwer Academic Publishers.

[50] Van Bemmel, J. and Musen, M. (1997), *Handbook of Medical Informatics*, Springer-Verlag, New York, USA.

[51] Rosenblatt, F. (1962), *Principles of Neurodynamics*, Spartan Books, New York.

[52] Minsky, M.L. and Papert, S. (1969), *Perceptrons: An Introduction to Computational Geometry*, MIT Press, Cambridge, MA.

[53] Rumelhart, D.E., Hinton, G.E., and Williams, R.J. (1986), "Learning internal representations by error propagation," in: Rumelhart, D.E. and McClelland, J.L. (Eds.), *Parallel Distributed Processing: Explorations in the Microstructure of Cognition*, vol. 1, MIT Press. Boston, USA.

[54] Broomhead, D.S. and Lowe, D. (1988), "Multivariable functional interpolation and adaptive networks," *Complex Systems*, vol. 2, pp. 321-355.

[55] Fahlman, S.E. and Lebiere, C. (1990), "The cascade-correlation learning architecture," in: Touretzky, D.S. (Ed.), *Advances in Neural Information Processing Systems 2*, pp. 524-532, Morgan Kaufman Publishers Inc., San Mateo, California.

[56] Bledsoe, W. and Browning, I. (1959), "Pattern recognition and reading by machine," *Proceedings of the Eastern Joint Computer Conference.*

[57] Aleksander, I. and Stonham, T. (1979), "Guide to pattern recognition using random-access memories," *IEE Proceedings on Computers and Digital Techniques*, vol. 2, pp. 29-40.

[58] Krishnaiah, P. and Kanal, L. (Eds.) (1982), *Classification, Pattern Recognition, and Reduction of Dimensionality*, vol. 2 of Handbook of Statistics, North Holland, Amsterdam.

[59] Kohonen, T. (1984), *Self-Organization and Associative Memory*, Springer-Verlag, Berlin.

[60] Witten, I. (1999), *Data Mining: Practical Machine Learning Tools and Techniques with Java Implementations*, Morgan Kaufmann.

[61] Billhardt, H., Maojo, V., Martín, F., Crespo, J., Pazos, A., Alamo, S., Rodríguez, J., and Sanandrés, J. (1998), "El proyecto armeda - acceso unificado a bases de datos medicas a través de la World Wide Web," *Actas del Congreso INFORSALUDNET 98*, Madrid.

[62] Wasson, J.H., Sox, H.C., Goldman. L., and Neff, R.K. (1985), "Clinical prediction rules; applications and methodologic standards," *New England Journal of Medicine*, vol. 313, pp. 793-799.

[63] Goldman, L., Cook, E.F., Brand, D.A., Lee, T.H., Rouan, G.W., Weisberg, M.C., Acampora, D., Stasiulewicz, C., Walshon, J., Terranova, G., et al. (1988), "A computer protocol to predict myocardial infarction in emergency department patients with chest pain," *The New England Journal of Medicine*, vol. 318, no. 13, pp. 797-803.

[64] Lorenzi, N. (1999), *Proceedings of the AMIA 99 Annual Symposium*, Hanley & Belfus, Inc, Philadelphia, USA.

[65] Cooper, G.F., Aliferis, C.F., Ambrosino, R., Aronis, J., Buchanan, B.G., Caruana, R., Fine, M.J., Glymour, C., Gordon, G., Hanusa, B.H., Janosky, J.E., Meek, C., Mitchell, T., Richardson, T.,

Spirtes, P. (1997), "An evaluation of machine-learning methods for predicting pneumonia mortality," *Artificial Intelligence in Medicine*, vol. 9, pp. 107-138.

[66] Tu, J.V., Weinstein, M.C., McNeil, B.J., Naylor, C.D., et al. (1998), "Predicting mortality after coronary artery bypass surgery," *Medical Decision Making*, vol. 18, pp. 229-235.

[67] Ciruelo, E., Crespo, J., Gómez, A., Maojo, V., Montes, C., and Sanandrés, J. (1997), "Predreuma: modelo de inducción constructiva en prognosis y clasificación en artritis reumatoide," *Actas del Congreso Inforsalud 97*, Madrid.

[68] Sanandrés, J., Maojo, V., Crespo, J., and Gomez, A. (2001), "A clustering-based constructive induction method and its application to rheumatoid arthritis,". in: Quaglini, S., Barahona, P., and Andreassen, S. (Eds.), *Proc. of Artificial Intelligence in Medicine.*

[69] Teach, R.L. and Shortliffe, E.H. (1981), "An analysis of physician attitudes regarding computer-based clinical computation systems," *Computers and Biomedical Research*, vol. 14, no. 6, pp. 542-558.

[70] Maojo, V. (2000), "A survey of data mining in medicine. methods and limitations," keynote speech at *ISMDA 2000*, Frankfurt.

[71] Cios, K. and Kacprzyk, J. (Eds.) (2001), *Medical Data Mining and Knowledge Discovery*, Springer Verlag, New York, USA.

Chapter 3

Internet-Based Decision Support for Evidence-Based Medicine

J. Simpson, J.K.C. Kingston, and N. Molony

The Protocol Assistant is a knowledge-based system, developed by the Department of Artificial Intelligence and AIAI at the University of Edinburgh, which advises on the treatment of parotid tumors. It was developed using a knowledge modeling technique named PRO*forma*, and implemented in HTML, using the knowledge models as the user interface. A set of rules was developed which were capable of "running" the protocol, using evidence-based reasoning to recommend decisions; however, the user is also supplied with access to the abstracts of all relevant published papers. The Protocol Assistant can thus be used either as a "wizard" which guides users through the decision making process, or as a "hypertext manual" which provides them with relevant information to make their own decisions. This dual-role capability is crucial for the acceptance of KBS in the real world.

Keywords: knowledge-based systems, Internet, clinical protocols

1 Introduction

Artificial Intelligence in Medicine was a field born during the 1970's amidst the euphoria surrounding the promises of artificial intelligence. At this time there was also an explosion in medical knowledge which was forcing health care professionals to become increasingly specialized. Because of this medicine seemed a logical field to apply the knowledge-based techniques that had been developed during the sixties for solving game playing, pattern recognition, and language understanding problems. Despite a number of early successes the dreams of the first ambitious researchers still remain little more than dreams and very few knowledge-based AI systems are actually in routine use in the medical world. This is not due to the technology failing; it has more to do with the poor integration of systems into the clinical working environment and the unwise marketing of expert systems as replacing their human counterparts.

Medical practice is, however, changing due to developments in clinical research and the recent enthusiasm for what has become known as "evidence-based" medicine. Back in 1960 randomized controlled trials were extremely rare and yet now it is accepted that practically no drugs are allowed to enter clinical practice without having been proved by a clinical trial. The swing towards basing clinical practice on the best evidence from clinical trials has been considerable and is evident in the sheer number of articles instructing clinicians on how to access, evaluate and interpret medical literature.

Yet evidence-based medicine is not without its problems. The major difficulties occur when published evidence is insufficient to use as a basis for clinical practice, when clinicians are unaware of the most recently published evidence, or (as often happens) different clinical studies produce inconsistent results. The doctor's dilemma can be summed up in the quotation below:

> "Those who have been in the profession of medicine, and especially surgery, for any length of time, know that basing every action on previously published *proof* is virtually impossible. Yet to speak against evidence-based medicine is akin to saying that the king has no clothes." [1 – italics added]

A solution is needed which finds a way of using all the evidence currently available to full effect, but also allows clinical judgment and experience to decide on the best practice when there is no clear evidence or when there is conflicting evidence. One of the most promising solutions to this problem is Protocol Assisted Care in which clinical protocols which detail the best-justified procedures for given clinical situations, are prepared by senior clinicians or public health organizations; the advantages of protocol assisted care are listed in [1] and [2], and some of these protocols are sufficiently well respected to be close to mandatory (e.g., the publications of the Scottish Inter-collegiate Guidelines Network – see [3]).

Computerized support for protocol-based care has been made available in the form of Internet-based publication [4], Internet-based libraries of protocols and abstracts of published clinical trials (see the website of the Cochrane collaboration [5]), and AI-based systems to provide decision support in following protocols (such as ONCOCIN [6] and EON [7]). However, none of these systems support clinicians in both following best practice protocols *and* in using clinical judgment; publications in text form lack any automated decision support, while ONCOCIN and EON follow protocols in a deterministic fashion without providing access to the supporting evidence for decisions.

The Protocol Assistant is a prototype knowledge-based system that has been developed by the Department of Artificial Intelligence and AIAI at the University of Edinburgh [8] to support both adherence to a protocol based on the latest evidence and the use of clinical judgment where the evidence is weak or inconsistent. It does this by representing the protocol using a simple yet expressive graphical notation; by providing a rule-based component which "runs" the protocol and asks the user for the necessary information; and by providing hypertext links from the protocol to both the abstracts and the full text of all published clinical trials relating to each decision point. The purpose of this chapter is to describe the techniques that were used to implement the system, and to describe the format of the system which was designed to be acceptable to clinical users.

2 The Protocol Assistant – Feasibility Assessment

There have been many attempts to build medical knowledge-based systems that have not resulted in systems entering routine use. Sometimes the reasons for this have been outside the control of the system developers; though on other occasions, the root causes of failure were present at the very start of the project, but were not highlighted. It is wise to start any such project with a detailed feasibility analysis to determine if there is a business case for developing such a system, how difficult it will be to construct the system, and what steps need to be taken to introduce it into routine use and to maintain its knowledge base.

The aim of this section is to present such a feasibility study, in the hope that it will serve as a model for future studies, and for identifying future opportunities for managing medical knowledge assets through the use of knowledge-based systems (KBS). The "knowledge asset" under examination is a clinical protocol for the treatment of parotid swellings; the study below considers the feasibility of building a knowledge-based system to provide decision support to clinical staff without specialist ENT training to help them use the protocol effectively.

2.1 Feasibility: Organizational Issues

The primary organizational issue for any innovation is its business case. There must be a justification for developing a KBS, whether it be in saving money through productivity gains, improving the accuracy of decision making, providing better management information, or even in 'archiving' the expertise of a senior staff member who is nearing retirement. The benefits must outweigh the associated costs of development, installation, and the ongoing cost of maintenance of the knowledge base as new knowledge appears.

For the Protocol Assistant, the heart of the business case lies in improving decision making. This would be achieved by encoding details of the studies and reports that justify each step, together with a measure of the reliability of each study. The reliability measure would be based on the "strength of evidence" as described above. The system could use this information to support more junior surgeons in choosing a protocol based on reliable studies, and (assuming the system was regularly updated) ensuring that all users of the system were made aware of new studies in the field which support or supersede old studies.

The system would also have associated benefits in providing on-the-job training. By presenting a list of cases supporting and opposing each protocol step, every user would become familiar with principles of good protocol design; also, if the system was able to provide access to the study reports, surgeons could study reports relevant to their current cases. It has been shown that presenting information to someone when they need it to solve a problem is a very effective method of teaching.

The analysis of costs and benefits is an important issue for any IT system. The financial benefits obtained by *faster* cures, *more effective cures*, or *longer lasting* cures can be estimated in terms of savings in salary, time and associated costs. If a new out-patient's appointment costs £70 without any investigations being undertaken, and a review £50, there are evidently huge savings to be made by minimizing reviews and reaching a decision at the first clinic visit. Traditionally consultants have seen new patients and ordered initial investigations, with sub-consultants performing reviews using the initial results; this is often the time when decision-making is most required. Many reviews have in the past been continued by juniors uncertain when a patient can safely be discharged; KBS protocols could guide junior surgeons in the light of results and provide well justified decisions. To halve the review appointments (around 12,000 per year in a typical department) would save £600,000; alternatively the routine new patient waiting time for an appointment could reduce from 8-10 weeks to around 2 weeks over around 4 months if referral rate remains steady. The positive effect of this on patient satisfaction would be enormous; there is also the possibility of a reduction in cases where errors are made, and the consequent exposure to claims for financial damages. The potential benefits of a system like this are therefore greater than the "bottom line" figure would suggest.

The costs of building such a system can be estimated on the basis of similar KBS projects. Building such a system to the stage of being a fully functional prototype would take roughly 6 man months at a cost of approximately £35,000[1]; presenting, revising, testing and installing the system into routine use would take roughly the same amount of time. In addition, knowledge base maintenance (adding in the details of the latest case studies to the system) might take up to one week every month. Hardware and software costs would be comparatively low;

[1] The DTI published a series of KBS case studies in 1989 that estimated 1 man year of effort from any professional as being worth £50,000. We have added 40% to allow for inflation.

perhaps £3,000 in total, with replacement every 3 years. The total estimated cost for the system is therefore £70,000, plus £13,500 per annum for knowledge base maintenance, software and hardware costs.

There are some further organizational features which should be considered before the business case for this system is declared to be sound. The system doesn't require any change to organizational responsibilities, unlike some pioneering expert systems which aimed to replace doctors rather than supporting them; the task of following surgical protocols is unlikely to be phased out in the near future; the only remaining risks relate to securing funding and preventing it from being cut before the project ends. For a relatively short IT project such as this one, the risk of funding being cut is comparatively low.

2.2 Feasibility: Technical Issues

When assessing the technical feasibility of a proposed application for KBS technology, one of the key factors to be considered is the *task type*. Tasks can be classified as analytic (analyzing an existing situation or artifact), synthetic (creating a new situation or artifact) or a combination of the two [9]. Analytic tasks (such as diagnosis, classification, assessment and monitoring) are normally easier than synthetic tasks (such as design, configuration, planning and scheduling); tasks which have both analytic and synthetic components (such as repair or control) are the hardest of all. In fact, almost 80% of KBS applications developed for commercial use perform either diagnostic or assessment tasks. It is clear that clinical protocols are used to carry out a diagnostic task, so KBS technology looks like a suitable approach.

The *form* of the knowledge is another important factor to consider. Knowledge-based systems can deal with symbolic knowledge (words, ideas or concepts) well, but are less appropriate for handling large amounts of numerical data, geometrical shapes, or perceptual information. A good rule of thumb is to ask whether the answer to the problem can be conveyed by telephone; if not, it is likely that a KBS will have difficulty in handling it. In this case, the protocols are currently described entirely in text, with few numbers or shapes, so it seems that the knowledge for this problem is indeed symbolic.

Knowledge-based systems are known to be good at representing certain types of *knowledge structures*. These include taxonomies, procedures, regulations, and heuristics ("rules of thumb"). KBS researchers have also developed techniques for representing uncertainty or assumptions within a reasoning process. For this task, a clinical protocol is a procedure, with some degree of uncertainty due to conflicts within the evidence base. Knowledge-based systems are less able to represent temporal information, spatial information, or simulation-based information, but none of these feature strongly in the Protocol Assistant; temporal information features in case histories but not in the procedure itself, spatial reasoning is limited to distinguishing "large" and "small" sizes, and there is no need for an explicit simulation of physiological processes. KBS technology may well be the most appropriate technology for solving this problem.

Other technical factors which need to be considered are whether explanations of reasoning are required (which may be useful for inexperienced users of the system), whether the knowledge is verifiable, and whether there is a clear definition of when the system is "finished". For this system, explanations are required and will be provided in the form of abstracts of clinical studies; recommendations should be verified by clinical trials; and completeness will be judged by closeness of match to existing protocols.

It's also important to consider the time required for a doctor to solve the problem. Tasks which take human experts more than an hour are generally too complex for a single KBS, whereas tasks which require less than 3 minutes are often not worth encoding because of the time required to input information into a computer. Given that a typical consultation lasts approximately 30 minutes, this task lies comfortably within these bounds.

Safety criticality is also important; applications which could be hazardous if they give wrong information (or omit necessary information) are generally poor choices for KBS, because the time required to verify that the system is totally complete, correct and consistent is significantly higher than for a KBS which merely aims to represent as much of an expert's knowledge as can be acquired in a reasonable time frame. Clearly the effective use of protocols is a safety-critical task; however, the protocols themselves represent the best available knowledge, and so completeness, correctness and consistency of the system can be judged by its conformance to the protocols.

The last technical feature to be considered is perhaps the least important from a "knowledge" viewpoint, but is often the single most important factor in determining the acceptance of the system. The user interface of the system must be designed so that the system is easy to use and yet gives adequate information. The aim of this project would be to implement a system that runs within an Internet browser, so that it can be used on an intranet or over the Internet. The user's opinions of the interface would be tested concurrently with the verification of the accuracy of the medical knowledge.

2.3 Feasibility: Project & Personnel Issues

It has often been said that without an expert, there can be no expert system. It is therefore essential that a surgeon or other medical expert is available to provide the expertise for this project. For the prototype Protocol Assistant, one protocol was considered sufficient, along with a locally available clinician (Dr. Molony) who was the author of the protocol and was available to answer questions regarding it.

In addition to the expert, it is important to secure commitment from three groups of people:

- Management: those who will be funding the system development, and who will be responsible for its introduction within an organization;
- Users: those who will be using the system;
- Developers: those who will be building it and keeping its knowledge up to date.

Management commitment can sometimes be difficult to obtain within a large organization where there are competing priorities; the problems are compounded if the system development takes a long time (more than a year). Feigenbaum et al. [10] discuss the development of a KBS into American Express which helped to decide if certain applications for credit on American Express cards should be granted; the system had to survive reductions in funding, re-prioritization by a new departmental head, accusations of over-generosity (from the finance department) and excessive strictness (from the sales department), inadequate communication from the database developers, and most significantly, resistance from middle management to a system which promised to improve decision-making accuracy but could not guarantee increased productivity. While the system proposed by this chapter is less ambitious than the million-dollar American Express system, it could still encounter some of the same problems. The solution chosen was to develop a prototype of the system, which could be used to demonstrate the potential of such a system to management, as well as providing an opportunity for users to give feedback on what they do and do not like about the system.

We have already discussed the importance of the user interface in obtaining user acceptance. The development of a prototype, fielded to a limited number of health care professionals for evaluation purposes, will give prospective users a chance to comment on all aspects of the system; its usability, its content, and its decision-making. It is also hoped that a medical evaluation will be possible, in which some patients are treated according to advice given by the system (and approved by the health care professionals), and the results are evaluated.

As for the developers of the system, it is important that they have the requisite skills in acquiring knowledge, developing the system in a methodological manner, understanding the users' needs and desires, and designing and writing programs to meet the need. For a knowledge-based system, it is equally important to consider the knowledge maintainers. Who will keep the knowledge in the system up to date? What skills will they require? Should a health care professional be trained to extend an existing program, or should an IT professional be trained in acquiring and understanding the latest medical expertise? Knowledge maintenance takes time (typically 1-3 days per month, depending on the frequency with which knowledge changes); who will fund this effort? If the system is taken beyond the prototype stage, such issues need to be considered.

3 Representing Clinical Protocols

The Protocol Assistant was developed based on knowledge from an experienced ear, nose & throat surgeon, and a terse draft of a text-based clinical protocol which he was developing. The protocol dealt with the diagnosis and treatment of parotid swellings (subcutaneous lumps which appear in the neck, below the ends of the jaw); this is an important aspect of otolaryngological work, since parotid swellings may be malignant, and so swift and accurate diagnosis is important.

3.1 Knowledge Acquisition and Modeling Using PRO*forma*

The first decision to be made when developing the Protocol Assistant was to decide how protocols would be represented in the system. Drawing on previous experience of knowledge modeling and analysis using techniques such as CommonKADS [9], [11] and IDEF3 [12], we decided to represent protocols using a node-and-arc-based knowledge modeling technique. Modeling knowledge provides an intermediate representation between an expert's knowledge and the final implemented system; if the modeling technique is good, the models should also be comprehensible to domain specialists, and can therefore be used to support further knowledge acquisition. The models resulting from such a technique resemble flow charts at first sight; however, they differ in some important respects. The models are hierarchically structured, so a single node in the top level model may be represented by one or more detailed sub-models; each node in a model represents a different type of knowledge (for example, IDEF3 differentiates activities, objects, and AND/OR junctions); and the nodes in a model may have descriptive attributes attached (for example, CommonKADS recommends that nodes which represent tasks or activities should have attributes describing their goal, their inputs and outputs, and the task specification).

The knowledge modeling technique which was chosen for this project was PRO*forma*, developed by the Imperial Cancer Research Fund [13]. The PRO*forma* language was developed specifically for the task of representing best practice guidelines. It assumes that three types of knowledge are required by a clinician when making patient management decisions. These are :

- General Medical Knowledge
- Specific Patient Knowledge
- Knowledge of Best-Practice Procedures.

Knowledge representation in PRO*forma* uses an ontology of four basic activities. These are shown in the diagram below:

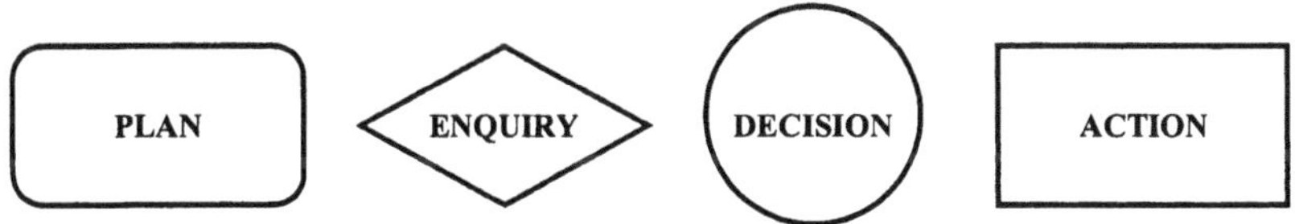

Figure 1. The ontology of activities within PRO*forma*.

These activities are defined in the following way:

- A *Plan* is a sequence of sub-tasks, or components, which need to be carried out to achieve a clinical objective, such as a therapeutic objective. Plan components are usually ordered, to reflect temporal, logical, resource or other constraints.
- An *Enquiry* is a task whose objective is to obtain an item of information which is needed in order to complete a procedure or take a decision. The specification of an enquiry includes a description of the information required (e.g. a lab result) and a method for getting it (e.g. by query on a local patient record, or a remote laboratory database).
- A *Decision* task occurs at any point in a guideline or protocol at which some sort of choice has to be made, such as a diagnostic, therapeutic or investigative choice.
- An *Action* is a procedure which is to be enacted outside the computer system, typically by clinical staff, such as the administration of an injection.

The PRO*forma* ontology specifies required attributes for each data type; these attributes proved helpful in specifying the knowledge which needed to be acquired in order to complete the model. It also permits Plans to be decomposed into lower-level models, thus allowing the total number of activities in the whole protocol (about 30 altogether) to be subdivided logically between 8 diagrams. This also simplified the representation of multiple paths to the same conclusion, since each path could be represented in a separate diagram, and then linked to another diagram representing the shared conclusion and its consequences.

Once initial knowledge acquisition had been performed, PRO*forma* diagrams were created using Hardy [14], a meta-CASE tool for creating node-and-arc diagrams of various types with additional hypertext facilities for linking between text, individual nodes, and whole diagrams. These diagrams constituted models of the clinical protocol, and the next stage of knowledge acquisition focused on verifying the accuracy of these models. The resulting models were then output by HARDY in a HTML-compatible format: the diagrams are converted into bitmaps and then into GIFs which can be displayed in a frame within a browser, the attributes of each node are stored in a *.htm* file which can be displayed in a separate frame by clicking on the node, and all the hyperlinks are preserved. This HTML representation of the models was used both for displaying the models to the clinical expert in order to facilitate further knowledge acquisition and knowledge refinement, and as a basis for the user interface of the Protocol Assistant. An example of the interface can be seen in Figure 2.

3.2 "Running" a Clinical Protocol Using JESS

Having obtained a representation of a clinical protocol in an HTML-compatible format, the next step in developing the Protocol Assistant was to provide a means of "running" the protocol. Running a protocol implies providing an automated "expert system" which would start at the beginning of the protocol, ask all (and only) the relevant questions, dynamically determine its path through the protocol based on the user's answers to previous questions, and finally reach a particular end point, thereby suggesting a diagnosis and recommending an approach to treatment. The requirement that the system should dynamically determine its next action based on previous input creates a preference for a production rule-based approach, since production rules take this approach by default. The chosen tool also needed to be able to obtain input from, and provide output to, an HTML-based user interface.

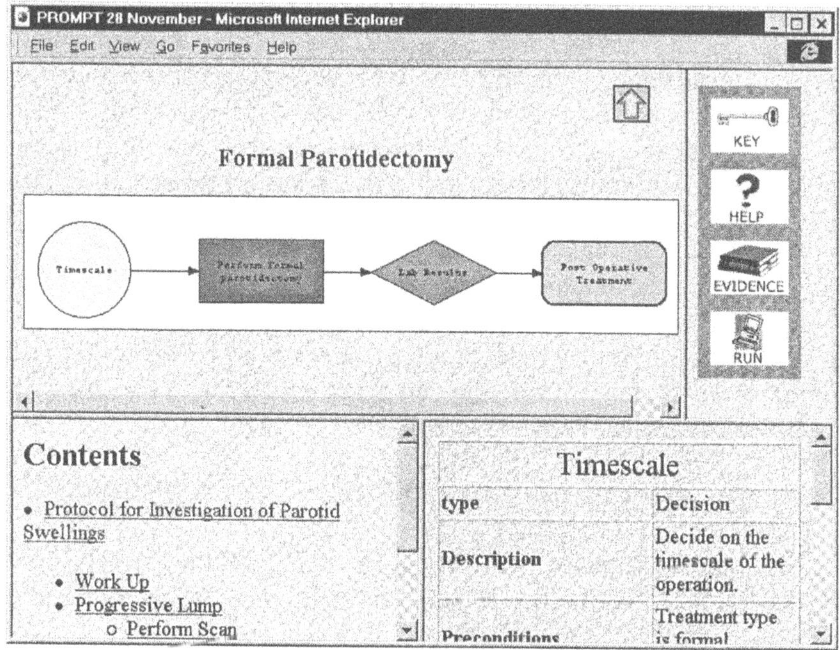

Figure 2. HTML-based representation of a clinical protocol.

The chosen tool was JESS (the Java Expert System Shell), a "clone" of CLIPS which was written in Java [15]. JESS is described as "essentially an interpreter for a rule language borrowed from CLIPS"; it therefore supports the development and execution of forward-chaining rules, which are compiled using a RETE algorithm. The major advantage of JESS for this project is that the rules can interact with Java code, thus fulfilling the requirement to be able to work with an HTML-based user interface. Details of the design and implementation of this "expert system" module are given later in the chapter.

3.3 Representing and Reasoning with Clinical Uncertainty

An innovative feature of the Protocol Assistant is its ability to represent clinical uncertainty. One of the main motivating factors for this project was to be able to represent protocols for which published clinical evidence was scarce or inconsistent – and yet hardly any other software for automating protocols supports this feature.* Uncertainty may arise at any Decision node in the protocol, and the Protocol Assistant is capable of representing evidence both for and against particular courses of action.

The assessments of clinical evidence are based on a ranking of the "goodness" of published evidence that was presented in [1]. Randomized control trials provide the "best" evidence, while unsupported opinions from respected authorities are considered the "weakest" evidence. The full ranking is as follows:

1. (a) Evidence obtained from meta-analysis of randomized control trials
   (b) Evidence obtained from at least one randomized control trial
2. (a) Evidence obtained from at least one well-designed controlled study without randomization
   (b) Evidence obtained from at least one other type of well-designed quasi-experimental study
3. Evidence obtained from well-designed non-experimental descriptive studies such as comparative studies, correlation studies and case-controlled studies
4. Evidence obtained from expert committee reports or opinions and/or clinical experience of respected authorities

* The only known package which claims to offer support for this feature is PROMPT and its more recent successor, Arezzo Composer, which are based on PRO*forma*. PROMPT was developed by the Imperial Cancer Research Fund [13]; Composer is available from Infermed Ltd. PROMPT was not used for this project for a number of reasons, including the desire for Internet-based delivery; Composer was not available when this work was carried out.

In order for the Protocol Assistant to reason about these types of evidence, a relative scoring system had to be devised: for example, do five expert committee reports outweigh a single randomized control trial? Since the "strength" of each type of evidence actually represents a level of certainty that the evidence is accurate, recognized AI methods of reasoning with measures of certainty were considered. Bayesian probability and Dempster-Shafer theory have the advantage of a strong theoretical basis, but were rejected because of the "loss of comprehensibility" which can arise when these theories are applied to real world situations (i.e. the propagated numerical certainty values can be hard for users to comprehend). MYCIN-style certainty factors provide a simple method for handling uncertainty, but they lack the theoretical weight of the other two numerical approaches. Cohen's theory of endorsements [16] is a particularly intuitive way of handling uncertainty, but it does not have any second order measure of uncertainty which makes combining evidence difficult. Fox's logic of argumentation [17] has a great deal of potential, but it is currently not easy to interpret how to use the method in a practical situation since most of the published work is theoretical.

It was decided that an approach based on endorsements would be used, utilizing some of the ideas proposed by the Logic of Argumentation. To address the weakness of endorsements, the ranking of evidence types described above has been used as a second order measure of uncertainty. All evidence relating to a particular proposition can then be combined to decide which advice the evidence indicates the system should give. The varying weights applied to different combinations of evidence are based on the results of a knowledge acquisition session with Dr. Molony; this session used a set of contrived "cases" to discover the relationships between the different quality of evidence. Once suitable weightings for each type of evidence have been established, it is then possible to combine the different strengths of evidence and recommend actions based on the results.

The "algorithm" for combining weights was simple addition and subtraction. This was used because it ought to be adequate if the weightings were calculated well; the number of items of evidence in each calculation was small (often less than 10, sometimes less than 5), reducing the utility of more complex calculations; and the project was viewed as an empirical test of a very simple approach, on the basis that it's often wise to use the simplest approach which works.

However, one of the major benefits of the Protocol Assistant for the user is that it does **not** enforce its choice of the "best" decision to take at each decision point. Instead, it suggests the best decision based on its certainty calculations, and then offers all the evidence to the users so that they can make up their own mind. By clicking on a Decision node, the users can view a list of relevant published articles (including conflicting evidence, if any), and can use the hypertext features of HTML to read the abstract of that article, or even to read the full paper. This feature allows a user to employ the Protocol Assistant either as a "wizard" which guides them through the decision making process, or as a "hypertext manual" which leads them to the information relevant to the decision they are making. Based on past experience of delivering AI systems, this dual-role capability is very important for the acceptance of KBS in the real world, since different users have very different requirements of fielded systems, and can quickly become irritated if the system does not meet their requirements.

An example in which the Protocol Assistant displays published evidence can be seen in Figure 3.

4 Design and Implementation of the Protocol Assistant

4.1 System Design

The design of the overall system was initiated by preparing a *use case diagram* (Figure 4), which describes the desired behavior of the system from the users' point of view. Use case diagrams consist of two elements; actors shown as a stick figure and use cases shown as a named oval. An actor is a "human user of the system in a particular role" or "an external system which in some way interacts with the system." A use case is defined as "a coherent work unit of the system which has value for an actor" [18].

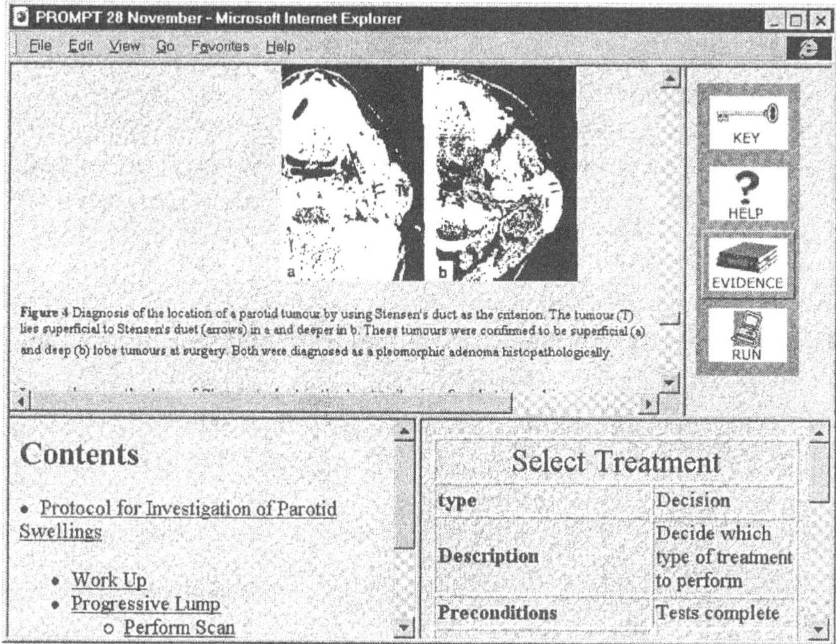

Figure 3. A published paper that gives evidence for a particular decision is presented by the Protocol Assistant.

From the use case analysis we can see that a typical doctor using the system would want to be able to enter patient data, receive advice on actions, view clinical protocols and view the evidence supporting the protocols. We can also see that the experts are responsible for defining the clinical protocols and that the World Wide Web supports a system with a user interface that can collect the evidence for a protocol. The final actor in the diagram is the patient who provides case specific details which would be entered and used to provide advice.

The main focus of this project – to generate advice on which action should be performed – is indicated by a note on the diagram. The storage of data to allow stop-start use was identified as a possible future requirement but due to complications with the legal aspects of storing patient data it was decided to leave this as a future extension.

The UML component diagram shown in Figure 5 represents the run-time dependencies of the five components of the system. Starting at the rule level the CLIPS rules are interpreted into Java using the Java

Expert System Shell (JESS). The user interface component provides an interface to the rules interpreted by JESS and allows the user to interact with the system from a Web browser. The Web browser interprets the HTML pages and is responsible for display the pages in the appropriate frames. The Web browser also interacts with the Java applet and receives requests from the applet to display particular HTML pages as the applet runs.

The HTML pages, the Java user interface and the CLIPS rules were all developed as part of this project whilst the Web browser and JESS are both externally developed packages.

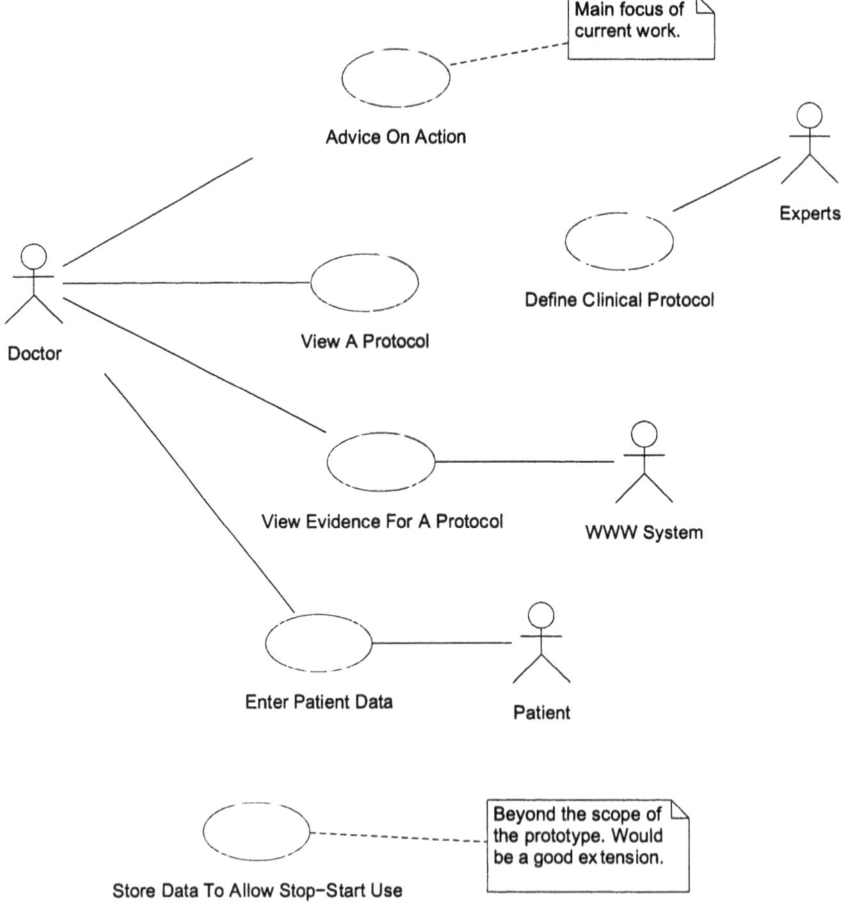

Figure 4. Use Case Analysis.

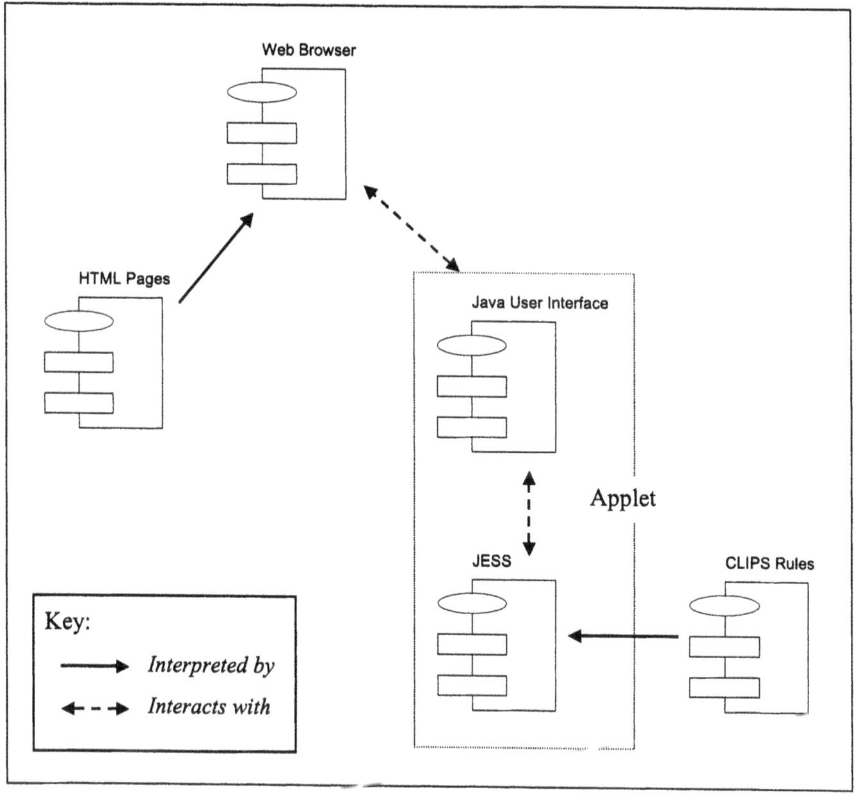

Figure 5. UML Component Diagram.

4.2 User Interface Design

One of the challenges when developing a medical expert system is getting the medical profession to accept and use the system. Because of this difficulty, considerable thought was given to the user interface so that the system would be as intuitive as possible for clinicians to use. To help maintain consistency throughout the interface a template was designed for the HTML pages, containing four frames. The top left-hand frame is used to show the protocol diagrams that were produced during the knowledge acquisition stage of the project. The top right-hand frame contains four control buttons that can be pressed at any time to take the user to a particular part of the system. The bottom left-hand frame is initially used to display a table of contents for the protocol and allow users to jump to examine any point in the protocol. Once the user has clicked on the "Run" button, the bottom left-hand frame is used as a notebook to record the information entered by the user so that they can check back over what they entered as the protocol proceeds. This recording is intended to capture the notes the clinician would normally take as a case is managed. The final frame is used as an information frame to display the online help menu and other information associated with particular steps in the protocol.

4.3 Implementation

For the prototype system, it was decided to implement an "all or nothing" approach to running the protocol; that is, the user must run the protocol from the beginning every time, rather than clicking on a node part way through the protocol and initiating the run from there. This avoided a number of problems, not least of which was deciding how to gather data that would normally have been obtained from the user earlier in the protocol. A set of CLIPS rules were therefore prepared, tested in CLIPS, and ported into JESS; a Java applet was then designed (using Java Development Kit 1.1) which could invoke JESS and also display input forms to a user when JESS requested data. The interaction between rules in JESS and the Java applet was less straightforward than expected; not only is JESS, as its web page says, "work in progress", but at the time when the Protocol Assistant was being developed, there were very few examples of how JESS could be used. The only working example that was available was the classic "monkey and bananas" program running as an applet, but this system ran from beginning to end without accepting any intermediate user input. This meant that the user interface classes and the method of linking them to JESS had to be developed from scratch without any examples to base them on.

With the assistance of an active JESS mailing list and some additional Java code, the rules were successfully linked to the user interface, so that activating certain rules not only performs reasoning, but also changes the display in the top left frame to reflect the stage of the protocol which is being carried out. The applet can be triggered by clicking on the "Run" button in the top right frame of the user interface; an example of the running system can be seen in Figure 6.

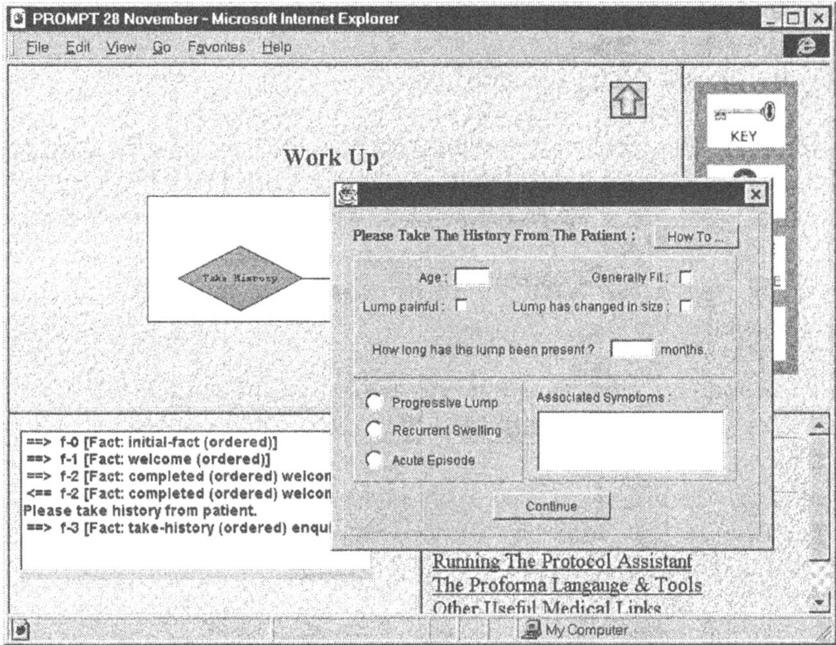

Figure 6. An input form appears during a "run" of the Protocol Assistant.

5 Evaluation and Future Work

When evaluating a system that is intended for real world use, one of the major criteria for evaluation must be how well the system meets the requirements of the users. In this section we will look at six user requirements and consider how well the Protocol Assistant meets them.

In a study of physicians' attitudes towards clinical consultation systems [19] the following six design features were identified as most important for consultation systems:

1. they should be able to explain their diagnostic and treatment decisions to physician users;
2. they should be portable and flexible so that clinicians can access them at any time and place;
3. they should display an understanding of their own medical knowledge;
4. they should improve the cost-efficiency of tests and therapies;

5. they should automatically learn new information when interacting with medical experts;
6. they should display common sense.

The first requirement is met within the Protocol Assistant by allowing the user access to the clinical protocol diagrams and the online evidence supporting the decisions recommended. By examining these, clinicians can easily find out why the system is recommending a particular decision and make an informed choice about following the recommendation.

The second user requirement of portability is also met by the design of the Protocol Assistant since all that is required to run the system is access to the World Wide Web. The system can be run using any operating system and can also easily be run locally on a portable laptop computer, as was used for demonstrating the system to the ENT department at the Royal Infirmary of Edinburgh.

The remaining requirements are more subjective and hence less easy to evaluate although it is possible to offer some opinions on how well these requirements have been met. From a purely AI perspective the system clearly has no conscious understanding of the recommendations it gives since it is simply following a path through a rule base. However, when using the system it can be observed that medics quickly become convinced that the system does have some inherent understanding of the reasons behind the advice it is giving and have been known to verbalize such beliefs by saying things like "that must be because it thinks X". So whilst in reality the system has no understanding of the advice it is giving it appears to convince some users that it does; it must therefore be displaying medical knowledge in a convincing manner.

To provide conclusive evidence that a system could improve cost efficiency would require a long-term study comparing the results before and after introducing the system. Since no such study has yet been carried out, it is difficult to justify suggestions that the system meets this criterion. However, since the system recommends the best justified practice it should reduce the number of non-essential tests that are performed and so may help to improve cost efficiency that way.

The next requirement, that the system should learn from interaction with expert users, is not met by the system. However, given that the system is based on the best knowledge supported by the current evidence, changing the knowledge base through interaction with users would seem like a bad idea. The system should of course be kept up to date with any new evidence that comes to light and by doing this it should be possible to ensure that the advice given is the best supported at any given time. A beneficial avenue of research would be to investigate the feasibility of establishing links to relevant publications on remote sites on the Internet (particularly the Cochrane Collaboration's material), thus providing a much wider collection of evidence for users to study.

The final user requirement identified is to show common sense. Again this is a fairly vague notion and is not something that was specifically addressed as part of this project. However, some of the decisions that are recommended by the system are based in some degree on common sense such as recommending against performing operations on very elderly patients because the shock of undergoing an operation may be more dangerous than the tumor.

Further evaluation was carried out by asking potential users with a range of expertise (ENT surgeons, junior doctors and medical students) to answer questions about the need for the system; the expertise level of the system; the usability of the system; the likely impact on patient management and well-being; and the cost-effectiveness of the system [20]. The results show high opinions of the usability, expertise level, and desirability of the system; cost effectiveness was more difficult to estimate. The usefulness of the dual capability as an expert system and a decision support system was not specifically asked about, but a number of favorable verbal comments were received. Perhaps the most compelling argument for the system, however, is the need for better availability and application of clinical protocols; at present,

- approx. 360,000 articles are published in medical journals each year;
- there are still very few protocols that have been published;
- protocols that have been published don't use the same way of describing procedures;

- clinical protocols are likely to change as new evidence comes to light;
- many of the advantages are nullified if an out of date protocol is used.

For these reasons, the Protocol Assistant is evaluated as supplying sufficient benefits to be worthy of further commercial design and development.

Acknowledgments

We would like to acknowledge the support of John Fox of the Imperial Cancer Research Fund in providing us with documents describing PRO*forma* and I.C.R.F.'s PROMPT tool [13], and the support of members of the Department of Otolaryngology at the Royal Infirmary of Edinburgh.

This work was carried out while Mr. Simpson was a student in the Department of Artificial Intelligence, University of Edinburgh. All correspondence regarding this work should be addressed to the second author.

References

[1] Maran, A.G., Molony, N.C., Armstrong, M.W.J., and Ah-See, K. (1996), "Is there an evidence base for the practice of ENT surgery?" *Clinical Otolaryngology*, vol. 22, pp. 152-157.

[2] Coiera, E., Baud, R., Console, L., Cruz, J., Durinck, J., Frutiger, P., Hucklenbroich, P., Rickards, A., and Spitzer, K. (1994), "The role of knowledge-based systems in clinical practice," in Barahona, P. and Christensen, J.P. (Eds.), *Knowledge and Decisions in Health Telematics – The Next Decade*, pp. 199-203.

[3] Scottish Intercollegiate Guidelines Network (1996), "Helicobacter pylori: eradication therapy," in *Scottish Intercollegiate Guidelines Network, Dyspeptic Disease*, Royal College of Physicians, Edinburgh.

[4] Detmer, W.M. and Shortliffe, E. (1997), "Using the Internet to improve knowledge diffusion in medicine," *Communications of the ACM.*

[5] The Cochrane Collaboration (1999), "Abstracts of cochrane reviews," http://www.hcn.net.au/cochrane/abstracts/intro.htm .

[6] Musen, M.A. and Johnson, P.D. (1997), "Development of a guideline authoring tool with PROTEGE II, based on the DILEMMA Generic Protocol and Guideline Model," Technical Report, Section on Medical Informatics, Stanford University School of Medicine.

[7] Shortliffe, E. (1984), "Studies to evaluate the ONCOCIN system," Stanford Heuristic Programming Project Report HPP 84-22, Stanford University.

[8] Simpson, J. (1998), "An expert system for "best practice" in medicine," Project Report, Department of Artificial Intelligence, University of Edinburgh.

[9] Schreiber, G., Akkermans, H., Anjewierden, A., de Hoog, R., Shadbolt, N., van de Velde, W., and Wielinga, B. (1999), "Knowledge engineering and management: the CommonKADS methodology," MIT Press, 2000.

[10] Feigenbaum, E., McCorduck, P., and Nii, P. (1988), "The rise of the expert company," MacMillan.

[11] Kingston, J.K.C. (1993), "Re-engineering IMPRESS and X-MATE into CommonKADS," in Bramer, M.A. and Macintosh, A.L. (Eds.), *Research and Development in Expert Systems X*, Proceedings of Expert Systems 93, St. John's College, Cambridge, 13-15 December 1993, pp. 17-42, Cambridge University Press.

[12] Kingston, J.K.C, Griffith, A., and Lydiard, T.J. (1997), "Multi-perspective modeling of air campaign planning," *Proceedings of the International Joint Conference on Artificial Intelligence (IJCAI '97),* Nagoya, Japan, 23-29 August 1997, pp. 668-677, AAAI Press.

[13] Fox, J., Johns, N., and Rahmanzadeh, A. (1997), "Protocols for medical procedures and therapies: a provisional description of the *PROforma* language and tools," *Proceedings of AIME '97*, Springer Berlin.

[14] Smart, J. (1993), "HARDY," *Airing*, AIAI, Edinburgh, April, pp. 3-7. See also http://www.aiai.ed.ac.uk/~hardy/hardy/hardy.html .

[15] Friedman-Hill, E. (1999), "JESS home page and JESS manual," http://www.herzberg.ca.sandia.gov/jess .

[16] Cohen, P.R. (1985), *Heuristic Reasoning about Uncertainty: an Artificial Intelligence Approach*, Pitmans Boston.

[17] Krause, P., Ambler, S., Elvang-Goransson, M., and Fox, J. (1995), "A logic of argumentation for reasoning under uncertainty," *Computational Intelligence*, vol. 11, p. 1.

[18] Stevens, P. (1997), *Software Engineering with Objects and Components*, Computer Science 4 Lecture Notes, University of Edinburgh.

[19] Teach, R. and Shortliffe, E. (1981), "An analysis of physician attitudes regarding computer-based clinical consultation systems," *Computers and Biomedical Research*, vol. 14, pp. 542-558.

[20] Shortliffe, E. and Davis, R. (1975), "Some considerations for the implementation of knowledge-based expert systems," *SIGART Newsletter*, vol. 55, pp. 9-12.

Chapter 4

Integrating Kernel Methods into a Knowledge-Based Approach to Evidence-Based Medicine

K. Morik, T. Joachims, M. Imhoff, P. Brockhausen, and S. Rüping

Operational protocols are a valuable means for quality control. However, developing operational protocols is a highly complex and costly task. We present an integrated approach involving both intelligent data analysis and knowledge acquisition from experts that supports the development and validation of operational protocols. The aim is to lower development cost through the use of machine learning and at the same time ensure high quality standards for the protocol through empirical validation. We demonstrate our approach of integrating expert knowledge with data driven techniques based on our effort to develop an operational protocol for the hemodynamic system.

1 Introduction

An abundance of information is generated during the process of critical care. Much of this information can now be captured and stored using clinical information systems (CIS) that have become commercially available for use in intensive care over the last years. These systems provide for a complete medical documentation at the bedside and their clinical usefulness and efficiency has been shown repeatedly [6], [7], [11]. While databases with more than 2,000 separate patient-related variables are now available for further analysis [8], the multitude of variables presented at the bedside even without a CIS precludes medical judgment by humans. A physician may be confronted with more than 200 variables in the critically ill during a typical morning round [21]. We know, however, that even an experienced physician is often not able to develop a systematic response to any problem involving more than seven variables [18]. Moreover, humans are limited in their ability to estimate the degree of relatedness between only two variables [12]. This problem is most pronounced in the evaluation of the measurable effect of a therapeutic intervention. Personal bias, experience, and a certain expectation toward the respective intervention may distort an objective judgment [4]. These arguments motivate the use of decision support systems.

Clinical decision support aims at providing health care professionals with therapy guidelines directly at the bed-side. This should enhance the quality of clinical care, since the guidelines sort out high value practices from those that have little or no value. The goal of decision support is to supply the best recommendation under all circumstances [22]. The computerized protocol of care can take into account more aspects of the patient than a physician can accommodate. It is not disturbed by circumstances or hospital constraints. It bridges the gap between low-level numerical measurements (the level of the equipment) and high-level qualitative principles (the level of medical reasoning). While knowledge-based systems have mostly been applied for diagnosis and therapy planning (e.g., [16], [25]), some systems also aim at on-line patient monitoring [5], [17], [22]. Methods that have proved their value in handling low-frequency patient data are not applicable for on-line monitoring [17]. Quantitative measurements and qualitative reasoning have to be integrated in a system that recommends interventions in real-time. The numerical measurements of the patients' vital signs have to be abstracted into qualitative terms of high abstraction. The aspect of time has to be handled both at the level of measurements and the level of expert knowledge [3], [14], [17], [25]. In the expert's reasoning, time becomes the relation between time intervals, abstracting from the exact duration of, e.g., an increasing heart rate, and focusing on tendencies of other parameters (e.g., cardiac output) within overlapping time intervals.

One of the big obstacles to the more frequent implementation of decision support systems is the tedious and time-consuming task of developing the knowledge base. The decision support system for respiratory care at the LDS Hospital, Salt Lake City, USA [22], for instance, has been developed in about 25 person years. The method of guideline development itself is not supported by a computer system.

Mechanisms of temporal abstraction and reasoning presuppose manually designed models or ontologies [3], [17], [25]. Why not use techniques of knowledge discovery and statistical time series analysis in order to ease the process of guideline generation? Machine learning and statistical analysis have been applied in building-up diagnostical systems successfully (e.g., [15]).

We now want to exploit the huge amount of data for the development of guidelines for on-line monitoring. Our task is to build a decision support system for on-line hemodynamic monitoring in the critically ill. We do not aim at modeling the actual physician's behavior. Imitating the actual interventions made by physicians is not the goal. Actual behavior is influenced by the overall hospital situation, e.g., how long is the physician on duty, how many patients require attention at the same time. Machine learning from patients' data could lead to a knowledge base that mirrors such disturbing effects. Therefore, the learned decision rules have to be checked by additional rules about effects of drug and fluid administration. Our approach is to combine statistics, knowledge acquisition, and machine learning. Our aim is to develop a method for guideline generation that is faster and more reliable than current methods.

Data for statistical evaluation and learning can be provided by the CIS. However, the nature of the data is different from that gathered in controlled experiments. While a CIS in modern intensive care can take numerous measurements every minute, the values of some vital signs are sometimes recorded only once every hour. Other vital signs are recorded only for a subset of the patients. Hence, the overall high dimensional data space is sparsely populated. Moreover, the average time difference between intervention as charted and estimated hemodynamic effect can show a wide variation [10]. Even the automatic measurements can be noisy due to manipulation of measurement equipment, flushing of pressure transducers, or technical artifacts. In some cases, relevant demographic and diagnostic parameters may even not be recorded at all. In summary, we have a large amount of high dimensional, numerical time series data that contains missing values and noise. Using this data already at the stage of development of the decision support system stave off surprises at the stage of clinical experience as has been reported in [17, p. 572]: "The huge number of measurements classified as invalid is quite astonishing although it reflects the real clinical environments."

In addition to problems of knowledge acquisition, we see a particular need for knowledge validation. It should be noted that many medical guidelines published today are neither evidence-based nor sufficiently validated against real patient data. The current procedure is to first develop the guideline, then represent it in a knowledge-based system, and finally to test it in clinical studies. In this "waterfall" process, unrealistic assumptions, mistakes, and flaws are recognized at a late stage. In contrast, our approach includes validation from the very beginning. Using a knowledge-based system early on supports the validation of the knowledge base at earlier stages. Inconsistencies within the knowledge base as well as a mismatch of rules and patient data are detected while developing the knowledge base. A mismatch may indicate that the model underlying the knowledge base is insufficient. Hence, applying the model to patient data helps to find errors in its design. A mismatch may also indicate a difference in the medical practices of the physician at the bed-side and the medical expert that helped to develop the knowledge base. Moreover, experts from different schools or countries can vary quite a bit in their behavior and knowledge. Matching the formally modeled guidelines with patient data facilitates and focuses the knowledge-acquisition process.

In order to test our approach to using real clinical data for building and validating a knowledge base for on-line monitoring, we have constructed a system. Its overall architecture is shown in Figure 1. The patients' measurements are used to recommend an intervention and are abstracted with respect to their course over time. The recommendation of interventions constitutes a model of physician behavior. This asks for further validation. Therefore, a recommended intervention is checked by calculating its expected effects on the basis of medical knowledge. In this way, a qualitative assessment of a statistical prediction enhances the model of physician behavior in order to obtain a model of best practice. The medical knowledge constitutes a model of the patients' hemodynamic system. This model is validated with respect to past patients' data. In detail, the processes we have designed are:

Data abstraction: Given series of measurements of one vital sign of the patient, detect and possibly eliminate outliers and find level changes by good statistical practice. This abstracts the measurements to qualitative propositions with respect to a time interval, e.g., within time point 12 and time point 63, the heart rate remained about equal, from time point 63 to time point 69 it was increasing. We used the statistical time series techniques of ARMA modeling and phase space embedding [1], [2], [9]

Data-driven acquisition of state-action rules: Given the numerical data describing signs of the patient, his or her current medication, find the appropriate intervention. An intervention is formalized as increasing, decreasing or not changing the dose of a drug. The decision is made every minute. These rules were learned by the Support Vector Machine [26].

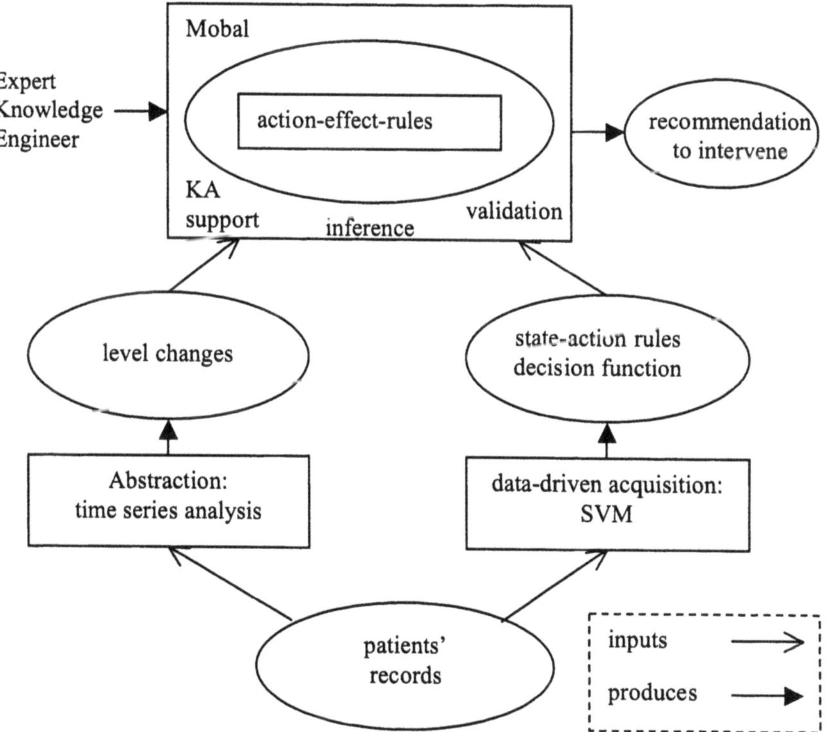

Figure 1. Overall system architecture.

Acquisition of medical knowledge: Given text book knowledge and explanations by an expert, represent the effects of substances in different dosages, relations between vital signs, and interrelations between different substances, and validate the knowledge on the basis of past patients' data. The knowledge acquisition and validation was supported by the MOBAL system [20].

validation of recommended interventions: Given

- the state of a patient described in qualitative terms,
- medical knowledge
- a sequence of interventions, and
- a current intervention,

find the effects of the current intervention on the patient. The derivation of effects is made for each intervention as forward inference within MOBAL. The effect should result in a stable state of the patient.

The outline of this chapter is as follows. Throughout the chapter we report on the continuous development of a decision support system for intensive care as performed at the city hospital and the university of Dortmund. We start with a description of the data acquisition process at the hospital and the resulting data set [11]. Section 3 shows, how we applied the support vector machine (SVM) to learn state-action rules. A short introduction to the MOBAL system [20] and its representation of medical knowledge leads to the issue of validation which is presented in Section 4.

2 Data Acquisition and Data Set

2.1 Data Acquisition

Most variables are entered by hand at the bedside. For entities such as clinical observations, nursing procedures, therapeutic measures, medications, or orders it appears very unlikely that entry of these variables can be automated in the foreseeable future. Only 5-10% of all variables in a CIS are acquired automatically. This includes the majority of bedside devices, e.g., physiologic monitors, ventilators, infusion devices. Additional data is interfaced from the hospital information system (HIS), the laboratory (LIS) or the microbiology information systems, where the LIS represents the clinically most relevant set of data among these centralized information systems. Although device data account for a comparatively small number of variables, they can, depending on the sampling rate, generate large amounts of data.

The data structure of a CIS shows a wide variety of different data types on different scales (nominal scales, e.g., sex, breathing sounds; ordinal scales, e.g., neurological scoring; absolute scales, e.g., vital signs), which are stored at different time intervals (ranging from seconds for vital signs to once during the length of stay for demographic data). Time intervals may also be regular or irregular.

For further analysis data must be structured, so that it can be subjected to statistical algorithms. Numeric data, e.g., vital signs, intake/output, is typically directly accessible for most applications. Free-text data, which traditionally makes up a large portion of medical documentation, cannot be statistically analyzed in any structured way. Therefore, free-text entries into a CIS should be avoided wherever possible. Qualitative information, such as clinical observations or interventions, should be documented in a strictly structured fashion with selection lists and menu items. This approach provides a consistent terminology throughout the entire medical institution. It is highly efficient and fast, especially for users not well trained in the use of computers and keyboards in particular. In clinical practice, with the stringent implementation of structured tabular documentation, it was possible to reduce the use of free-text notes by more than 90%. Structured qualitative data can, in contrast to free-text information, be directly exported for statistical analysis.

These general propositions also hold for the city hospital of Dortmund, a 1,900-bed tertiary referral center. There, all medication data of the 16-bed surgical intensive care unit was charted with a CIS, allowing the user one minute time resolution for all data. Moreover, data from bedside devices, e.g., patient monitors, is gathered automatically every minute.

2.2 Data Set

The entire database of intensive care patient records at the city hospital of Dortmund comprises about 2,000 different variables (attributes). Data from the CIS is selected through customizable data filters and copied into a standard relational database where it is accessible for further data analysis.

For this investigation, data was acquired from 148 consecutive critically ill patients (53 female, 95 male, mean age 64.1 years), who had pulmonary artery catheters for extended hemodynamic monitoring. Recording in one minute intervals, this amounts to 679,817 sets of observations.

From the original database 118 attributes in 9 groups were taken for learning state-action rules (Table 1).

Table 1. Overall attribute set for learning state-effect rules.

16 demographic attributes	5 intensive care diagnoses	6 continuously infused drugs
11 vital signs	9 derived parameters	14 respiratory variables
37 intake/output variables	10 bolus drugs	10 laboratory tests

Categorical attributes are broken down into a number of binary attributes, each taking the values {0,1}. Real valued parameters are either scaled so that all measurements lie in the interval [0,1], or they are normalized by empirical mean and variance:

$$norm(X) = (X - means(X)) / \sqrt{\mathrm{var}(X)} \quad (1)$$

We systematically evaluated a large number of plausible attribute sets using a train/test scheme on the learning task described in Section 3.2. The set with the best performance is given in Table 2. These attributes are actually the most important parameters of the patient according to expert judgment. Only the relevant attributes "Cardiac Output" and "Net Intake/Output" are missing, but they cannot be used as they are not continuously available.

We also experimented with different ways of incorporating the history of the patient. We tried:

Table 2. Best feature set for learning state-action rules using SVM.

Vital signs (measured every minute)	Continuously given drugs (changes charted at 1-min-resolution)	Demographic attributes (charted once at admission)
Diastolic Arterial Pressure	Dobutamine	Broca-Index
Systolic Arterial Pressure	Adrenaline	Age
Mean Arterial Pressure	Glyceroltrinitrate	Body Surface Area
Heart Rate	Noradrenaline	Emergency Surgery y/n
Central Venous Pressure	Dopamine	
Diastolic Pulmonary Pressure	Nifedipine	
Systolic Pulmonary Pressure		
Mean Pulmonary Pressure		

- using only the last minute before the intervention
- using the last up to 10 minutes before the intervention
- using the averages of up to 60 minutes before the intervention
- combinations of these
- the state of the patient at the previous intervention

None of the more complex approaches gave significantly better results on the learning task in Section 3.2 than just using the measurements from one minute before the intervention. All the feature selection experiments were done on the training set, leaving a separate test set to measure the results presented in this chapter.

Since each patient record covers several interventions, data from 148 patients gives us sufficiently large sets of examples. For learning state-action rules, we used a total of 1319 training and 473 test examples. For the rule validation we analyzed 8200 interventions corresponding to 27400 intervention-effect pairs.

2.3 Statistical Preprocessing

Given series of measurements of one vital sign of the patient, the goal of statistical data abstraction is to detect and possibly eliminate outliers and find level changes by good statistical practice. This abstracts the measurements to qualitative propositions with respect to a time interval, e.g., within time point 12 and time point 63, the heart rate remained about equal, from time point 63 to time point 69 it was increasing. We used an approach based on statistical time series analysis. Classical

ARMA (autoregressive moving average) modeling [2] is applied with corresponding outlier- and level shift detection procedures using the new tool of a phase space embedding [1], [9].

3 Data-Driven Acquisition of State-Action Rules

3.1 Support Vector Machine

Support vector machines (SVMs) [26] represent a method to learn either binary classifiers or function approximators from examples. For a set of training examples they find the classification rule for which they can guarantee the lowest error rate on new observations. Each example consists of a vector (describing, e.g., the state of a patient represented by the current measurements of blood pressures, heart rate, etc.) and its label (classification or functional value).

In their basic form, SVMs learn linear decision rules $h(\vec{o}) = sign(\vec{w} \cdot \vec{o} + b)$. The weight vector $\vec{w}$ and the threshold b are the result of learning and describe a hyperplane. Observations are classified according to which side of the hyperplane they are located. A typical decision rule is given in Figure 2. During training, the SVM calculates the hyperplane so that it classifies most training examples correctly while keeping a large "margin" around the hyperplane. If the training data can be separated without error, the margin is the distance from the hyperplane to the closest training examples.

Since we will be dealing with very unbalanced numbers of positive and negative examples in the following, we introduce cost factors to be able to adjust the cost of false positives vs. false negatives. Training an SVM can now be translated into the following optimization problem:

$$\text{Minimize:} \quad J(\vec{w}, b, \vec{\xi}) = \tfrac{1}{2}\vec{w} \cdot \vec{w} + C_{+} \sum_{i:y_i=1} \xi_i + C_{-} \sum_{j:y_j=-1} \xi_j \tag{2}$$

$$\text{subject to:} \quad \forall t \in \{1 \ldots n\}: y_t[\vec{w} \cdot \vec{o}_t + b] \geq 1 - \xi_t \wedge \xi_t \geq 0 \tag{3}$$

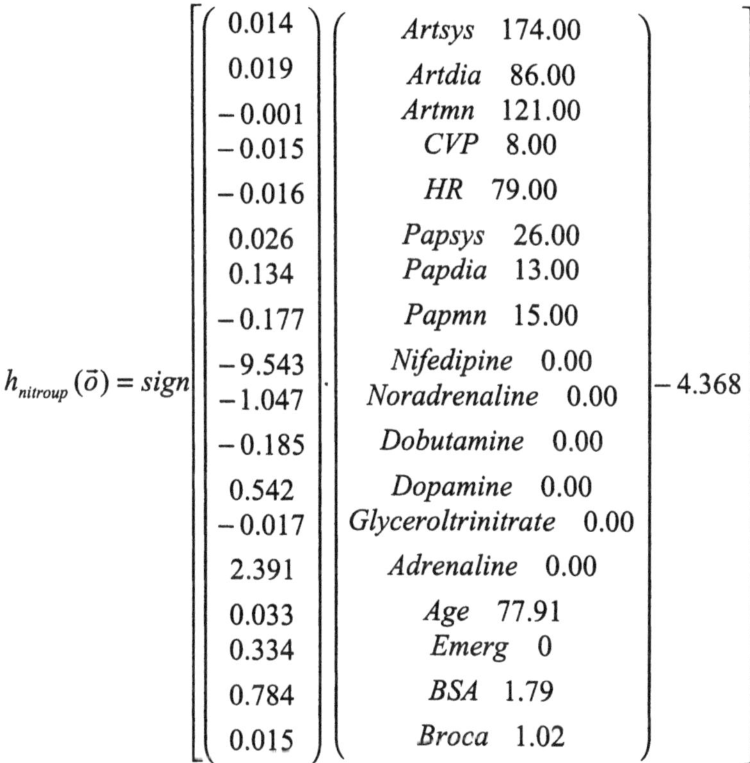

Figure 2. Decision rule and an instantiation for predicting an intervention that increases the dosage of Glyceroltrinitrate.

Training error is represented by the variables ξ_i, ξ_j, while the margin is measured by $\vec{w} \cdot \vec{w}$. We solve this optimization problem in its dual formulation using SVMlight [13], extended to handle unsymmetrical cost-factors.

3.2 Learning the Directions of Interventions

The first question we asked ourselves was: Given that we know the physician changed the dosage of some drug, can we learn when he increased the dosage and when he decreased the dosage based on the state of the patient? For each drug, examples are taken from the points in time where, in fact, the dosage changed. For all drugs, linear SVMs are trained on the problem "increase of dosage" ($y_t = 1$) vs. "decrease of dosage" ($y_t = -1$) using the attributes in Table 2 for describing the state of the patient. The performance of the respective SVM on a previously untouched test set is given in Table 3.

Table 3. Accuracy in predicting the right direction of an intervention.

Drug	Accuracy	StdErr
Dobutamine	83.6%	2.6%
Adrenaline	81.3%	3.7%
Glyceroltrinitrate	85.5%	3.0%
Noradrenaline	86.0%	5.2%
Dopamine	84.0%	7.3%
Nifedipine	86.8%	7.0%

To get an impression about how good these prediction accuracies are, we conducted an experiment with a physician. On a subset of 41 test examples we asked an expert to do the same task as the SVM for Dobutamine, given the same information about the state of the patient. In a blind test the physician predicted the same direction of dosage change as actually performed in 32 out of the 41 cases. On the same examples the SVM predicted the same direction of dosage change as actually performed in 34 cases, resulting in an essentially equivalent accuracy.

3.3 Learning When to Intervene

The previous experiment shows that SVMs can learn in how far drugs should be changed given the state the patient is in. In reality, the physician also has to decide when to intervene or just keep a dosage constant. This leads to the following three class learning problem. Given the state of the patient, should the dosage of a drug be increased, decreased or kept constant? Generating examples for this task from the data is difficult. The particular minute a dosage is changed depends to a large extend on external conditions (e.g., an emergency involving a different patient). So interventions can be delayed and the optimal minute an intervention should be performed is unknown. To make sure that we generate examples only when a physician was closely monitoring the patient, we consider only those minutes where some drug was changed. This leads to 1319 training and 473 test examples.

For each drug we trained two binary SVMs. One is trained on the problem "increase dosage" vs. "do not increase dosage (i.e., lower or keep dosage equal)", the other one is trained on the problem "lower dosage" vs. "do not lower dosage (i.e., increase or keep dosage equal)". An intervention is predicted if exactly one such decision rule recommends a change. As an example, Figure 2 shows the decision rule that the SVM learned for increasing the dosage of Glyceroltrinitrate. Since the class distribution is very skewed towards the "do not ... dosage" class, we use a cost model. The cost-factors are chosen so that the potential total cost of the false positives equals the potential total cost of the false negatives. This means that the parameters of the SVM are chosen to conform to the ratio

$$\frac{C_+}{C_-} = \frac{\textit{number of negative training examples}}{\textit{number of positive training examples}} \quad (4)$$

Table 4 shows the test results for Dobutamine and Adrenaline. The confusion matrices give insight into the class distributions and the type of errors that occur. The diagonal contains the test cases, where the prediction of the SVM was the same as the actual intervention of the physician. This accounts for 63% of the test cases for Dobutamine and for 79% of the test cases for Adrenaline. The SVM suggests the opposite intervention in about 1.5% for both drugs.

Table 4. Confusion matrix for predicting time and direction of Dobutamine and Adrenaline interventions.

	actual intervention		
Dobutamine	up	equal	down
predicted up	46	32	3
predicted equal	50	197	54
predicted down	5	30	56
	actual intervention		
Adrenaline	up	equal	down
predicted up	23	22	3
predicted equal	21	310	15
predicted down	4	34	41

Again, we would like to put these numbers into relation to the performance of an expert when given the same information. For a subsample of 95 examples from the test set, we asked a physician to perform the same task as the SVM. The results for Dobutamine and Adrenaline are given in Table 5. The results of the SVM on this subsample are followed by the performance of the human expert in brackets. Both are aligned remarkably well. Again, the learned functions of the SVM are comparable in terms of accuracy with a human expert. This also holds for the other drugs.

Table 5. Confusion matrix for predicting time and direction of Dobutamine and Adrenaline interventions in comparison to human performance (results from an experienced intensivist in brackets).

	actual intervention		
Dobutamine	up	equal	down
predicted up	10 (9)	12 (8)	0 (1)
predicted equal	7 (9)	35 (31)	9 (9)
predicted down	2 (1)	7 (15)	13 (12)
	actual intervention		
Adrenaline	up	equal	down
predicted up	4 (2)	3 (1)	0 (0)
predicted equal	4 (6)	65 (66)	2 (2)
predicted down	1 (1)	8 (9)	8 (8)

3.4 SVM Rules in Evidence Based Medicine

To use the SVM decision functions in a bigger learning environment the binary decisions of the SVM often do not offer enough information to decide for the appropriate action. For example a decision to increase a drug may have been triggered by random effects in the data or different decision rules may advise two or more contradicting actions. Hence, a measure of evidence of the SVM decisions would be very useful. The numerical value of the SVM function $f(\vec{o}) = \vec{w} \cdot \vec{o} + b$ (remember that the SVM decision function is given by $h(\vec{o}) = sign(f(\vec{o}))$) can be used as such a measure [23].

As an example, Figure 3 shows the actual dosage of adrenaline of a patient over a period of 110 minutes (upper line) and compares this to the output of the SVM that was trained to the task of classifying whether or not to increase the dose of adrenaline (lower line). It can be seen that the SVM did recommend to increase the dosage for some time before the intervention took place, but the evidence to intervene rapidly increases a few minutes before the actual intervention. Shortly after the intervention the recommendation of the SVM quickly changes to "do not increase the dosage."

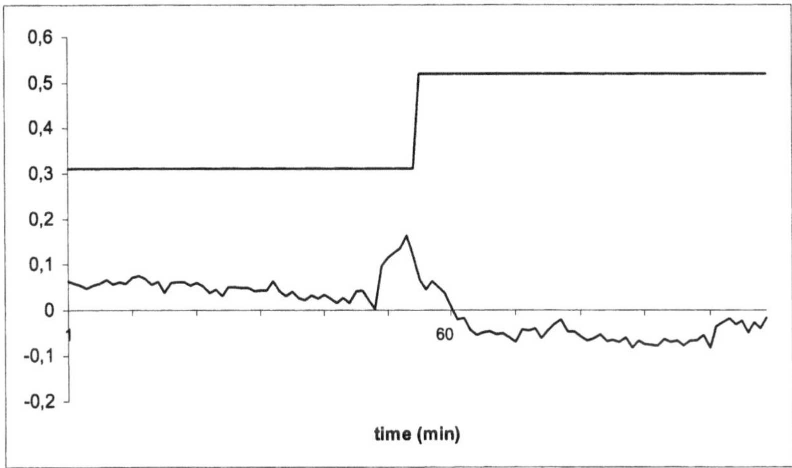

Figure 3. Actual dose of adrenaline (upper line) and evidence of SVM for "increase dosage" (lower line).

From the viewpoint of quality control in medicine, the question whether the intervention should have been taken some time earlier, as the output of the SVM indicates, deserves further investigation. This might be an example of a situation where a more sophisticated alarm system would have alerted the intensivist on duty much earlier.

3.5 More Learning Tasks

Let us now reason about the appropriate learning tasks for our goals. One may ask whether learning the appropriate direction of an intervention is justified at all, or whether the real task is to find the optimal dosage of a drug. In other words: should medical interventions be modeled as a classification or a regression problem? This is how we try to answer the question: For every drug, medical reasoning gives a value δ which a dosage change has to exceed in order to be considered to have a significant effect. We found that for all drugs at least 84% of the changes (96% at the average) lie within the range of $\pm\delta$. This justifies our approach. A higher dosage change can be realized by re-evaluating the decision to increase / decrease a drug a few minutes after the intervention.

Another interesting learning task would be to predict a trend in the vital signs of a patient. Discovering life-threatening situations as early as possible is a major key for optimal medical treatment. Moreover, this would also bring important advantages from the viewpoint of quality control and knowledge revision: In the validation of the effects of medical interventions, both of human experts and computer systems, one often finds cases where the expected effect of an intervention cannot be found in the data (e.g., medical knowledge says that the application of a drug will increase the blood pressure but the blood pressure stays stable). Confronted with this contradiction, a frequent explanation of experts is that the intervention anticipated an imminent change of the patients state into the opposite direction. As it is impossible to do a controlled experiment where the reaction of the patient with and without the intervention can be compared, the prediction of the patient's state based on examples of time periods without an intervention could offer a possibility to validate the success of an intervention.

Unfortunately, our experiments to predict vital signs of a patient in the nearer future (5 to 30 minutes) failed. For each vital sign, a regression version of the SVM [26] learned how much the parameter would increase or decrease. The learning results failed to predict these changes with more than default accuracy. As we tried many different representations of the patient's state (with and without history, learning an individual predictor per patient vs. learning on all patients, using Fourier transforms of the measurements of vital signs), we feel that this learning task is ill-posed. At the level of numerical measurements, i.e., disregarding the qualitative knowledge about physiological processes, the prediction cannot become more precise. Hence, we combine data-driven numerical methods with a knowledge-based approach (see Section 4).

Another learning task could aim at characterizing a stable state by the observed measurements. Instead of judging the patients state in term of necessary interventions, a learning algorithm could find a description of regions in the high-dimensional attribute space that can be considered safe. When the patients state leaves the safe regions, an alarm can be generated. There exists an extension of the SVM algorithm to estimate the support of high-dimensional data [24] that seems to be promising for this learning tasks.

4 Medical Knowledge Base

Decision rules learned by the SVM reflect the average behavior of a physician, not the "gold standard." As argued above, they have to be checked against medical knowledge about the effects of drugs. This section presents an approach to building a knowledge base that helps accomplish this task automatically and that makes decision support transparent.

Knowledge acquisition from experts is performed according to the current state of the art: first, knowledge is elicited from the expert, second, a knowledge base is modeled, third, the model is inspected, validated, and enhanced in collaboration with the expert. These steps form a cycle, i.e., the third step actually leads to obtain more expert knowledge, which is then modeled, etc. [19]. This expert knowledge augments and validates the data-driven knowledge acquisition using machine learning.

4.1 Knowledge Acquisition and Representation

The knowledge base of action-effect rules serves three purposes. First, it is used in order to model a protocol of care. Second, it is used to base learned decision functions on explicit and qualitative knowledge. Third, it is used for the validation of predictions. Let us describe the knowledge acquisition from experts before we show how this knowledge is integrated with the learned decision functions (Section 4.3) and how it is used for validating predictions (Section 5).

A medical expert defined the necessary knowledge. This knowledge is medical textbook knowledge for the cardiovascular system. It reflects direct pharmacological effects of a selected list of medical interventions on the basic hemodynamic variables. Any interaction of these interventions with other organ systems or of other organ systems with the cardiovascular system were ignored. An excerpt of intervention-effect relations is shown in Table 6. The dosage intervals indicated for each drug are not shown in the table, but modeled in the knowledge base. Also parameter dependencies have been modeled. It should be noted that the knowledge is qualitative with intervals of dosages, trends of changes, and implicit time intervals.

Table 6. Medical Knowledge base for hemodynamic effects: + = increase of the respective variable or intervention; - = decrease; 0 = no change.

Intervention		Effect on hemodynamic variable				
		Heart Rate	Mean Arterial Pressure	Mean Pulmonary Artery Pressure	Central Venous Pressure	Cardiac Output
Dobutamine	+	+	+	+	o	+
	-	-	-	-	o	-
Adrenaline	+	+	+	+	o	+
	-	-	-	-	o	-
Noradrenaline	+	-	+	+	o	-
	-	+	-	-	o	+
Nitroglycerin	+	+	-	-	-	+
	-	-	+	+	+	-
Fluid intake /	+	-	+	+	+	+
output	-	+	-	-	-	-

For the representation of qualitative medical knowledge we chose the MOBAL system [20]. MOBAL is a knowledge acquisition and maintenance system. Several tools facilitate the construction and inspection of a knowledge base. Its representation formalism is a restricted many-sorted first-order logic with explicit negation. A four-valued logic is used in order to allow for unknown and contradictory facts in addition to true and false facts. The inference engine derives new facts on the basis of rules and given facts. Due to the expressive power of first-order logic, compact models can be built. What would be a rule in propositional logic, can be expressed by a mere fact in first-order logic. For instance, using a propositional logic, explicitly stating that up is the opposite of down requires the rule

heart_rate_trend=up --> not (heart_rate_trend=down)

and its dual form for all parameters. Using first-order logic, the fact

opposite(up, down)

is stated and can be used for any parameter. The pharmacological knowledge from Table 6 is expressed by facts of the form

effect(adrenaline, 0.01, 0.03, art, up)

stating that Adrenaline in a dosage between 0.01 and 0.03 mg/kg/min has the effect up on mean arterial pressure. Effects are modeled for substances. Additional facts indicate the particular drugs in which the substance is contained.

Patient records are also expressed by facts. The time is indicated by minutes, starting with the first measurement of a patient and ending with his or her discharge from intensive care.

intervention(pat4711, 10, 62, supra,0.02)

means that the patient 4711 from the tenth minute to minute 62 received Suprarenin (a drug containing Adrenaline) in a dosage of 0.02 mg/kg/min. Given the abstractions described in Section 2, the values of hemodynamic parameters are stated in terms of level changes.

level(pat4711, 11, 62, hr, up)

states that the heart rate of patient 4711 had an upward level change at minute 11 and then remained almost stable until minute 62. In addition to this abstract description of a vital sign in a time interval, its deviation from the stable state is calculated. For each vital sign, the desired range of values is given, e.g., [60, 100] for the heart rate. For a patient's parameter values within a time interval, the standard deviation is calculated and added to (subtracted from) the upper (lower) value of the desired range. If the patient's actual value does not lie within this enlarged interval, a fact stating a deviation is entered. For instance, the following fact states that arterial mean pressure of patient 4999 is beyond the desired range:

deviation(pat4999, 0, 31, art, up)

We now want to use the pharmacological knowledge for deriving expected effects of an intervention on a particular patient. This is done by rules. The advantage of first-order logic is particularly important for modeling relations between intervals. For instance, stating that two time intervals are immediately succeeding, can be expressed by simply unifying the end point of one time interval with the start point of the other time interval. The following statement states, for instance, that two interventions were directly succeeding each other:

intervention(Patient, T1, T2, M, D1)
intervention(Patient, T2, T3, M, D2)

This statement can be instantiated by all patients, points in time, parameters and dosages as long as the same argument variable (e.g., Patient) is instantiated by the same value (e.g., pat4711). Different argument variables (e.g., D1, D2) can be instantiated by different values.

```
intervention(pat4711, 73, 83, supra,0.05)
intervention(pat4711, 83, 177, supra, 0.02)
```

Intervals of dosages are handled in a similar manner. We can distinguish between major and minor changes of a dosage. A minor change is one within the same interval for which an effect has been stated by pharmacological facts. The rule and an actual instantiation is the following:

```
intervention(Patient, T1, T2, M,D1),
intervention(Patient, T2, T3, M,D2),
contains(M, S),
effect(S, FromD1, ToD1, Param, Trend),
FromD1=< D1 <ToD1, FromD1=< D2 <ToD1
-->
interv_effect(Patient,T2,T3,M,Param,Trend,minor)

intervention(pat4711,441,968,nitro,1.9),
intervention(pat4711,968,1081,nitro,2.38),
contains(nitro, glyceroltrinitrat),
effect(glyceroltrinitrat, 1, 10, hr, up),
1 =< 1.9 < 10, 1 =< 2.38 < 10
-->
interv_effect(pat4711,968,1081,nitro,hr,up,minor)
```

Changing into another such interval is a major change. The actual dosage of a drug given to a patient is compared with the dosage interval of effect facts. The following rule expresses the enforcement of an effect because of a major change of dosage.

```
intervention(Patient, T1, T2, M,D1),
intervention(Patient, T2, T3, M,D2),
contains(M, S),
effect(S, FromD1, ToD1, Param, Trend),
effect(S, FromD2, ToD2, Param, Trend),
FromD1 =< D1 < ToD1, FromD2 =< D2 < ToD2,
```

ToD1 < FromD2
-->
interv_effect(Patient, T2,T3, M, Param, Trend, major)

Note, that if the substance S of drug M has a decreasing effect on a parameter of the patient, the rule predicts a further decrease of that vital sign. The variable Trend is then instantiated by down. Another rule states that decreasing a substance with an increasing effect on a parameter will decrease the parameter's value. We use such rules in order to predict effects of interventions. The prediction of intervention effects is used to check interventions that are proposed by the learned decision rules. Not counting the patient records, the knowledge base consists of 39 rules and 88 facts.

4.2 Validating Action-Effect Rules

In order to validate the knowledge base we applied it to the data of 148 patients. The data contain 8,200 interventions. The validation is easy, since rules can directly be applied to patient data. MOBAL's inference engine derived 27,400 effects of the interventions using forward chaining. For 22,599 effects the actual effects in terms of level changes could be computed by the time series analysis (see Section 2). When matching the derived effects with the actual ones, the system detected:

- 13,364 effects (i.e., 59.14%) took place in the restricted sense, that the patient's state remained stable. E.g., a drug with an increasing effect on a patient's vital sign does not lead to a significant level change of this parameter. This is not in conflict with medical knowledge, but shows best therapeutical practice. Smooth medication keeps the patient's state stable and does not lead to oscillating reactions of the patient.
- 5,165 effects (i.e., 22.85%) took place in the sense, that increasing or decreasing effects of drugs on vital signs match corresponding level changes.
- 4,070 contradictions (i.e., 18.01%) were detected. The observed level change of a vital sign went into the opposite direction of the knowledge-based prediction.

The ratio of 83.56 percent correct predictions of effects is quite positive. Some decisive features are not present in the data. Particularly the lack of data about cardiac arrhythmias and cardiac output could possibly explain many deviations of observed from predicted effects.

4.3 Integrating Learned Decision Functions with the Knowledge Base

Since the goal of our work is an integrated system for intensive care monitoring, the numerical approach using the SVM has to be incorporated into the logic of MOBAL. While training SVM classifiers can take place offline in a separate program, MOBAL needs to be able to evaluate SVM decision rules and access the results online. We achieve this by introducing the special predicate svm_calc/6 with the following semantic. The first two arguments indicate the patient and the drug. The third argument is either "up" or "down" depending on whether the svm_calc fact belongs to the SVM predicting dose increase or decrease (compare Section 3.3). The fourth argument is the time and the fifth is the current dosage of the drug. The last argument finally contains the value of that particular SVM rule for the measurements at that time. Calculation can be done very efficiently, since it mainly consists of computing a dot product between the SVM weight vector and the measurement vector. From each pair of decision rules (i.e., up and down) an intervention for the respective drug is recommended, if exactly one decision rule has a value larger than a confidence threshold of 0.8.

The decision rule for an increase of Glyceroltrinitrat (nitro) together with the actual parameter values of patient 4999 at time 32 is shown in Figure 2. The dot product plus –4.368 (the value of b) is 1.85598. The fact entered into the fact base for patient 4999 is *svm_calc(pat4999, nitro, up, 32, 0.0, 1.85598)*. An intervention to increase nitro is derived. The dose is calculated on the basis of the former dose. The SVM actually only decides whether to increase, to decrease, or not to change the dose. For each drug, a level of granularity is defined. For instance, the granularity of Glyceroltrinitrat is 1, whereas that of Suprarenin (containing adrenaline) is 0.01. The dose is changed by just one step. In our example, the proposed intervention is:

pred_intervention(pat4999, 32, nitro, 1.0).

5 Using the Knowledge Base of Effects to Validate Interventions

Medical knowledge is used for validation in two different ways. On the one hand, learned decision rules are validated on patient data by comparing the effects of their recommended interventions with the effects of actual physicians` interventions. This validation means to incorporate an evaluation step already into the knowledge acquisition phase. On the other hand, we believe that even an evaluated decision support system should check its decisions by considering their effects.

5.1 Validating Learned Decision Rules

There are usually several different combinations of drugs that achieve the same goal of keeping the patient in a stable state. And indeed, different physicians, depending on their experience in the ICU, do use different mixtures and follow different strategies to reach this goal. For comparing treatment strategies, the real criterion is whether the recommendations have the same effect as the actual interventions. Therefore, we apply the action-effect rules from the knowledge base to both the proposed intervention of the SVM classifiers and to the intervention actually performed by the physician. If the derived effects are equal, then the proposed decision of the SVM classifiers can be considered as "equivalent" to the intervention executed by the physician. The results of this comparison for 473 interventions are shown in Table 7. The right-most column indicates the accuracy, i.e., in how many cases the classification of SVM and physician were identical (same behavior of SVM and physician). The other columns state how often the SVMs' intervention leads to the same effects as the intervention of the physician. The first two columns show, how many of interventions had the same effect on arterial blood pressure or heart rate, respectively. The third column gives a more concise evaluation. Here it is stated, how many interventions recommended by the SVM had the same effects on all vital signs as the actual intervention. For instance, the SVM correctly classifies 299 test cases for Dobutamine (63%). If we compare the resulting effects of the predicted interventions concerning Dobutamine with the effects of the actual physician's interventions, we find that in 383 cases (81%) the deduced effects will be equal. Thus, in 84 cases the recommendation of the SVM does not match the physician's behavior, but the derived effects are the same, since the physician has chosen an "equivalent" drug or combination of drugs. An inspection of these cases helps to clarify issues of best practice and thus supports knowledge acquisition.

Table 7. Equivalence of decisions regarding effects.

Interventions	Mean arterial pressure	Heart rate	Same effect all parameters	Same behavior
Dobutamine	403	395	383	299
Adrenaline	407	406	393	374
Glyceroltrinitrate	437	388	380	342
Noradrenaline	436	428	424	420
Nifedipine	457	457	455	438

5.2 Validating Proposed Interventions

As depicted in the overall architecture (cf. Figure 1), we have chosen a design which allows us to use the action-effect rules in the knowledge base for validating predicted interventions. The underlying argument is that accuracy measures only reflect how well the SVMs' learning results fit actual behavior of the physician. However, we aim at best practice. Hence, we validate a proposed intervention with respect to its effects on the patient. If the effects push vital signs in the direction of the desired value range, the recommendation is considered sound, otherwise it is rejected. An example may clarify this. Patient 4999 is older than 75 years and stays at the ICU after a surgical operation. He suffers from high arterial mean pressure (around 124), where the heart rate is normal (around 80). Using its decision rules, the SVM recommends to increase Glyceroltrinitrat (see Figure 2). This proposed intervention is checked by the medical knowledge about effects. The derived effects are an increase of the heart rate and a decrease of arterial mean pressure as well as left ventricular stroke work index (lvswi) and systemic vascular resistance (svr): *interv_effect(pat4999,32, T, art, down)*. The observed deviation is *deviation(pat4999, 0,31, art, up)*. Since down is the opposite of up, the proposed intervention is considered sound. In this way, the prescriptive medical knowledge (action-effect rules) is used to control the knowledge that is learned from actual therapies (state-action rules).

6 Comparison with Related Work

Using data from the most comprehensive singular clinical data repository at the LDS Hospital, Salt Lake City, Utah, USA, the group of Morris [22] developed a rule-based decision support system (DSS) for respiratory care in acute respiratory distress syndrome. Time is handled by introducing time points into the rules where a certain parameter value needs to be obtained. The development of this highly specialized system required more than 25 person years. It is a propositional rule base without a mechanism for consistency checking or matching rules and data. All validation efforts started only after the knowledge base had been completed.

Temporal reasoning is taken seriously in other developments [3], [5], [17], [25]. The Stanford approach uses an explicit time ontology for low-frequency data [25]. This approach is not feasible for our application. The VIE-VENT system is comparable with our approach in that it combines numerical data and a knowledge base [17]. Qualitative abstractions are derived for deviations of measurements from the target range. Time intervals refer to the validity of a measurement. The detection of outliers (data validation) is handled by a trend-based component. The validated measurements are used by the therapy planning component which aims at pushing vital signs into the value ranges of a stable state. Similar to our approach, therapy planning is divided into state-action rules (therapeutic actions based on status interpretation) and verifying the effectiveness of interventions. However, the system was developed without using actual patient data. Hence, the observation that parameter values oscillate considerably was made as late as the first clinical experience. In contrast, this observation has motivated our use of the phase space procedure for abstracting from numerical time series. Temporal correlations can also be included in trend templates, which are used by Haimowitz and Kohane [5]. Trend templates consist of sets of low order polynomial regression models describing qualitative characteristics. Pattern abstraction is done based on the fit of these templates to the observed data. The major drawbacks of this method are the demand for predefined expected behavior and absolute value thresholds. However, time series in intensive care often show irregular behavior like patchy outliers, or outliers and level changes occurring in short time lags. Such behavior is difficult to specify in advance. Moreover, thresholds should be depending dynamically on the patient's status in the past. This has already been included in our approach, which does not need prespecified patterns either. Altogether, statistical time series analysis seems to be the most sophisticated method to model and investigate dynamical data since other approaches capture only parts of the time dependent structure of the data.

Our goals of easing the development of guidelines and validating the knowledge early on is shared by the two-step approach by Mani and coworkers [16]. They use machine learning in order to first characterize scores of dementia with respect to six categories (e.g., memory, orientation). These learning results are then used to learn the global clinical dementia rating. After a two years effort an efficient and effective system was accomplished. While the goals are the same, the application characteristics and, hence, the methods are completely different. The clinical rating is a classification task and the patient data is of qualitative nature, whereas our task is on-line monitoring and the patient data are time series of numerical measurements.

7 Conclusions

We presented an approach towards integrating learning and knowledge-based methods for the development of decision support algorithms in critical care. The SVM was chosen for learning state-action rules due to its ability to handle multiple features. For modeling medical knowledge in terms of action-effect rules we chose a first-order logic representation using MOBAL. This allowed a compact representation of medical knowledge with a small number of rules, fulfilling the real-world demand for a knowledge base to be understandable by humans and accessible for expert validation.

The validation issue has been treated with special care. Each process has been validated in the standard way, i.e., tested on data not used for training. In addition, the results of state-action rules were compared with the results of a human expert who classified the same data. Moreover, recommended interventions of state-action rules are validated by formalized medical knowledge. On the one hand, the effect of a recommended intervention is compared with the effect of an actual intervention. Of course, this comparison can only be made for past cases. In case of conflict, the expert inspects the particular cases. This may lead to the generation of explicit additional knowledge. On the other hand, the formalized effects of interventions are applied to current cases and evaluated with respect to the target ranges of vital signs.

Our new approach combines modeling of expert knowledge with data-driven methods. This eases the task of building operational protocols. Moreover, the data-driven method allows for an ongoing enhancement of the knowledge base on the basis of current practice. The knowledge base is validated against existing patient data. This approach is meant to be significantly more effective than the tedious, time-consuming, and costly process of traditional development of on-line operational decision support systems. The effect of this is an improvement in both the extensibility of an existing knowledge base and the control of the quality of medical treatment.

Acknowledgements

This work has been funded in part by the Deutsche Forschungsgemeinschaft (SFB475, "Reduction of Complexity for Multivariate Data Structures").

References

[1] Bauer, M., Gather, U., and Imhoff, M. (1999), "The identification of multiple outliers in online monitoring data," *Technical Report 29, SFB 475*, University of Dortmund.

[2] Box, G.E.P., Jenkins, G.M., and Reinsel, G.C. (1994), *Time Series Analysis. Forecasting and Control*, 3rd edition, Prentice Hall, Englewood Cliffs.

[3] Dojat, M. and Sayettat, C. (1995), "A realistic model for temporal reasoning in real-time patient monitoring," *Applied Artificial Intelligence*, vol. 10, no. 2, pp. 121-143.

[4] Guyatt, G., Drummund, M., Feeny, D., Tugwell, P., Stoddart, G., Haynes, R., Bennett, K., and LaBelle, R. (1986), "Guidelines for the clinical and economic evaluation of health care technologies," *Soc Sci Med*, vol. 22, pp. 393-408.

[5] Haimowitz, I.J. and Kohane, I.S. (1996), "Managing temporal worlds for medical trend diagnosis," *Artificial Intelligence in Medicine*, vol. 8, pp. 299-321.

[6] Imhoff, M. (1995), "A clinical information system on the intensive care unit: dream or night mare?" *Medicina Intensiva 1995, XXX. Congreso SEMIUC*, Murcia, pp. 17-22.

[7] Imhoff, M. (1996), "3 Years clinical use of the Siemens Emtek System 2000: efforts and benefits," *Clinical Intensive Care 7 (Suppl.)*, pp. 43-44.

[8] Imhoff, M. (1998), "Clinical data acquisition: what and how?" *Journal für Anästhesie und Intensivmedizin*, vol. 5, pp. 85-86.

[9] Imhoff, M., Bauer, M., Gather, U., and Löhlein, D. (1998), "Statistical pattern detection in univariate time series of intensive care on-line monitoring data," *Intensive Care Med*, vol. 24, pp. 1305-1314.

[10] Imhoff, M., Bauer, M., and Gather, U. (1999), "Time-effect relations of medical interventions in a clinical information system," *KI-99: Advances in Artificial Intelligence*, LNAI vol. 1701, Springer-Verlag, pp. 307-310.

[11] Imhoff, M., Lehner, J.H., and Löhlein, D. (1994), "2 Years clinical experience with a clinical information system on a surgical ICU," *7th European Congress on Intensive Care Medicine*, pp. 163-166.

[12] Jennings, D., Amabile, T., and Ross, L. (1982), "Informal covariation assessments: data-based versus theory-based judgements," in: *Judgment under Uncertainty: Heuristics and Biases*, Cambridge University Press, Cambridge, pp. 211-230.

[13] Joachims, T. (1999), "Making large-scale SVM learning practical," in: *Advances in Kernel Methods – Support Vector Learning*, MIT Press, Cambridge, pp. 169-184.

[14] Keravnou, E.T. (1996), "Temporal diagnostic reasoning based on time-objects," *Artificial Intelligence in Medicine*, vol. 8, pp. 235-265.

[15] Kukar, M., Kononenko, I., Groselj, C., Kralj, K., and Fettich, J. (1999), "Analyzing and improving the diagnosis of ischaemic heart disease with machine learning," *Artificial Intelligence in Medicine*, vol. 16, pp. 25-50.

[16] Mani, S., Shankle, W.R., Dick, M.B., and Pazzani, M.J. (1999), "Two-stage machine learning for guideline development," *Artificial Intelligence in Medicine*, vol. 16, pp. 51-71.

[17] Miksch, S., Horn, W., Popow, C., and Paky, F. (1996), "Utilizing temporal abstraction for data validation and therapy planning for artificially ventilated newborn infants," *Artificial Intelligence in Medicine*, vol. 8, pp. 543-576.

[18] Miller, G. (1956), "The magical number seven, plus or minus two: some limits to our capacity for processing information," *Psychol Rev*, vol. 63, pp. 81-97.

[19] Morik, K. (1994), "Balanced cooperative modeling," *Machine Learning – a Multistrategy Approach*, Morgan Kaufmann, pp. 295-318.

[20] Morik, K., Wrobel, S., Kietz, J.-U., and Emde, W. (1993), *Knowledge Acquisition and Machine Learning – Theory, Methods, and Applications,* Academic Press, London.

[21] Morris, A. and Gardner, R. (1992), "Computer applications," in: *Principles of Critical Care*, McGraw-Hill, New York, pp. 500-514.

[22] Morris, A. (1998), "Algorithm-based decision-making," in: *Principles and Practice of Intensive Care Monitoring*, McGraw-Hill, New York, pp. 1355-1381.

[23] Platt, C. (1999), "Probabilistic outputs for support vector machines and comparisons to regularized likelihood methods," in: *Advances in Large Margin Classifiers*, MIT Press.

[24] Schölkopf, B., Williamson, R., Smola, A., and Shawe-Taylor, J. (1999), "SV-estimation of a distribution's support," *NIPS 99*.

[25] Shahar, Y. and Musen, M. (1996), "Knowledge-based temporal abstraction in clinical domains," *Artificial Intelligence in Medicine*, vol. 8, pp. 267-298.

[26] Vapnik, V. (1998), *Statistical Learning Theory*, Wiley, New York.

Chapter 5

Case-Based Reasoning Prognosis for Temporal Courses

R. Schmidt and L. Gierl

Since clinical management of patients and clinical research are essentially time-oriented endeavors, reasoning about time has recently become a hot topic in medical informatics. Here we present a method for prognosis of temporal courses, which combines temporal abstractions with Case-Based Reasoning. The method was originally generated for multiparametric time course prognosis of the kidney function. Recently, we have started to apply the same ideas for the prognosis of the temporal spread of diseases. In this chapter, we mainly describe both applications and subsequently present a generalization of our method.

1 Introduction

Since clinical management of patients and clinical research are essentially time-oriented endeavors, reasoning about time has recently become a hot topic in medical informatics. Here we present a method for prognosis of temporal courses based on numeric values, e.g., for organ functions or for incidences of the spread of diseases.

Since traditional time series techniques [1] work well with known periodicity, but do not fit in domains characterized by possibilities of abrupt changes, much research has been performed in the field of medical temporal course analysis in the recent years. However, the methods developed so far either require a complete domain theory or well-known standards (e.g., course pattern or periodicity).

An ability of RÉSUMÉ [2] is the abstraction of many parameters into one single parameter and to analyze courses of this abstracted parameter. However, interpretation of these courses requires complete domain knowledge. Haimowitz and Kohane [3] compare many parameters of current courses with well-known standards (trend templates). In VIE-VENT [4] both ideas are combined: Courses of single quantitative measured parameters are abstracted into qualitative course descriptions that are matched with well-known standards.

When we started building a system for course analysis and prediction of the kidney function, we were confronted with a domain where the domain theory is extremely incomplete and no standards are yet known. So we had to design our own method. Because of our good experience with Case-Based Reasoning (CBR) methods for medical applications (e.g., antibiotics therapy advice [5], post-operative management for liver-transplanted patients [6], diagnosis of dismorphic syndromes [7]), we decided to use Case-Based Reasoning again. For temporal courses, our general idea is to search with CBR retrieval methods [8], [9] for former patients with similar courses and to consider their course continuations as possible prognosis for the current patient.

So far, we have successfully applied our method in the kidney function domain [10] and have recently started to apply it again for the prognosis of the spread of diseases (especially of influenza) [11].

2 Methods

2.1 Case-Based Reasoning

Case-Based Reasoning means using previous experience represented as cases to understand and solve new problems. A case-based reasoner remembers former cases similar to the current problem and attempts to modify solutions of former cases to fit for the current problem. Figure 1 shows the Case-Based Reasoning cycle developed by Aamodt and Plaza [12], which consists of four steps: retrieving former similar cases, adapting their solutions to the current problem, revising a proposed solution, and retaining new learned cases. However, there are two main subtasks in Case-Based Reasoning [12], [13]: the retrieval, the search for a similar case, and the adaptation, the modification of solutions of retrieved cases. Since differences between two cases are sometimes very complex, especially in medical domains, many case-based systems are so called retrieval-only systems. They just perform the retrieval task, visualize current and similar cases, and sometimes additionally point out the important differences between them [14].

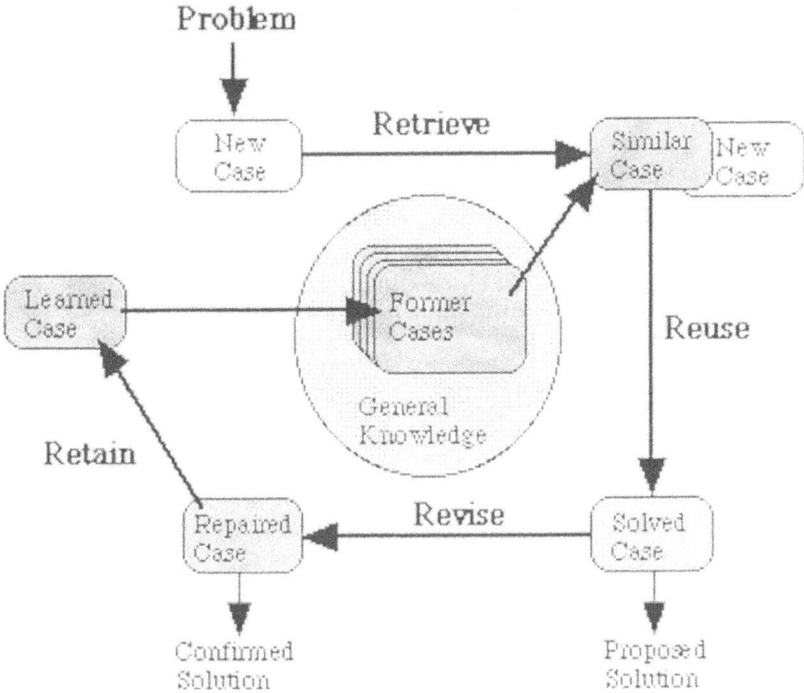

Figure 1. The Case-Based Reasoning cycle developed by Aamodt.

To make CBR applicable an appropriate case representation has to be found. Usually, a list of attribute-value-pairs that contains all case attributes is sufficient. However, for multiparametric time courses the choice of suitable attributes is not obvious. Firstly, not complete courses (they may differ in their length, they may go much further back than it is relevant for the current situation), but only patients current developments of a certain length should be compared with parts of former patients courses, which should have about the same length. Secondly, each course consists of a sequence of measured or calculated parameter sets. It cannot be assumed that all parameters are of the same importance, especially the more current parameter sets are usually more important than those further back.

And even the importance of parameters within the same set may extremely differ. One idea is to look for appropriate weightings for the parameters. However, hundreds of parameters might be involved, much domain knowledge may be required, and weights can be very subjective. Furthermore, it seems to be impossible to visualize a sequence of parameter sets in such a way that a user can rapidly discern the important characteristics.

In the kidney function domain, we chose a different alternative. With the help of medical experts we defined kidney function states based on the most important parameters and subsequently we abstracted each daily parameter set into such a function state. So, courses are represented as a sequence of function states.

2.2 Prognostic Model

As a spin-off from our work on the analysis and prognosis of kidney function courses, we have developed a prognostic model for multi-parametric time courses (Figure 2). It combines two abstraction steps with Case-Based Reasoning.

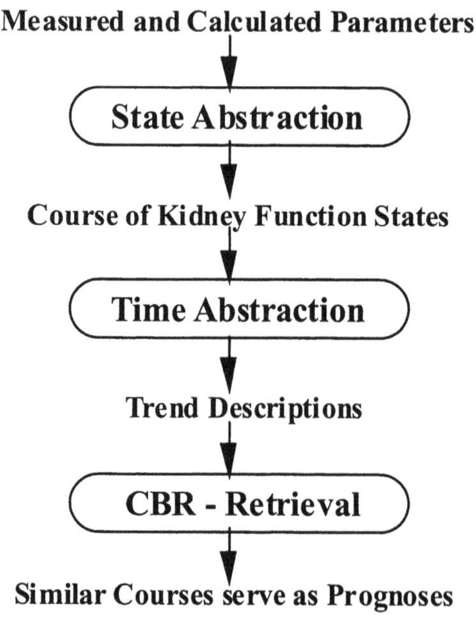

Figure 2. The prognostic model for ICONS.

2.2.1 State Abstraction

The first step is an abstraction from a set of parameter values to a single function state. Therefore few requirements have to be met. Meaningful states to describe the parameter sets and a hierarchy of these states must exist. Furthermore, knowledge to define the states must be available. These definitions may consist of obligatory or optional conditions on the parameter values. Of course, all obligatory conditions should be met, while for the optional ones some alternatives exist how to determine the appropriate state. One simple idea is to count the met conditions. Additionally, the quality of meeting graduated conditions may be considered (e.g., fuzzy methods may be applied). An alternative is to leave at least a part of this state determination process to the user, e.g., if more than one state is under consideration, the user decides which one is the most appropriate one.

2.2.2 Temporal Abstraction

The second abstraction means to describe a course of states. An often-realized idea is to use different trend descriptions for different periods of time, e.g., short-term or long-term trend descriptions etc. (see, e.g., [4]). The lengths of each trend description can be fixed or they may depend on concrete values (e.g., successive equivalent states may be concatenated).

However, concrete definitions of the trend descriptions depend on characteristics of the application domain:

(1) On the number of states and on their hierarchy,
(2) On the lengths of the considered courses, and
(3) On what has to be detected, e.g., long-term developments or short-term changes.

However the trend descriptions may be defined, they can be expressed by four parameters: the length, the first and last state, and an assessment. The lengths and the assessments of the descriptions can vary with domain-dependent demands, while the state definitions and their hierarchy are domain dependent anyway.

2.2.3 CBR Retrieval

To determine the most similar cases sophisticated similarity measures provide the best results, but they consider all stored cases sequentially. Especially for large case bases a sequential process is too time consuming. So, a few non-sequential retrieval algorithms have been developed in the CBR community. Some more have been incorporated from related fields; e.g., classification algorithms like ID3 [15] or Nearest Neighbor classification [16] from the Machine Learning community. Most of the retrieval algorithms can handle various sorts of attributes, but usually they only work well for those sorts of attributes or problems they have been developed for, e.g., the Tree-Hash-Retrieval indexing algorithm [17] works very well for huge case bases and for cases with non-ordered nominal attribute values.

This means, the choice of the retrieval algorithm should mainly depend on the sort of values of case attributes and sometimes additionally on application characteristics like the size of the case base. The question arises: Of which sort are the four parameters that describe a trend? The states are obviously nominal valued ordered according to their hierarchy. The assessments should have ordered nominal values too, e.g., steady, decreasing etc. Only the lengths should have numeric values. If the time points of the parameter measurements are few integers, they can be treated as ordered nominal values. The proposed retrieval algorithms for ordered nominal valued attributes are CBR-Retrieval-Nets [8], which are based on Spreading Activation [18]. So, if all four parameters have ordered nominal values, the choice of the retrieval algorithm should obviously be CBR-Retrieval-Nets.

However, we made some assumptions that may not necessarily be met in every domain. For example, the lengths may not be transformable into nominal values, the trend assessments may not be just simple nominal values, but more sophisticated descriptions, and there are of course alternatives to describe trends, e.g., even a computed real value might somehow express a trend.

3 Applications

Since the prognosis of temporal courses is not limited to medical problems, prognostic applications which use Case-Based Reasoning techniques have been developed in a few domains, for example for weather prediction [19], for predicting air pollution levels [20], and for predicting the physical structure around a sea going vessel [21]. In this chapter we show how we have applied CBR for the prediction of kidney function courses and how we are going to apply the same method for predicting the temporal spread of diseases.

3.1 Kidney Function Courses

3.1.1 Objectives

Up to 60% of the body mass of an adult person consists of water. The electrolytes dissolved in body water are of great importance for an adequate cell function. The human body tends to balance the fluid and electrolyte situation. But intensive care patients are often no longer able to maintain adequate fluid and electrolyte balances themselves due to impaired organ functions, e.g., renal failure, or medical treatment, e.g., parenteral nutrition of mechanically ventilated patients. Therefore physicians need objective criteria for the monitoring of fluid and electrolyte balances and for choosing therapeutic interventions as necessary.

At our ICU, physicians daily get a printed renal report from the monitoring system NIMON [22] which consists of 13 measured and 33 calculated parameters of those patients where renal function monitoring is applied. For example, the urine osmolality and the plasma osmolality are measured parameters that are used to calculate the osmolar clearance and the osmolar excretion. The interpretation of all reported parameters is quite complex and needs special knowledge of the renal physiology.

The aim of our knowledge based system ICONS is to give an automatic interpretation of the renal state to elicit impairments of the kidney function on time and to give early warnings against forthcoming kidney failures. That means, we need a time course analysis of many parameters without any well-defined standards.

However, in the domain of fluid and electrolyte balance, neither a prototypical approach in ICU settings is known nor exists complete knowledge about the kidney function. Especially, knowledge about the behavior of the various parameters on time is yet incomplete. So, we combined the idea of RÉSUMÉ [2] to abstract many parameters into one single parameter with the idea of Haimowitz and Kohane [3] to compare many parameters of current courses with well-known standards. Since well-known standards were not available, we used former similar cases instead.

3.1.2 Methods

Our procedure to interpret kidney function courses can be seen in Figure 2. First, the monitoring system NIMON gets 13 measured parameters from the clinical chemistry and calculates 33 meaningful kidney function parameters. To elicit the relationships among these parameters a three-dimensional presentation was implemented inside the renal monitoring system NIMON. However, complex relations among all parameters were not visible.

So, we decided to abstract these parameters. For this data abstraction we use states of the renal function, which determine states of increasing severity beginning with a normal renal function and ending with a renal failure. Based on these state definitions, we determine the appropriate state of the kidney function per day. Therefore, we present the possible states to the user sorted according to their probability. The physician has to accept one of them. Based on the transitions of the states of one day to the state of the respectively next day, we generate four different trends. These trends, that are abstractions of time, describe the courses of the states. Subsequently we use Case-Based Reasoning retrieval methods [8, 9] to search for similar courses. We present similar courses together with the current one as comparisons to the user, the course continuations of the similar courses serve as prognoses. When presenting such a comparison, ICONS supplies the user with the ability to access additional renal syndromes and to access courses of single parameter values.

State abstraction. Based on the kidney function states (e.g., in Figure 3), characterized by obligatory and optional conditions for selected renal parameters, we first check the obligatory conditions. For each state that satisfies the obligatory conditions we calculate a similarity value concerning the optional conditions. We use a variation of Tversky's [23] measure of dissimilarity between concepts. Only if two or more states are under consideration, ICONS presents them to the user sorted according to their similarity values together with information about the satisfied and not satisfied optional conditions.

Reduced Kidney Function

<u>Obligatory Condition:</u>	c_kreat	40 - 80
<u>Optional Conditions:</u>		
Retention Rates:	p_kreat_se	< 2
	p_urea_se	< 150
Tubular Function:	u_osmol	320 - 600
	u_p_osmol	1.1 - 1.8
	u_kreat	10 - 40
	u_p_kreat	20 - 50
Urine Volume:	urine volume	0.7 - 3.0
	osmol_ex	800 - 3000

Figure 3. Definition of the reduced kidney function state.
Abbreviations: c = clearance, p = plasma, u = urine, kreat = kreatinin, osmol = osmolality, se = serum, ex = excretion.

The user can accept or reject a presented state. When a suggested state has been rejected, ICONS selects another one. Finally, we determine the central state of occasionally more than one states the user has accepted. This central state is the closest one towards a kidney failure. Our intention is to find the state indicating the most profound impairment of the kidney function.

Temporal abstraction. First, we have fixed five assessment definitions for the transition of the kidney function state of one day to the state of the respectively next day. These assessment definitions are related to the grade of renal impairment:

steady: both states have the same severity value.
increasing: exactly one severity step in the direction towards a normal function.
sharply increasing: at least two severity steps in the direction towards a normal function.

decreasing: exactly one severity step in the direction towards a kidney failure.

sharply decreasing: at least two severity steps in the direction towards a kidney failure.

These assessment definitions are used to determine the state transitions from one qualitative value to another. Based on these state transitions, we generate three trend descriptions. Two trend descriptions especially consider the current state transitions.

short-term trend	:=	current state transition; abbreviation: T1
medium-term trend	:=	looks recursively back from the current state transition to the one before and unites them if they are both of the same direction or one of them has a "steady" assessment; abbreviation: T2
long-term trend	:=	characterizes the considered course of at most seven days; abbreviation: T3

For the long-term trend description we additionally introduced four new assessment definitions. If none of the five former assessments fits the complete considered course, we attempt to fit one of these four definitions in the following order:

alternating: at least two up and two down transitions and all local minima are equal.

oscillating: at least two up and two down transitions.

fluctuating: the distance of the highest to the lowest severity state value is greater than 1.

nearly steady: the distance of the highest to the lowest severity state value equals one.

Only if there are several courses with the same trend descriptions, we use a minor fourth trend description T4 to find the most similar among them. We assess the considered course by adding up the state transition values inversely weighted by the distances to the current day. Together with the current kidney function state these four trend descriptions form a course depiction, that abstracts the sequence of the kidney function states.

Looking back from a time point t, these four trend descriptions form a pattern of the immediate course history of the kidney function considering qualitative and quantitative assessments.

Why these four trend descriptions? There are domain specific reasons for defining the short-, medium- and long-term trend descriptions T1, T2 and T3. If physicians evaluate courses of the kidney function, they consider at most one week prior to the current date. Earlier renal function states are irrelevant for the current situation of a patient. Most relevant information is derived from the current function state, the current development and sometimes a current development within a slightly longer time period. That means, very long term trends are of no interest in this domain. In fact, very often only the current state transition or short continuous developments are crucial.

The short-term trend description T1 expresses the current development. For longer time periods, we have defined the medium- and long-term trend descriptions T2 and T3, because there are two different phenomena to discover and for each, a special technique is needed. T2 can be used for detecting a continuous trend independent of its length, because equal or steady state transitions are united recursively beginning with the current one. As the long-term trend description T3 describes a well-defined time period, it is especially useful for detecting fluctuating trends.

Since every abstraction loses some specific information, information about the daily kidney function states is lost in the second abstraction step. The course description contains only information about the current and the start states of the three trend descriptions. The intermediate states are abstracted into trend description assessments.

Example. The following kidney function states may be observed in this temporal sequence (Figure 5):

selective tubular damage, reduced kidney function, reduced kidney function, selective tubular damage, reduced kidney function, reduced kidney function, sharply reduced kidney function.

So we get these six state transitions:

decreasing, steady, increasing, decreasing, steady, decreasing.

with these trend descriptions:

current state: sharply reduced kidney function
T1: decreasing, reduced kidney function, one transition
T2: decreasing, selective tubular damage, three transitions
T3: fluctuating, selective tubular damage, six transitions
T4: 1.23

In this example, the short-term trend description T1 assesses the current state transition as "decreasing" from a "reduced kidney function" to a "sharply reduced kidney function." Since the medium-term trend description T2 accumulates steady state transitions, T2 determines a "decrease" in the last four days from a "selective tubular damage" to a "sharply reduced kidney function." The long-term trend description T3 assesses the entire course of seven days as "fluctuating," because there is only one increasing state transition and the difference between the severity values of a "selective tubular damage" and a "sharply reduced kidney function" equals two.

Retrieval. We use the parameters of the four trend descriptions and the current kidney function state to search for similar courses. As the aim is to develop an early warning system, we need a prognosis. For this reason and to avoid a sequential runtime search along the entire cases, we store a course of the previous seven days and a maximal projection of three days for each day a patient spent on the intensive care unit.

Since there are many different possible continuations for the same previous course, it is necessary to search for similar courses and for different projections. Therefore, we divided the search space into nine parts corresponding to the possible continuation directions. Each direction forms an own part of the search space. During the retrieval these parts are searched separately and each part may provide at most one similar case. The similar cases of these parts together are presented in the order of their computed similarity values.

Before the main retrieval, we search for a prototypical case (see Section 3.1.3) that matches most of the trend descriptions. Below this prototype the main retrieval starts (Figure 4). It consists of two steps for each part. First we search with an activation algorithm concerning qualitative features. Our algorithm differs from the common spreading activation algorithm [18] mainly due to the fact that we do not use a net for the similarity relations. Instead, we have defined explicit activation values for each possible feature value. This is possible, because on this abstraction level there are only ten dimensions (see the left column of Table 1) with at most six values. The right column of Table 1 shows the possible activation values for the description parameters. For example there are four activation values for the current kidney function state. Courses with the same current state as the query course get the value 15, while those cases whose distance to the current state of the query course is one step in the severity hierarchy get 7 and so forth.

Subsequently, we check the retrieved cases with a similarity criterion [9] that looks for sufficient similarity, because even the most similar course may differ from the current one significantly. This may happen at the beginning of the use of ICONS, when there are only a few cases known to ICONS, or when the query course is rather exceptional.

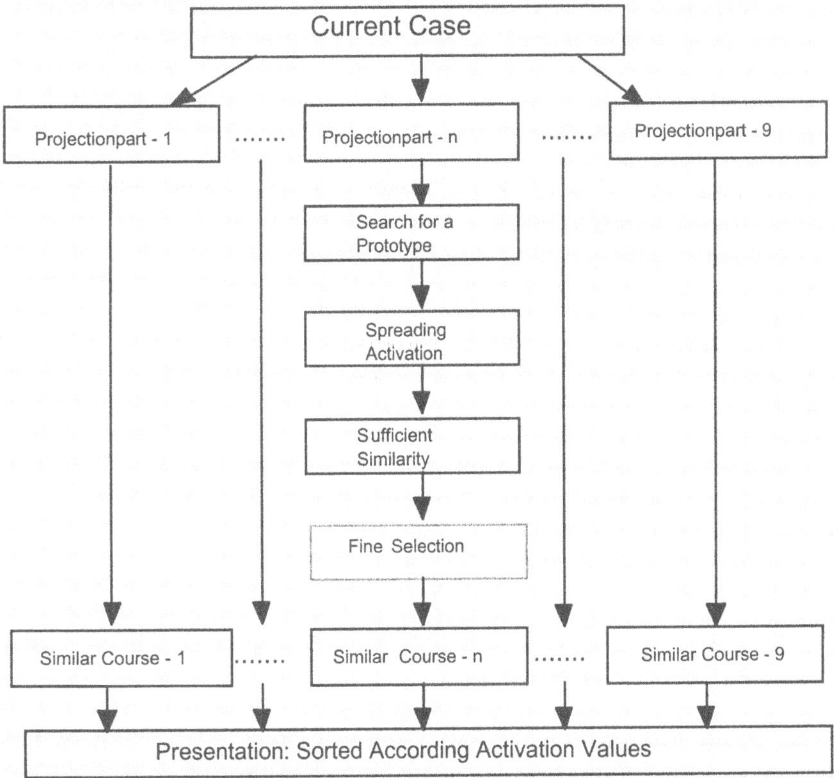

Figure 4. The retrieval procedure.

Table 1. Retrieval dimensions and their activation values.

Dimensions	Activation values
Current state	15, 7, 5, 2
Assessment T1	10, 5, 2
Assessment T2	4, 2, 1
Assessment T3	6, 5, 4, 3, 2, 1
Length T1	10, 5, 3, 1,
Length T2	3, 1
Length T3	2, 1
Start state T1	4, 2
Start state T2	4, 2
Start state T3	2, 1

If two or more courses are selected in the same projection part, we use the sequential similarity measure of TSCALE [24], which goes back to Tversky [23], concerning the quantitative features in a second step.

Continuation of the example. For the example above, the following similar course (Figure 5) with these transitions is retrieved:

decreasing, increasing, decreasing, steady, steady, decreasing

with these trend descriptions:

current state: sharply reduced kidney function
T1: decreasing, reduced kidney function, one transition
T2: decreasing, selective tubular damage, four transitions
T3: fluctuating, selective tubular damage, six transitions
T4: 1.17

T1 describes a "decrease" from a "reduced kidney function" and T2 describes a "decrease" from a "selective tubular damage" to a "sharply reduced kidney function" in the last five days. T3 assesses the considered course as "fluctuating." For T4, a slightly lower value in comparison to the current course has been calculated, because the change from a "selective tubular damage" to a "reduced kidney function" state occurs earlier.

After another day with a "sharply reduced kidney function" the patient belonging to the similar course had a kidney failure. The physician may notice this as a warning and it is up to him to interpret it.

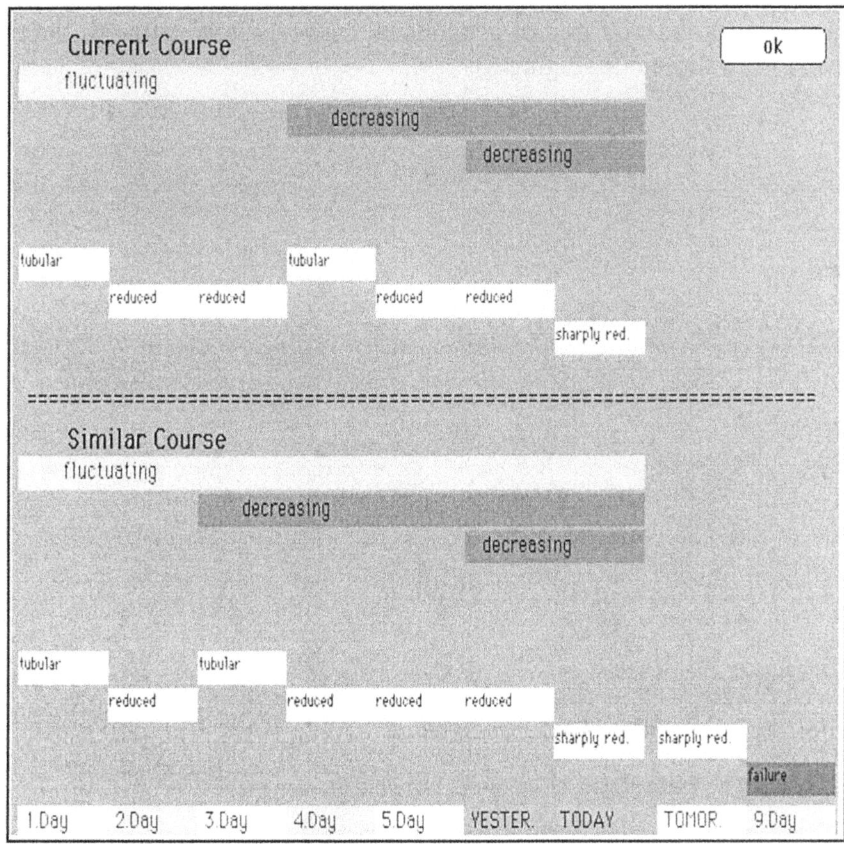

Figure 5. Comparative presentation of a current and a similar course. In the lower part of each course the (abbreviated) kidney function states are depicted. The upper part of each course shows the deduced trend descriptions.

This former course was retrieved, because especially the features with the highest weights (the current state and all assessments) equal the features of the query course. As there is no significant difference between both courses, there is no reason for the sufficient similarity criterion to reject this similar course.

3.1.3 Learning a Tree of Prototypes

Prognosis of multiparametric courses of the kidney function for ICU patients is a domain without a medical theory. Moreover, we can not expect such a theory to be formulated in the near future. So we attempt to learn prototypical course pattern. Therefore, knowledge on this domain is stored as a tree of generalized cases (prototypes) with three levels and a root node (Figure 6).

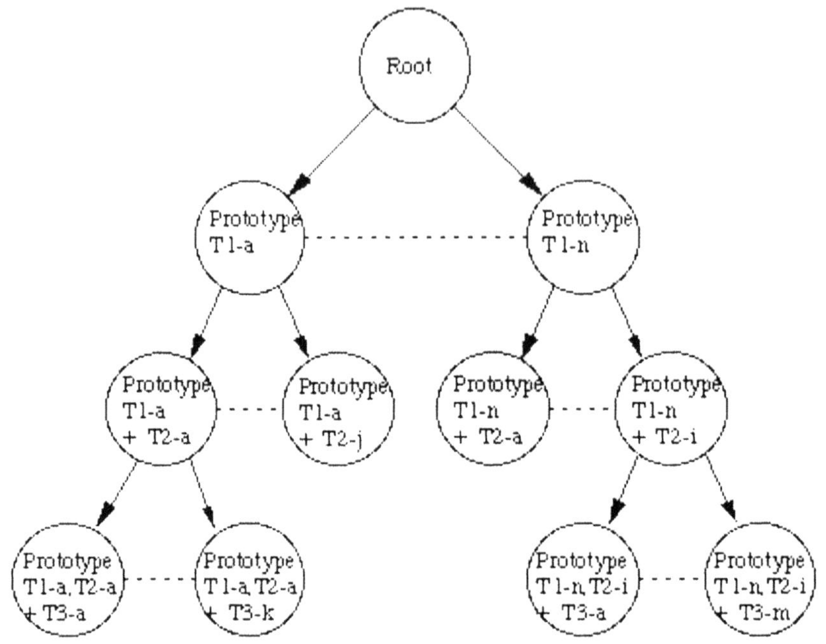

Figure 6. Prototype architecture for the trend descriptions T1, T2, and T3.

Except for the root, where all not yet clustered courses are stored, each level corresponds to one of the trend descriptions T1, T2 or T3. As soon as enough courses that share another trend description are stored at a prototype, we create a new prototype with this trend. At a prototype at level 1, we cluster courses that share T1, at level 2, courses that share T1 and T2 and at level 3, courses that share all trend descriptions T1, T2 and T3.

We can do this, because regarding their importance, the short-, medium- and long-term trend descriptions T1, T2 and T3 refer to hierarchically related time periods. T1 is more important than T2 and T3, and so forth.

We start the retrieval with a search for a prototype that has most of the trend descriptions in common with the query course. The search begins at the root node with a check for a prototype with the same short-term trend description T1. If such a prototype can be found, the search goes on below this prototype for a prototype that has the same trend descriptions T1 and T2, and so forth. If no prototype with a further trend in common can be found, we search for a course at the last accepted prototype. If no prototype exists that has the same T1 as the query course, we search at the root node, where links to all courses in the case base exist.

Continuation of the example. In the example above, we can create just one prototype at level one, because at the second level the query course and the similar one, called "similar-1" differ in their length. Although the long-term trend description T3 is equal for both courses, we can not create a prototype at level three because of the strictly hierarchical organization of the prototype tree. However, learning a prototypical description "fluctuating in seven days from a selective tubular damage to sharply reduced kidney function" which does not consider any more similarities or deviations within this time period would be too general and therefore too impracticable.

Assuming we find another similar course, called "similar-2," for the current case of the example above with the following kidney function states:

reduced kidney function, reduced kidney function, selective tubular damage, selective tubular damage, reduced kidney function, reduced kidney function, sharply reduced kidney function

with these trend descriptions:

current state: sharply reduced kidney function
T1: decreasing, reduced kidney function, one transition
T2: decreasing, selective tubular damage, four transitions
T3: oscillating, reduced kidney function, six transitions
T4: 1.33

The current query course, "similar-1", and "similar-2" will be clustered at level 1 to prototype T1-a, defined by T1 as "decreasing, reduced kidney function, one transition." Afterwards at level 2 the current course and "similar-2" will be clustered to a prototype T1-a + T2-a, defined by T1 as "decreasing, reduced kidney function, one transition" plus by T2 as "decreasing, selective tubular damage, four transitions." The attempt to create another prototype at level 3 fails, because the trend descriptions T3 have different assessments and different start states. The result, a tree of prototypes learned from the three courses is shown in Figure 7.

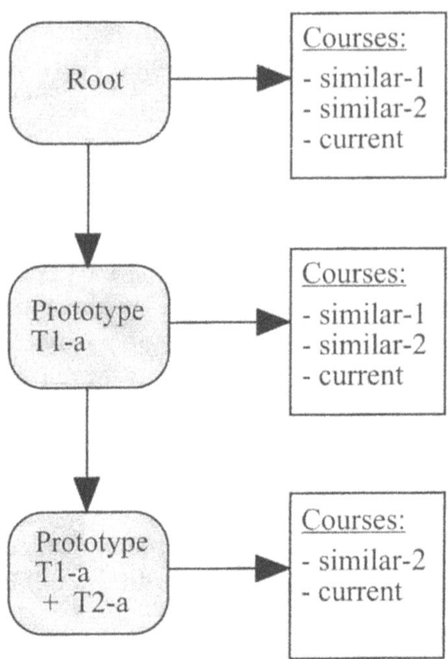

Figure 7. Generated prototype tree from three example courses.

3.1.4 Evaluation

Retrieval experiments. Since we wished to be convinced that CBR-retrieval-nets really are the appropriate retrieval algorithm for our prognostic model, we compared them with an indexing algorithm, which had been developed for non-ordered nominal values [13]. The results of this comparison are as follows: The indexing algorithm works faster (Table 2), but provides worse results, because stored cases get only activation values for attribute values that exactly match the query case values. The CBR retrieval nets additionally send smaller activation values to cases with attribute values similar to query case values. Hence, courses can be determined to be similar which have attribute values that slightly deviate from the query case values.

Table 2. Retrieval times (in seconds) for the retrieval algorithms "CBR retrieval nets" and "Indexing" with and without using prototypes.

Courses	Retrieval-Nets	Retrieval-Nets, Use of Prototypes	Indexing	Indexing, Use of Prototypes
No. 1	0.163	0.163	0.155	0.155
No. 2	0.284	0.281	0.214	0.218
No. 3	0.316	0.366	0.165	0.213
No. 4	0.455	0.513	0.404	0.452
No. 5	0.514	0.544	0.428	0.506
No. 6	1.328	0.759	0.600	0.717
No. 7	0.649	0.401	0.246	0.347
No. 8	0.685	0.642	0.376	0.469
No. 9	0.550	0.617	0.444	0.551
No. 10	0.386	0.476	0.257	0.394
No. 11	0.537	0.553	0.234	0.367
No. 12	1.396	0.870	0.743	0.890
No. 13	0.577	0.607	0.244	0.332
No. 14	0.518	0.425	0.340	0.494

Since one idea of using prototypes is to speed-up the retrieval by structuring the case base, we additionally compared both algorithms with and without using prototypes. To decide when a prototype should be generated, a threshold parameter is required. We set this parameter to the value of 2, which means, that already 2 cases with the same trend description are sufficient to generate a prototype. Hence many prototypes were generated. At first glance the results (Table 2) are not very encouraging for using prototypes. However, for the CBR retrieval nets the time differences between with and without prototypes are very small except for those two courses where the retrieval worked noticeably slower (No. 6 and No. 12): Here, using prototypes reduces the retrieval by at least a third.

However, so far the determination of the appropriate prototype occurred by sequentially matching the trend description parameters. So, most of the time gained by reducing the number of cases worth to consider is used up to determine the appropriate prototype. This indicates that not only the retrieval algorithm for cases, but also the determination of appropriate prototypes should be organized in a non-sequential way.

Evaluation of the knowledge base. To verify the knowledge base, we selected 100 data sets from the NIMON database. The selection happened only partly randomly, because we wanted an adequate representation of all kidney function states. Two physicians experienced with the kidney function were asked to classify the selected data sets according to the concepts, but without knowing ICONS's obligatory and optional conditions of the kidney function states. We compared the results of the physicians with ICONS's classifications of the same data sets. The comparison was mostly satisfactory. For 83 parameter sets the classifications of ICONS corresponded to those of the physicians. The 17 deviations are shown in Table 3. In 16 cases ICONS tended more towards the direction of kidney failures. This indicates that the intention of the experts who had defined the states had been to be more on the saver side. Just once ICONS classified the parameter set as a "reduced kidney function," while the most experienced physician assessed it as a "kidney failure" and the other physician as a "filtration rate reduction due to renal impairment." However, as a result of the evaluation we slightly modified the state definition of the "reduced kidney function."

Table 3. Classification deviations between the most experienced physicians and ICONS.

Physicians	ICONS	Quantity
normal kidney function	reduced kidney function	7
normal kidney function	selective tubular damage	3
sharply reduced kidney function	kidney failure	3
selective tubular damage	reduced kidney function	1
selective tubular damage	sharply reduced kidney function	1
reduced kidney function	sharply reduced kidney function	1
kidney failure	reduced kidney function	1

3.2 Prognosis of the Spread of Diseases

Recently we have started the TeCoMed project. The aim of this project is to send early warnings against forthcoming waves or even epidemics of infectious diseases, especially of influenza, to interested practitioners, pharmacists, etc., in the German federal state Mecklenburg-Western Pomerania. Available data are written confirmations of unfitness for work, which have to be sent by affected employees to their employers and their health insurance schemes. These confirmations contain the diagnoses made by their doctors. Since 1997 we receive data from the main German health insurance scheme.

Influenza waves are complicated to predict, because they are cyclic, but not regular [25]. Usually, each winter one influenza wave can be observed in Germany. However, the intensities of these waves vary very much. In some years they are nearly unnoticeable, while in other years doctors and pharmacists even run out of vaccine. And a wave may occur in December, in March or sometime in between. Because of the irregular cyclic behavior it is insufficient to determine average values based on former years and to give warnings as soon as such values are noticeably overstepped. So, again we apply temporal abstraction and use Case-Based Reasoning to search for similar developments in the past.

However, there are some differences in comparison to the kidney function domain. Here, a state abstraction is unnecessary and impossible, because now we have got just one parameter, namely weekly incidences of a disease. So we have to deal with courses of integer values instead of nominal states related to a hierarchy.

And the data are not measured daily, but weekly. Since we believe that courses should reflect the development of four weeks, courses consist of 4 integer values.

Again, we use three trend descriptions. Here they are simply the assessments of the developments from last week to this week (T1), from last but one week to this week (T2) and so forth. For retrieval we use these three assessments (nominal valued) plus the four weekly data (integers). We use these two sorts of parameters, because we want to ensure that the current and the similar course are on the same level (similar weekly data) and that they have similar changes on time (similar assessments).

3.2.1 Searching for Similar Courses

So far, we sequentially compute the distances between a query course and all courses stored in the case base. In the future we hope to develop a more efficient retrieval algorithm.

The considered attributes are the trend assessments and the weekly incidences. For each trend we have defined separate assessments based on the percentage of change between the weekly data, e.g., we assess the third trend T3 as "threatening increase" if the data of the current week is at least 50% higher than three weeks ago. When comparing a current course with a former one, equal assessments are valued as 1.0 and related ones as 0.5. Again the current trend (T1) is weighted higher than longer ones (T2, T3).

For the weekly data, we simply compute differences between a query and a former course and weight them with the number of the week within the four weeks course (e.g., the first week gets the weight 1.0, the current week gets 4.0).

To bring both sorts of parameters together on equal terms we multiply the computed assessment similarity with the doubled mean value of the weekly data. The result of this similarity computation is a list of all stored 4-weeks courses sorted according to their distances with respect to the query course.

As we have already done in the kidney function domain, we reduce this list by checking for sufficient similarity. For the sum of the three assessments we use a distance threshold which should not be overstepped. Concerning the four weekly incidences we have defined individual constraints that allow specific percentage deviations from the query case values. At present we are attempting to learn good settings for the parameters of these similarity constraints by using former courses in retrospect.

3.2.2 Adaptation

For adaptation, we apply a variation of Compositional Adaptation [26]. Our aim is not to present the most similar former courses – as we did in the kidney function domain, but to send a warning – when appropriate – e.g., against a forthcoming influenza wave to interested people (practitioners, pharmacists etc.). We first marked those points of the former courses where we thought a warning was appropriate. So, we can split the reduced list of similar courses in two lists, namely split concerning the question if a warning was appropriate or not. For both of these new lists we compute a sum of the reciprocal distances of their courses. Subsequently, the decision about the appropriateness of a warning depends on the question, which of the two sums is bigger.

However, so far the definitions for retrieval and for adaptation are just based on few experiments and have to be funded or modified by further experiments and experiences. For adaptation, further information like the spatial spread of a disease and the local vaccination situation should be considered in the future. For retrieval, we again intend to generalize from single courses to prototypical ones, but a method to do this for mainly integer valued parameters still has to be generated.

4 Generalization of Our Prognostic Method

We have adapted the well-known Case-Based Reasoning cycle developed by Aamodt and Plaza (Figure 1) to our medical temporal abstraction method (Figure 8).

Since the idea of both of our applications is to give information about a specific development and its probable continuation, we do not generate a solution that should be revised by a user. So, in comparison to the original CBR cycle (Figure 1) our method (Figure 8) does not contain a revision step.

In our applications the adaptation mainly consists of a sufficient similarity criterion. However, in the TeCoMed project we additionally apply a variation of Compositional Adaptation and we intend to broaden the adaptation to further criteria and information sources.

On the other hand, we have added some steps to the original CBR cycle. For multiple parameters (as in ICONS) we propose a state abstraction. For single parameters (as in the TeCoMed project) this step is unnecessary. The second step, a temporal abstraction, should provide some trend descriptions, which should not only help to analyze current courses, but the description parameters should also be used for retrieving former, similar courses. A domain dependent similarity has to be defined for retrieval and additionally a sufficient similarity criterion has to be determined, which can be viewed as part of the adaptation step.

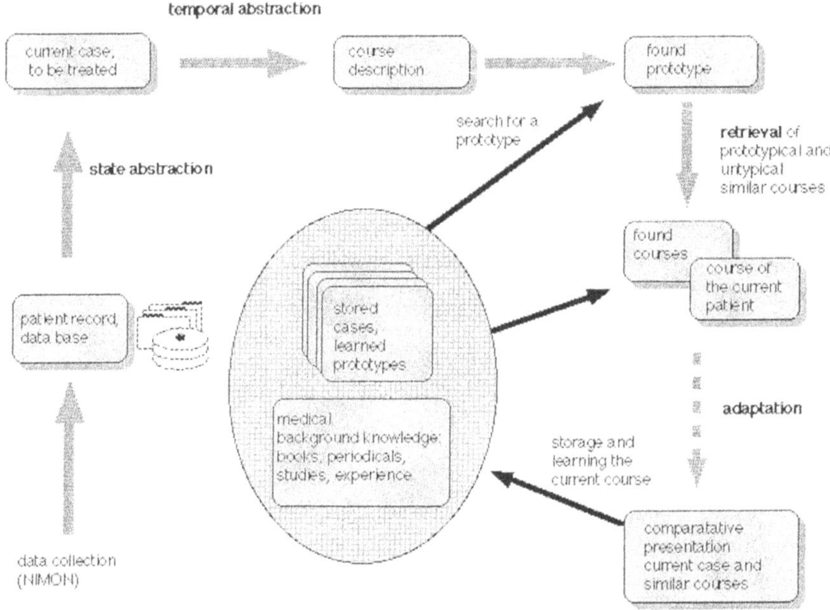

Figure 8. The Case-Based Reasoning cycle adapted to medical temporal abstraction.

We believe that – at least in the medical domain – prototypes are an appropriate knowledge representation form to generalize from single cases [27]. They help to structure the case base, to guide and to speed up the retrieval, and to get rid of redundant cases [28]. Especially for course prognoses they are very useful, because otherwise too many very similar, former courses would remain in the case base. So, the search of the fitting prototype becomes a sort of preliminary selection, before the main retrieval just has to take those cases into account that belong to the determined prototype. As our evaluation experiments in the kidney function domain have shown the fitting prototype should be searched non-sequentially.

5 Summary

In this chapter we have proposed a method for the prognoses of temporal courses, especially for application domains without well-known standards and without a complete domain theory. The idea of our method is to combine temporal abstractions with Case-Based Reasoning. We have presented the application of this method for the prognosis of kidney functions courses. Additionally, we have outlined first ideas for applying the same method for the prognosis of the spread of diseases in the TeCoMed project. Though there are some differences between both applications the main principles are the same. Finally, based on our experiences and considerations, we have adapted the well-known Case-Based Reasoning cycle to temporal course prognoses.

References

[1] Robeson, S.M. and Steyn, D.G. (1990), "Evaluation and comparison of statistical forecast models for daily maximum ozone concentrations," *Atmospheric Environment*, vol. 24 B, no. 2, pp. 303-312.

[2] Shahar,Y. (1999), "Timing is everything: temporal reasoning and temporal data maintenance in medicine," *Proceedings of AIMDM'99 Aalborg*, Springer-Verlag, Berlin, pp. 30-46.

[3] Haimowitz, I.J. and Kohane, I.S. (1993), "Automated trend detection with alternate temporal hypotheses," *Proceedings of the 13th International Joint Conference on Artificial Intelligence*, Morgan Kaufmann Publishers, San Mateo, pp. 146-151.

[4] Miksch, S., Horn, W., Popow, C., and Paky, F. (1995), "Therapy planning using qualitative trend descriptions," *Proceedings of the Fifth AIME Pavia*, Springer Verlag, Berlin, pp. 197-208.

[5] Schmidt, R., Boscher, L., Heindl, B., et al. (1995), "Adaptation and abstraction in a case-based antibiotics therapy adviser," *Proceedings of the Fifth AIME Pavia*, Springer-Verlag, Berlin, pp. 209-217.

[6] Swoboda, W., Zwiebel, F.M., Spitz, R., and Gierl, L. (1994), "A case-based consultation system for postoperative management of liver-transplanted patients," *Proceedings of the 12th MIE Lisbon*, IOS Press, Amsterdam, pp. 191-195.

[7] Gierl, L. and Stengel-Rutkowski, S. (1994), "Integrating consultation and semi-automatic knowledge acquisition in a prototype-based architecture: experiences with dysmorphic syndromes," *Artificial Intelligence in Medicine*, vol. 6, pp. 29-49.

[8] Lenz, M., Auriol, E., and Manago, M. (1998), "Diagnosis and decision support," in: Lenz, M., Bartsch-Spörl, B., Burkhard, H.-D., Wess, S. (Eds.), *Case-Based Reasoning Technology, from Foundations to Applications*, Springer-Verlag, Berlin, pp. 51-90.

[9] Smyth, B. and Keane, M.T.(1998), "Adaptation-guided retrieval: questioning the similarity assumption in reasoning," *Artificial Intelligence*, vol. 102, pp. 249-293.

[10] Schmidt, R., Pollwein, B., and Gierl, L. (1999), "Medical multiparametric time course prognoses applied to kidney function assessments," *International Journal in Medical Informatics*, vol. 53, no. 2-3, pp. 253-264.

[11] Schmidt, R. and Gierl, L. (2001), "Temporal abstractions and Case-Based Reasoning for medical course data: two prognostic applications," *Proceedings of Second MLDM Leipzig*, Springer-Verlag Berlin. (To appear.)

[12] Aamodt, A. and Plaza, E. (1994), "Case-Based Reasoning: foundation issues. Methodological variation- and system approaches," *AI Communications*, vol. 7, no. 1, pp. 39-59.

[13] Kolodner, J. (1993), "*Case-Based Reasoning*," Morgan Kaufmann Publishers, San Mateo.

[14] Macura, R. and Macura, K. (1995), "MacRad: radiology image resources with a case-based retrieval system," *Proceedings of the First ICCBR Sesimbra*, Springer-Verlag, Berlin, pp. 43-54.

[15] Quinlan, J.R. (1993), *C4.5, Programs for Machine Learning*, Morgan Kaufmann Publishers, San Mateo.

[16] Broder, A.J. (1990), "Strategies for efficient incremental nearest neighbor search," *Pattern Recognition*, vol. 23, pp. 171-178.

[17] Stottler, R.H., Henke, A.L., and King, J.A. (1989), "Rapid retrieval algorithms for Case-Based Reasoning," *Proceedings of the 11th IJCAI Detroit*, Morgan Kaufmann Publishers, San Mateo, pp. 233-237.

[18] Anderson, J.R. (1989), "A theory of the origins of human knowledge," *Artificial Intelligence*, vol. 40, Special Volume on Machine Learning, pp. 313-351.

[19] Jones, E.K. and Roydhouse, A. (1994), "Iterative design for case retrieval systems," *Proceedings of AAAI Workshop on Case-Based Reasoning, Seattle*, AAAI Press.

[20] Lekkas, G.P., Arouris, N.M., and Viras, L.L. (1994), "Case-Based Reasoning in environmental monitoring applications," *Applied Artificial Intelligence*, vol. 8, pp. 349-376.

[21] Lees, B. and Corchado, J. (1997), "Case based reasoning in a hybrid agent-oriented system," *Proceedings of the 5th GWCBR Kaiserslautern*, University Press Kaiserslautern, pp. 139-144.

[22] Wenkebach, U., Pollwein, B., and Finsterer, U. (1992), "Visualization of large datasets in intensive care," *Proc Annu Symp Comput Appl Med Care*, pp. 18-22.

[23] Tversky, A. (1977), "Features of similarity," *Psychological Review*, vol. 84, pp. 327-352.

[24] DeSarbo, W.S. et al. (1992), "TSCALE: a new multidimensional scaling procedure based on Tversky's contrast model," *Psychometrika*, vol. 57, pp. 43-69.

[25] Farrington, C.P. and Beale, A.D. (1977), "The detection of outbreaks of infectious diseases," in: Gierl, L. et al. (Eds.),

International Workshop on Geomedical Systems Rostock, Teubner-Verlag, Stuttgart Leipzig, pp. 97-117.

[26] Wilke, W., Smyth, B., and Cunningham, P. (1998), "Using configuration techniques for adaptation," in: Lenz, M., Bartsch-Spörl, B., Burkhard, H.-D., Wess, S. (Eds.), *Case-Based Reasoning Technology, from Foundations to Applications*, Springer-Verlag, Berlin, pp. 139-168.

[27] Schmidt, R. and Gierl, L. (2000), "Case-Based Reasoning for medical knowledge-based systems," *Proceedings of MIE and GMDS Hannover*, IOS Press, Amsterdam, pp. 720-725.

[28] Schmidt, R., Pollwein, B., and Gierl, L. (1999), "Experiences with Case-Based Reasoning methods and prototypes for medical knowledge-based systems," *Proceedings of AIMDM'99 Aalborg*, Springer-Verlag, Berlin, pp. 124-132.

Chapter 6

Pattern Recognition in Intensive Care Online Monitoring

R. Fried, U. Gather, and M. Imhoff

Clinical information systems can record numerous variables describing the patient's state at high sampling frequencies. Intelligent alarm systems and suitable bedside decision support are needed to cope with this flood of information. A basic task here is the fast and correct detection of important patterns of change such as level shifts and trends in the data. We present approaches for automated pattern detection in online-monitoring data. Several methods based on curve fitting and statistical time series analysis are described. Median filtering can be used as a preliminary step to reduce the noise and to remove clinically irrelevant short term fluctuations.

Our special focus is the potential of these methods for online-monitoring in intensive care. The strengths and weaknesses of the methods are discussed in this special context. The best approach may well be a suitable combination of the methods for achieving reliable results. Further investigations are needed to further improve the methods and their performance should be compared extensively in simulation studies and applications to real data.

1 Introduction

In the last three decades there has been a rapid development of the equipment used for monitoring of critically ill patients in intensive care. Clinical information systems (CIS) allow a comprehensive acquisition and storage of this data with high sampling frequencies. Up to 2000 physiological variables, laboratory data, device parameters, medication parameters etc. are recorded in the course of time. The volume of these data frequently exceeds the human ability to assimilate, identify and conceptually relate the observations [1]. While a physician can be confronted with more than 200 variables of the critically ill patient during a typical morning round [2], it is well-known that human beings are not able to develop a systematic response to any problem involving more than seven variables [3]. Moreover, humans are limited in their ability to judge the degree of relatedness between only two variables [4]. Another serious problem is the detection of slow trends in the data. Additionally, personal bias, subjective experience, and a certain expectation toward the respective intervention may distort an objective judgment [5]. Thus electronic bedside decision support offers large potential benefit.

Clinical decision support aims at providing physicians with therapy guidelines directly at the bedside. The best recommendation possible should be supplied under all circumstances [6]. To achieve this goal quantitative measurements and qualitative reasoning have to be integrated in a system that recommends interventions in real time. One step to reach this aim is to abstract the numerical measurements of the patient's vital signs into qualitative patterns of change which are clinically relevant. Such patterns provide information on whether, e.g., an intervention is successful or a clinical complication occurs.

Currently, in intensive care only critical patterns are searched for by comparing each observation individually to fixed thresholds which have to be chosen by the health care professional. These automatic alarms systems produce a huge number of false alarms due to measurement artifacts, patient movements or transient fluctuations past the set alarm limit. Coughing, turning of and movements of the patient, therapy, blood sampling and flushing of the catheters cause transient artifacts. Most of the alarms, about 90%, are irrelevant in terms of patient care [1], [7], [8]. O'Carrol [9] reports that in a study of 1455 alarms that occurred during three weeks only eight cases were actually life-threatening. While the large number of false alarms could be reduced by choosing sufficiently wide thresholds, this precludes the detection of intervention effects and the problems in detecting slow monotone trends increase even more. The unreliability of fixed threshold alarms may even lead to critical or life-threatening situations.

Usually changes of a variable over time are more important than a single pathological value at the time of observation. The online detection of qualitative patterns such as artifacts, level changes and trends is important for assessing the patient's state. Several approaches for qualitative abstraction of the data into such patterns have been suggested. Haimowitz and Kohane [10], [11] fit trend templates to the data, which are predefined functional forms of relevant patterns. Miksch et al. [12] propose to measure deviations of measurements from a given target range. Mäkivirta [1] suggests to preprocess the data by a median filter, which is common practice in signal processing. Statistical intervention analysis is useful for retrospective assessment of the effectiveness of therapeutical actions. In statistical process control, Shewhart, CUSUM- and EWMA-charts are applied. Smith and West [13] and Gather et al. [14] use methods based on statistical time series analysis.

In the following sections we describe several approaches to pattern recognition in a sequence $y_1, \ldots, y_n$ of subsequent observations of a physiological variable measured at equidistant time points with a high sampling frequency such as one observation per minute. We neglect approaches like RESUME [15], [16] which are designed for low-frequency data.

2 Curve Fitting

Several curve fitting methods have been developed and applied to clinical problems, the most obvious being the first order approximation with a straight line. Although higher order polynomials have also been used, their applications are limited to cases where such a relation is assumed to exist.

Haimowitz and coworkers [10], [11] developed TrenDx using the concept of trend templates for diagnosing pediatric growth disorders and detecting clinically significant trends in hemodynamics and blood gases in intensive care units. A trend template denotes a time-varying pattern in multiple variables common to a diagnostic population. Predefined patterns of normal and abnormal trends represent disorders as typical patterns of relevant parameters. Trends are diagnosed by matching the observed data to the trend templates.

Each pattern contains representations of landmark events and a set of phases. The trend templates are temporally linked to the patient history. The anchor points used for the trend templates are not necessarily identical to particular events in the patient's history, but may lay within a certain time range around the time point of an event. The phases are represented by a partially ordered set of time intervals with variable endpoints to consider uncertainties. Low-order polynomials constrain the observed values of the variables during each time interval. Hence, these value constraints are parameterized linear regression models describing variation in data assigned to an interval. Haimowitz and coworkers use seven qualitatively distinct elementary regression models corresponding to constant, linear and quadratic models. They consider these seven models to be sufficient to roughly distinguish between different behaviors.

The trend templates are organized in monitor sets which belong to a certain clinical state ("context") of the patient, i.e., all trend templates within a monitor set are assumed to belong to the same patient's state described by the context. One of the monitor sets is the expected or normal model, while the others should warrant the attention of the physician. All members of a monitor set are concurrently matched to the observed data and a matching score is calculated for each template. This score is defined to be the mean absolute percentage error between the observations and regression model estimations. An overall score is obtained for each monitor set using a weighted average of these error scores.

If multiple hypothesized diseases and disease chronologies have to be matched to the data and compared to one another this repeated calculation can become prohibitively expensive. Therefore some simplifying strategies are implemented in the system.

The major disadvantages of TrenDx are the necessary predefinition of the expected normal behavior of the variables during the whole time considered and the usage of absolute value thresholds matching a trend template, which do not take into account the different degrees of parameters' abnormalities. Moreover, the thresholds should be dynamically derived according to the patient's status in the past [12]. TrenDx is designed for trend detection only and does not cover the activation of therapeutic actions and the assessment of the effectiveness of therapeutic actions.

Miksch et al. [12] developed VIE-VENT. In this system knowledge-based monitoring and therapy planning for artificially ventilated newborn infants are integrated. First, the incoming data are validated to arrive at reliable values. Then these values are abstracted into qualitative descriptions, i.e., temporal patterns. The procedure for temporal data-abstraction consists of transformation of quantitative measurements to qualitative values, smoothing of data oscillating near thresholds, smoothing of schemata for data-point transformation, context-sensitive adjustment of qualitative values and transformation of interval data.

The measurements are transformed into qualitative values by dividing the numerical range of each variable into regions of interest and attainable goals. Each of these regions corresponds to a qualitative value. For blood-gas measurements, e.g., they use seven qualitative categories of blood-gas abnormalities. The corresponding regions are smaller the nearer the target range since they represent different degrees of abnormalities. The definition of these regions requires specific predefined target values depending on different attainable goals.

Since oscillation of the measurements near a threshold causes rapid oscillation of the qualitative categories smoothing is applied to keep the categories stable. Miksch et al. predefine neighborhoods of the regions of interest and a maximum smoothing activation time period. The size of the neighborhood depends on the size of the region. Smoothing starts if the category for the incoming observation y_t is not the same as the category for the previous measurement and y_t lies within the neighborhood of the previously observed category. In this case the qualitative values of the new observations are set to the category before the change. Smoothing stops if the new observation leaves the neighborhood or the maximal smoothing time is reached.

For different contexts, i.e., different clinical states, different target values are used for the categories. Changing the context would result in a sudden shift of the category resulting possibly in a recommendation for drastic therapeutic intervention. Therefore the thresholds of the schemata are smoothed within a predefined time period, which may be three to eight hours, after changing the context.

Schemata for trend-curve fitting are defined describing the dynamics of different degrees of parameter abnormalities. The qualitative categories are combined with qualitative trend descriptions expected by the physician. In this way qualitative descriptions resulting in an explicit categorization of the change of the variables over time are obtained. Changes of the parameters are modeled by exponential functions, which are piecewise linearized to reduce the complexity. For instance, Miksch et al. consider four different durations of trends. In this way, e.g., the results of therapeutic actions can be assessed.

Miksch et al. tested their approach on simulated and real data of blood-gas measurements. They found the therapeutic recommendations to be consistent and reasonable, except when invalid measurements occurred, but they also considered some limitations of their approach. Information about the frequency of therapeutic interactions in the past would be useful for future reasoning. Data abstraction should include a memory which weights the data since more recently observed data are more important for the reasoning process than data observed in older time periods. Moreover, they found more noise in the real data than expected because of measurement errors, online transmission problems and input from different people in different environments and in different experimental settings.

We use two sequences of heart rate measurements of critically ill patients to illustrate the presented techniques. In example 1, shown in Figure 1, an experienced physician found a clinically relevant increase of the heart rate. For qualitative data abstraction as suggested by Miksch et al. we divide the range of the measurements into categories corresponding to the target range $[60, 80]$, slight deviation $[50, 60) \cup (80, 100]$ from this target range, strong deviation $[40, 50) \cup (100, 130]$ and extreme deviation $[0, 40) \cup (130, \infty)$. The neighborhood of the target range is defined to be $[55, 60) \cup (80, 90]$, while the neighborhood of slight deviation is $[45, 50) \cup [60, 65] \cup [75, 80] \cup (100, 110]$, compare the reference lines shown in the Figure. It should be noted that ideally these regions should depend on the physical constitution of the individual and thus they are hard to specify properly in advance. The maximal smoothing time is set to three subsequent observations. Abstracting the quantitative measurements into qualitative information at an early stage as is done here means a substantial loss of information for the further analysis. If we only use these qualitative categories we can detect a shift only if the categorical value changes. This may prevent the detection of clinically interesting shifts towards the boundary of the current category. On the other hand, in spite of the smoothing applied we may detect meaningless changes occurring when the time series oscillates at the border of two categories. This effect may be diminished by increasing the maximal smoothing time, but this also implies a larger delay of the detection of clinically relevant changes. The category obtained for the current observation may even be random and depend on the actual beginning of the data recording. This becomes obvious if we consider the simple example $77, 77, 77, 83, 83, 83, 77, 77, 77, 83, 83, 83 \ldots$ and the regions defined above. The qualitative classification of this sequence depends on whether we start the analysis at time point one or at time point four. Of course such behavior can be detected by the physician if he takes a look at the quantitative measurements, but this information is lost for a further automatic analysis.

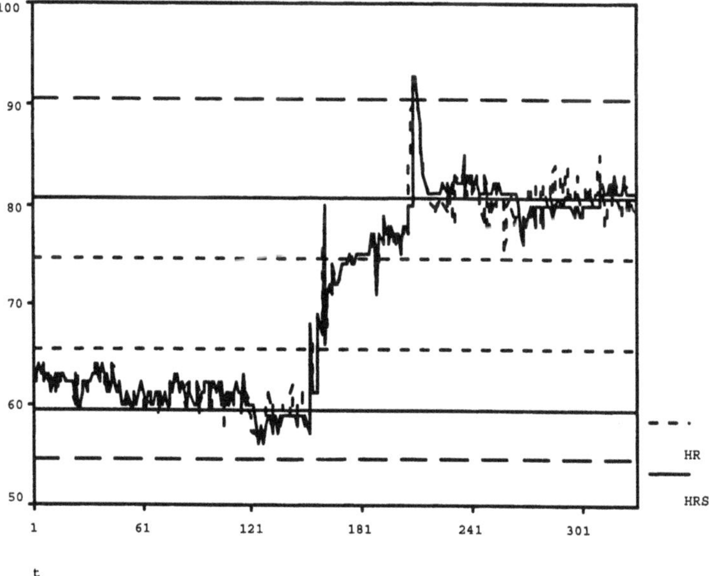

Figure 1. Example for data abstraction of a sequence of heart rate measurements (HR) into qualitative categories. The data are smoothed within certain neighborhoods of the interesting regions (HRS).

In example 2, shown in Figure 2, a slow monotone decrease (negative trend) of the heart rate occurs. We use this example to illustrate some basic weaknesses of curve fitting methods. We fit a linear trend to each time window of 60 subsequent observations and test whether a linear trend has occurred in this time period. This is accomplished by testing the significance of the slope parameter in this linear model. In this example, the test statistic is well between -2.0 and 2.0 and thus it is not significant at any time point if we compare it to the critical values ±3.0, which are used since we do multiple testing. Curve fitting methods behave poorly in the presence of large noise and short term transients. These problems, which occur frequently in intensive care monitoring [1], can be overcome by median filtering, which is discussed in the next section.

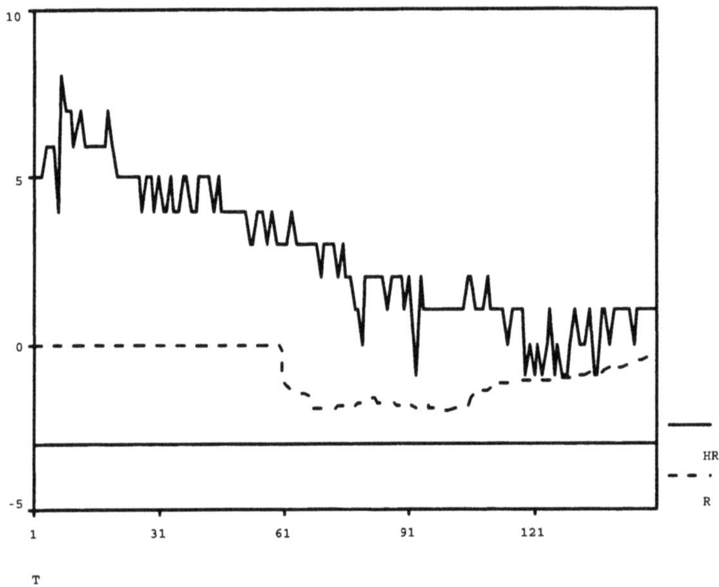

Figure 2. Sequence of test statistics (R) for the slope in a simple linear model fitted to a moving time window of 60 subsequent observations of the heart rate (HR). We subtracted 70 from the measurements for the reason of illustration. The test statistic never crosses the reference line at -3.0 although an experienced physician found a negative trend in this sequence.

Both methods presented in this section do not consider correlations within the data and they demand predefinition of expected behavior, which may be hard to specify in advance in critical care.

3 Median Filtering

Mäkivirta and coworkers [1], [17] suggest the use of median filters for preprocessing the observations. Median filtering is a basic non-linear method for signal processing which helps to remove noise and transients from the signal without distortion of the signal's baseline. For subsequent observations $y_1, \ldots, y_N$ the median filter of length $2l + 1$, $l \in \mathbb{N}$, is defined by

$$\tilde{y}_t = median\{y_{t-l}, \ldots, y_{t+l}\}, \quad t = l + 1, \ldots, N - l .$$

This filter responds to a change in the signal with a time delay of l observations. On the other hand, this filter tolerates up to l artifacts within a time span of $2l + 1$ observations without breaking down completely. Therefore, there is a trade-off between noise attenuation and the time delay. Particularly, these filters have excellent attenuation of impulsive noise since their response to an impulse is zero. Median filters preserve sudden level changes in the signal and diminish the strong noise-like variability with frequent transients. The bias of the filter is related to the variance of the noise and the height of the edge. Therefore, in the presence of excessive noise the ability of a median filter to preserve edges deteriorates.

Mäkivirta [1] noted that increasing the length of the median filter causes a radical decrease of the incidence of both false and true alarms. Therefore, a dual limit alarmer was proposed [17] based on a short and a long median filter. For the short median filter, wide control margins are used to detect sudden critical changes in the patient's state while for the long median filter narrower margins are used to assess therapy effects. Trends can be detected by fitting a linear regression model to the filtered data. In a clinical study, a decrease of about 64 % of the false alarms could be achieved in comparison to usual threshold alarming performed from the unfiltered data.

Figure 3 illustrates the idea of a dual limit alarmer. Filtered series are obtained by applying two median filters with l=2 and l=12 respectively to examples 1 and 2. There is a couple of possibilities for calculating control margins based on these filtered series. We used a deviation of 10% from the last value of the series filtered by the short filter and of 5%

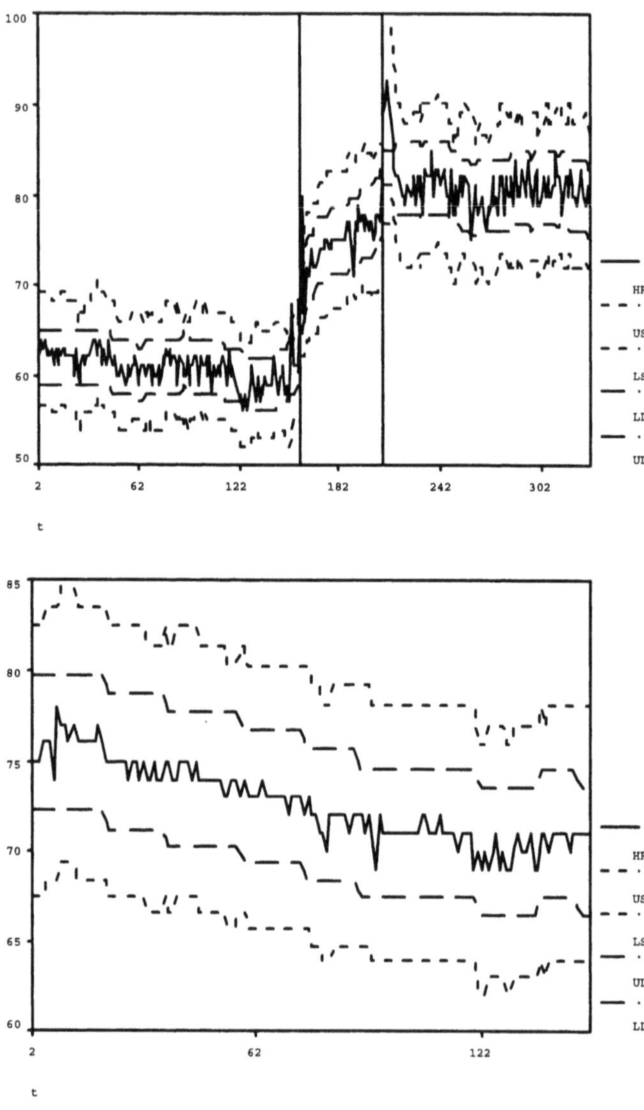

Figure 3. Pairs of control limits based on a short (l=2, LS and US) and a long (l=12, LL and UL) median filter with wide and narrow margins respectively. In the first example, alarms are triggered at $t = 158$ and $t = 207$ since the filtered series exceeds the control margins for the short median filter. In the second example it becomes obvious that additional methods such as curve fitting are needed to detect a slow trend, but filtering the series makes this task easier because of the reduced nonsystematic variability. The smoother behavior of the narrow control margins is due to the larger length of the underlying median filter.

from the last value of the series filtered by the large filter for calculating the control margins. Changes of these magnitudes have to be regarded as clinically relevant and clinically interesting respectively.

Several variations of the basic principle of median filtering can be found in the literature. Justusson [18], [19] suggests a weighted median filter giving more weight to the current observation. Nieminen et al. [20], [21] develop median filter based algorithms for real-time trend detection. Their methods are based on so-called FIR median hybrid (FMH) filters which are constructed from several linear subfilters using median operations. These algorithms gradually refine trends when new data becomes available. A particular concern is the detection of slope changes in a trend. This is important because of the frequent transients and sudden changes in the physiological variables of a critically ill patient. Nieminen et al. state that their methods have excellent noise attenuation properties with respect to the delay of the system. However, all these approaches do not include correlations within the data into the analysis.

4 Statistical Time Series Analysis

Endresen and Hill [22] consider methods which are based on the independence of the observations to be not appropriate for the analysis of an observed series from variables such as the heart rate and the blood pressures. When analyzing such sequences of observations (time series) they found large positive correlations between subsequent measurements. Woodward and Gray [23] point out that high positive correlations within subsequent observations of a time series do often lead to the false detection of a trend. This supports the demand to incorporate autocorrelations within a time series into the analysis as stated by Endresen and Hill [22].

Statistical time series analysis allows to consider autocorrelations, leads to interpretable descriptions of complex underlying dynamics, provides forecasts, gives confidence bounds and allows to assess the clinical effects of therapeutic interventions [24], [25]. It has also been shown to be useful for online detection of characteristic patterns in univariate time series. For pattern detection in single variables dynamic linear models

[26], ARIMA-models [27], [28] and models based on a phase space type approach [29] have been applied.

In this section we give a brief, but general introduction into statistical time series analysis. Here, the measured sequence of observations (time series) $y_1, \ldots, y_N$ is assumed to be generated from a stochastic process $\{Y_t : t \in \mathbb{Z}\}$, i.e., y_t is regarded as a realization of a random variable $Y_t, t = 1, \ldots, N$. We assume mean and variance to exist for the random law describing the distribution of Y_t and denote them with μ_t and σ_t^2 respectively. Often joint normality of the random variables is assumed. Under this assumption all information about the dependencies between the variables is contained in their correlations. Statistical time series analysis incorporates these dependencies into the reasoning process and uses them to improve decision making based on the observed values. Therefore, the main focus of basic time series models are usually the correlations between the variables while the mean of the random variables Y_t is assumed to be constant over time, i.e., $\mu_t = \mu$ for all $t \in \mathbb{Z}$. This property is called mean-stationarity, while we call the series variance stationary if the variances of the random variables are constant over time, i.e., $\sigma_t^2 = \sigma^2, t \in \mathbb{Z}$. If not only the variances $\sigma_t^2 = Cov(Y_t, Y_t)$, but all covariances $Cov(Y_t, Y_{t+h})$ of the stochastic process are independent of t for all time lags $h \in \mathbb{N}$ we call the series second order stationary and define the autocorrelation function $\rho : \mathbb{N} \to \mathbb{R}$ by

$$\rho(h) = \frac{Cov(Y_t, Y_{t+h})}{Var(Y_t)}.$$

A series which is both mean stationary and second order stationary is called weakly stationary. For the ease of notation, the constant mean μ is usually assumed to be zero. In a practical application we can estimate the mean by the sample mean and subtract this estimate from the series if we assume the underlying stochastic process to be at least mean stationary.

Autoregressive integrated moving average (ARIMA) models are perhaps the most often used time series models, see [30] and [31]. An ARIMA(p, d, q) model assumes that the d-times differenced time series follows a weakly stationary ARMA(p, q) model. Differencing means to calculate the increments, i.e., the differences $d_t = y_t - y_{t-1}$ between subsequent observations.

An ARMA(p, q) model assumes that Y_t can be represented as

$$Y_t - \phi_1 Y_{t-1} - \ldots - \phi_p Y_{t-p} = a_t - \theta_1 a_{t-1} - \ldots - \theta_q a_{t-q},\ t \in \mathbb{Z},$$

where $\{a_t : t \in \mathbb{Z}\}$ is a white-noise process of independent identically $N(0, \sigma_a^2)$-distributed random variables ("shocks") and $\phi_1, \ldots, \phi_p$, $\theta_1, \ldots, \theta_q$ are unknown weights measuring the influence of past observations and past shocks on the current observation. For $q = 0$ we simply call the process AR(p) process, while it is called MA(q) process if $p = 0$. The statistical properties of these models can be exploited using the algebraical theory for polynomials. For this reason the backshift operator B is defined by $By_t = y_{t-1}$. Then $\Phi(B) = 1 - \phi_1 B - \cdots - \phi_p B^p$ and $\Theta(B) = 1 - \theta_1 B - \cdots - \theta_q B^q$ are polynomials in B of degrees p and q respectively. In ARMA models it is assumed that both polynomials have all their roots outside the unit circle. The former demand follows the assumption of (weak) stationarity of the time series. For an AR(1) model for instance this means that the absolute value $|\phi_1|$ must be smaller than one. An AR(1) model with $\phi_1 = 1$, which has a root located at 1, is a non-stationary ARIMA(0,1,0) model. In this simple model, which is called a random walk, the variance of the observation Y_t increases over time.

Any ARMA model which fulfills the assumptions stated above can also be written in the so-called $AR(\infty)$ form as

$$\Pi(B)Y_t = a_t$$

where $\Pi(B) = \Theta(B)^{-1}\Phi(B) = 1 - \pi_1 B - \pi_2 B^2 - \ldots - \ldots$, or in the $MA(\infty)$ form as

$$Y_t = \Psi(B)a_t$$

where $\Psi(B) = \Phi(B)^{-1}\Theta(B) = 1 - \psi_1 B - \psi_2 B^2 - \ldots$.

ARMA models for a given time series $y_1, \ldots, y_N$ are often constructed using an iterative procedure described in [31]. In this procedure the steps identification, estimation and diagnosis are repeated. An investigator using this procedure should have some experience with it since active interaction is needed.

In the identification step, the orders p and q of the ARMA model are determined. For this reason the sample autocorrelation function SACF

$$\hat{\rho}(h) = \frac{\sum_{t=1}^{N-h}(y_t - \overline{y})(y_{t+h} - \overline{y})}{\sum_{t=1}^{N}(y_t - \overline{y})(y_t - \overline{y})}, \quad \overline{y} = \frac{1}{N}\sum_{t=1}^{N} y_t,$$

and the sample partial autocorrelation function SPACF are analyzed. The former estimates the unknown autocorrelation function of the process, while the latter estimates the autocorrelations between Y_t and Y_{t+h} after elimination of the linear influences of $Y_{t+1}, \ldots, Y_{t+h-1}$ on Y_t and Y_{t+h}. The SPACF can be calculated from the SACF using some recursions [30], [31]. For an AR(p) process the SACF should be exponentially declining and the SPACF should be about zero at all time lags $h > p$. Conversely, for an MA(q) process the SACF should be about zero from time lag $q + 1$ on, while the SPACF should be exponentially declining. For mixed ARMA(p, q) processes the SACF and SPACF show more complex patterns. Their identification often takes several cycles of the procedure.

In the estimation step, the unknown parameters $\phi_1, \ldots, \phi_p, \theta_1, \ldots, \theta_q$ are estimated from the data. While usage of exact maximum likelihood estimators is preferable for short time series with $N \leq 50$ observations, for long time series with $N > 100$ simpler techniques such as conditional maximum likelihood, which is equivalent to conditional least squares under the assumption of normality, usually provide almost identical results. A detailed description of these algorithms can be found in [30] and [31].

In the diagnosis step we check whether the model provides an adequate description of the time series. Commonly the estimated residuals are compared to white noise. The hypothesis of white noise can be tested with the Box-Ljung Q-statistics for instance, while the Durbin-Watson test gives information about possible non-stationarities of the time series. If the model turns out to be satisfactory we can use it for further analysis, otherwise we should modify it according to the impressions gained in the diagnosis stage and then iterate the three steps of the procedure until a satisfactory model is derived.

5 Intervention Analysis

The basic assumption of (weak) stationarity is not fulfilled for the physiological time series observed in intensive care. In the contrary, the detection of changes in the patient's state, which may be caused by possibly life-threatening complications or by changes in medicamentation, are the main reason for analyzing these data. Moreover, biorhythm causes systematic long-term oscillations, and many measurement artifacts occur in clinical time series. Intervention analysis has been proposed to incorporate patterns of change in statistical time series analysis, and appropriate methods for the retrospective detection of such extraordinary events have been developed. The effects of such events are modeled by using deterministic functions of time for describing a time-varying mean structure. Although these methods cannot be applied online since past and future observations of the variable are used in the analysis, their retrospective application can be useful to assess the effectiveness of therapeutical methods and to construct a knowledge base for future bedside decision support.

Intervention analysis is accomplished via an iterative procedure for detection and removement of patterns of change. Dummy variables are used to model changes in the mean of a time series [32]. Chang et al. [33] propose an iterative procedure based on repeated likelihood ratio tests for outlier detection and parameter estimation which is generalized in Tsay [34] to include level shifts and temporary changes. Nowadays, the procedure of Chen and Liu [35] for outlier detection and parameter estimation in ARIMA models seems to be widely used. Commonly four patterns of change in the mean are considered in statistical time series analysis. These are additive outliers AO, innovational outliers IO, level shifts LS and transient changes TC [36].

An additive outlier AO represents an isolated spike in the time series, which can be caused by an external error, e.g., a measurement artifact, changing the observed value at one particular time point τ without further effects on the future values of the time series. In this case, we observe a modified series $z_1, \ldots, z_N$ instead of $y_1, \ldots, y_N$, which is related to the latter by

$$z_t = \begin{cases} y_t, & t \neq \tau \\ y_t + \omega_A, & t = \tau. \end{cases}$$

Here, ω_A is the unknown effect of the AO. Using the $AR(\infty)$-representation of the ARMA model, we can write the AO model as

$$\Pi(B)(z_t - \omega_A I_t^{(\tau)}) = a_t .$$

The dummy variable $I_t^{(\tau)}$ is identical to zero for all t with the only exception $t = \tau$ where $I_\tau^{(\tau)} = 1$. An additive outlier can have serious effects on the estimated residuals of the future observations after time point τ and on the estimated model parameters. The residuals e_t of the observed process after time point τ are related to the residuals of the underlying ARMA process via

$$e_{\tau+j} = a_{\tau+j} - \pi_j \omega_A, \quad j \geq 0 ,$$

where $\pi_0 = -1$. Moreover, it can be proved that a single additive outlier pushes all sample autocorrelations toward zero [36].

An innovational outlier IO can be due to an internal change or endogenous effect which causes an extraordinary shock at some time point τ. The model for an IO is

$$z_t = y_t + \Psi(B)\omega_I I_t^{(\tau)} ,$$

where ω_I is the unknown size of the IO. Hence, the effects of an IO on the observed time series depend on the ARIMA model. The IO model can be written equivalently as

$$\Pi(B) z_t = a_t + \omega_I I_t^{(\tau)} .$$

In case of known model parameters, an IO only affects the residual of the observed process at the time point τ

$$e_\tau = a_\tau + \omega_I ,$$

and $e_t = a_t$ otherwise. It is well-known that IOs usually have minor effects on the sample autocorrelations and on the parameter estimates

[37]. For a time series which forms white noise AO and IO are equivalent, while an IO in a random walk is equivalent to a level shift.

A level shift LS corresponds to a step change in the mean level of the process at a time point τ. The model for a LS is

$$z_t = \begin{cases} y_t, & t < \tau \\ y_t + \omega_L, & t \geq \tau, \end{cases}$$

where ω_L is the size of the step. This model can also be written as [36]

$$\Pi(B)z_t = a_t + \omega_L \Pi(B)(1-B)^{-1} I_t^{(\tau)} .$$

A LS affects all residuals after the time point when the shift occurs:

$$e_{\tau+j} = a_{\tau+j} + \ell_j \omega_L, \quad j \geq 0,$$

where $\ell_j, j = 0, 1, \ldots$ are the coefficients of $\ell(B) = \Pi(B)(1-B)^{-1}$. A LS pushes the sample autocorrelations at all time lags to one if there are many observations before and after the shift.

A transitory change TC is a temporary LS that dies out exponentially with rate δ and initial impact ω_T, i.e., it is a level shift that decreases with time and fades to zero, see Tsay [34]. Alternatively we could describe it as a spike that takes a few time periods to disappear. The model for a TC is

$$\Pi(B)z_t = a_t + \omega_T \Pi(B)(1-\delta B)^{-1} I_t^{(\tau)}.$$

For $\delta = 1$, a TC is identical to a LS, while for $\delta = 0$, a TC is an AO. Therefore, a TC with $0 < \delta < 1$ can be seen as an intervention in between a LS and an AO. Generally, the parameter δ specifies how fast the effect of a TC decreases. Tsay [34] chose the values $\delta = 0.8$ and $\delta = 0.6$ and stated that the results were only slightly different in his applications. In case of high positive autocorrelations within the time series, the TC model is close to the IO model. This can result in some misclassifications [34].

Furthermore, in the physiological variables observed in intensive care slow monotone trends can be found which lead to a modified level of the process. A linear trend can be modeled by a ramp shift outlier RS [36]

$$z_t = \Psi(B)a_t + \omega_R R_t^{(\tau)} \quad (1)$$

$$R_t^{(\tau)} = \begin{cases} 0, & t < \tau \\ t-\tau, & t \geq \tau. \end{cases} \quad (2)$$

$R_t^{(\tau)}$ is called a ramp effect. This model means a temporary LS over the first differences after time point τ

$$(1-B)z_t = \Psi(B)(1-B)a_t + \omega_R(1-B)^{-1}I_t^{(\tau)}.$$

Piecewise linear trends, i.e., trends with a slope which changes occasionally, can be modeled by several subsequent ramp shift outliers.

For retrospective estimation and testing of intervention effects we first assume that the parameters of the ARMA model for y_t are known. Let $e_t = \Psi(B)z_t$ be the residuals from the observed series given the true model parameters. For an AO, IO, LS and TC the model for the residuals can be written in regression form as

$$e_t = \omega x_t + a_t$$

where we have $\omega = \omega_A$ and $x_t = \Pi(B)I_t^{(\tau)}$ for AO, $\omega = \omega_I$ and $x_t = I_t^{(\tau)}$ for IO, $\omega = \omega_L$ and $x_t = \Pi(B)(1-B)^{-1}$ for LS, and $\omega = \omega_T$ and $x_t = \Pi(B)(1-\delta B)^{-1}$ for TC. As the model parameters and therefore x_t are assumed to be known and the residuals a_t are independent, ω can be estimated by ordinary least squares, leading to $\hat{\omega} = \sum_t e_t x_t / \sum_t x_t^2$, with variance $\sigma_a^2(\sum x_t^2)^{-1}$ [36]. Estimation and testing of an RS can be done by applying the methods designed for an LS to the differenced residuals.

In the AO case this approach leads to

$$\hat{\omega}_A = \rho_A^2 \Pi(F) e_\tau,$$

where F is the forward shift operator defined by $Fz_t = z_{t+1}$, and $\rho_A^2 = (1 + \pi_1^2 + \ldots + \pi_{N-\tau}^2)^{-1}$. This result corresponds to the fact that all residuals after τ are affected by an AO and therefore all of them carry some information. The variance reads $Var(\hat{\omega}_A) = \rho_A^2 \sigma_a^2$.

For an IO only the residual at time τ carries information on ω_I. The estimate is

$$\hat{\omega}_I = e_\tau$$

and the variance is $Var(\hat{\omega}_I) = \sigma_a^2$.

In case of an LS, all residuals after the change point τ are informative. The estimate

$$\hat{\omega}_L = \rho_L^2 \ell(F) e_\tau$$

with $\rho_L^2 = (1 + \ell_1^2 + \ldots + \ell_{N-\tau}^2)^{-1}$ combines this information in a linear way. It can be shown that this statistic measures the difference of the levels before and after time point τ [36]. The variance of the estimate is $Var(\hat{\omega}_L) = \rho_L^2 \sigma_a^2$.

For a TC we have

$$\hat{\omega}_T = \rho_\tau^2 \beta(F) e_\tau$$

where $\rho_T^2 = (1 + \beta_1^2 + \ldots + \beta_{N-\tau}^2)^{-1}$ and the β_i are the coefficients of $\beta(B) = \Pi(B)(1 - \delta B)^{-1}$. The variance is $Var(\hat{\omega}_T) = \rho_T^2 \sigma_a^2$.

In order to test whether an outlier of known type $j \in \{AO, IO, LS, TC\}$ has occurred at time point τ, one typically tests

$$H_0 : \omega_j = 0 \text{ versus } H_1 : \omega_j \neq 0$$

using the likelihood ratio method. This criterion leads to comparing the parameter estimate to its standard error. The resulting test statistic

$$\lambda_{j,\tau} = \frac{\hat{\omega}_{j,\tau}}{\rho_{j,\tau} \sigma_a},$$

can be compared with a percentile of the student-t-distribution. We add the index τ to stress that the corresponding statistics are calculated for a particular time point τ.

On the other hand, if the type, but not the location of an outlier is known, a test statistic based on the maximized likelihood ratio $\lambda_j = \max_\tau \{\lambda_{j,\tau}\}$ can be used. The distribution of this maximum is complicated because of the correlations between the $\lambda_{j,\tau}$. Approximations of these distributions based on simulations can be found in Chang et al. [33] and Ljung [38].

For the realistic case that neither the model parameters nor the location or the type of the outliers are known, Tsay [34] suggested an iterative procedure for outlier detection. This procedure consists of specification, estimation, detection and removal cycles. Simulation studies revealed that this procedure seems to work very well in case of isolated outliers [34]. It can be summarized as follows:

1. An ARMA model is fitted to the observed time series using maximum likelihood under the assumption that no outlier has occurred; the model residuals are calculated.

2. The maximum of all likelihood ratio statistics $\lambda = \max_j \lambda_j$ is calculated. If λ is larger than a predetermined constant c we assume that an outlier of the corresponding type has occurred at this particular time point. We define new residuals by subtracting the estimated effects of this outlier from the model residuals.

3. Using these new residuals, a new estimate of the residual variance is obtained, and the likelihood ratios are computed again using the new residuals from step 2. Steps 2 and 3 are repeated until no further outliers are found.

4. In the second stage the sizes of the identified outliers and the parameters of the time series model are estimated jointly by fitting the joint model including the interventions effects to the observed time series. Using the resulting residuals steps 1 to 4 are repeated until no further outliers are found.

Imhoff et al. [39] employed intervention analysis to assess the effects of therapeutic interventions at known times within short time series of pulmonary target variables. They only considered level shifts since these describe relevant, long-term therapeutic effects measuring the change in mean level caused by an intervention. As a result, they could separate effective from non-effective interventions. This allows to reconsider therapeutic strategies.

Applying intervention analysis to example 1 results in the detection of level shifts at $t = 160$, $t = 186$ and $t = 227$ with sizes 6.5, 1.8 and 1.4 respectively, while a transitory change with size 14.4 was detected at $t = 208$ for $\delta = 0.8$, cf. Figure 4. For the mean corrected series, an AR(2) model with $\phi_1 = 0.21$ and $\phi_2 = 0.30$ was found to be satisfactory. Intervention analysis has some problems just like any other method if many disturbances occur within short time intervals. In this example, dummy variables for both a transitory change and a level shift at $t = 208$

were added during the iterations, but the dummy for the level shift turned out to be not significant in the final analysis if the transitory change was included in the model.

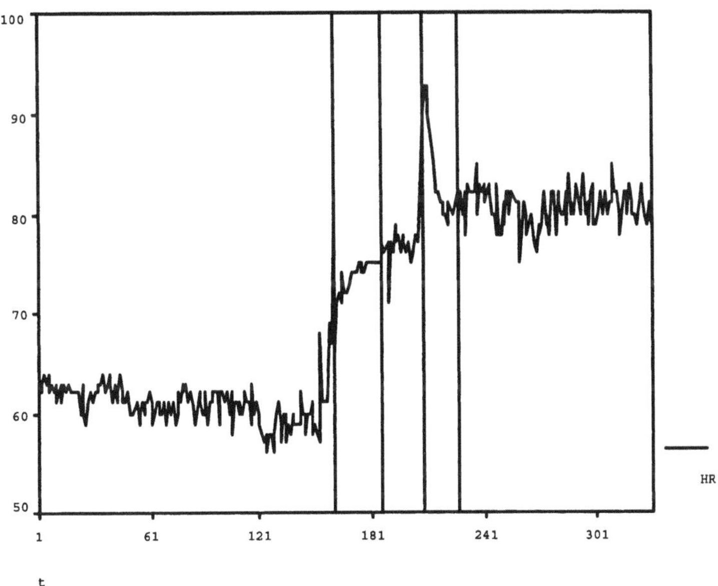

Figure 4. Application of intervention analysis to example 1 results in the detection of four disturbances at the time points marked by vertical reference lines.

6 Statistical Process Control

Statistical process control is a widely applied methodology for controlling industrial manufacturing processes for instance. Control charts such as the Shewhart-, the CUSUM- and the EWMA-chart aim at fast detection of systematic changes, particularly level shifts, in a process. The standard measure of the performance of a control chart is the average run length ARL, which is the expected time until an alarm is given. One has to consider both the in-control and the out-of-control ARL of a chart. The former is the ARL of a process which is under control, i.e., if no shift occurs within the time period considered. The latter is the ARL in the case that a shift by a certain amount occurs. It is a function of the size of this shift. For simplicity the shift is usually assumed to occur at the first observation considered. Obviously, large values of the in-control ARL and small values of the out-of-control ARL are good. One has to find a balance between these goals since increasing the in-control ARL by choosing larger control limits for a special chart means to increase the out-of-control ARL, too. To compare two control charts, usually the in-control ARL of both charts is set to a certain value by adjusting the control limits and then the out-of control ARLs of the charts are considered.

The performance of classical control charts suffers seriously from autocorrelations [40]-[43]. Two approaches have been proposed to overcome this problem [42]: Modified control charts compare the original measurements to control limits which are appropriately changed to consider autocorrelations. Forecast-based monitoring schemes consist of a two-stage procedure. First a time series model is fitted to the data and then the one-step ahead forecast residuals are monitored using traditional control charts. If the time series model is correctly specified, the forecast residuals are independent and identically distributed with mean zero during the steady-state, and thus the traditional control chart used in the second step should be reliable. Forecast-based monitoring schemes often employ autoregressive AR(p)-models since they constitute a quite flexible model class describing a wide variety of autocorrelation functions, and since simple computational formulae for prediction and confidence bounds estimation exist. Simulation studies show that the properties of forecast-based monitoring schemes strongly depend on the parameter estimates [42]-[45].

The individuals-chart is a special form of the Shewhart-chart for which the individual observations are compared to predetermined upper and lower control limits UCL and LCL. An observation outside the control limits results in an alarm suggesting that some action is required. These control limits are based on a certain allowable deviation from a target value, which may be an estimate of the mean during the steady state, for instance. The allowable deviation is often based on normal theory resulting in $UCL = \mu_0 + ks$ and $LCL = \mu_0 - ks$, where μ_0 is the target value, s is an estimate of the variability found in the undisturbed process, e.g., the standard deviation, and k can be chosen as a certain percentile of the standard normal distribution. Fixed threshold alarms can be regarded as classical individuals-charts where the original observations are monitored. In case of autocorrelated observations, the individuals chart can be applied to the forecast errors of an AR model [46]. Both the classical and the forecast-based individuals charts are not robust against measurement artifacts since an alarm is triggered in case of a single extreme value. Sometimes it is suggested to add a run rule to an individual chart, e.g., "if two out of the last three observations are between two and three standard deviations away from the centerline and on the same side of it, an alarm should be given" [45].

EWMA-charts use exponential weighting methods for suppressing large variability in the monitored variable. The EWMA-chart with weighting factor λ monitors the weighted sum of the observations

$$EM_t = \lambda Y_t + (1 - \lambda) EM_{t-1} \, .$$

An initialization EM_0 is needed which can be set to a target value or a mean value obtained from past data. Then EM_t is compared to appropriately chosen control limits. It is well-known that EM_t describes the best predictor if the observed variable follows an ARIMA(0,1,1) process with parameter $\theta_1 = 1 - \lambda$. Although this property could be used to choose the value of λ by fitting an ARIMA(0,1,1) model to the data [47], usually a value $\lambda \leq 0.2$ is chosen [48], [49] since λ defines the influence of the latest measurement. An EWMA-chart with $\lambda = 1$ is an individuals chart. For small values of λ, EM_t is a low-pass filter which diminishes artifact noise and most of the fast fluctuations of the measured signal. Generally, EWMA-charts with large λ perform better for large level shifts, while EWMA-charts with small λ do better for small shifts [42]. Zhang [49] proposed an EWMA-chart with control limits modified to consider correlations within the data, while Montgomery and Mastrangelo [50] developed an EWMA-chart based on the forecast errors of an AR(1) model. EWMA-charts can also detect slow trends [48]. Nevertheless, Mäkivirta [1] stated that there is no evidence that exponential weighting is preferable to any other weighting method, apart from the fact that exponentially weighting allows to use a very convenient recursive structure.

Cumulative sum (CUSUM-)charts [22] accumulate differences between the actual and the expected values of a monitored variable. Timmer, Pignatiello and Longnecker [51] among others constructed a CUSUM-chart for an AR(1) process using a likelihood ratio approach. Two recursive filters, an upper and a lower one, are used for the detection of upward and downward shifts in the mean of an AR(1) process with mean μ_0 during the steady state. Choosing a certain out-of-control value μ_A of the mean, which should be detected with high probability, the upper filter is given by

$$U_t = \max[0, U_{t-1} + (W_t - m_{W_t} - k_t \sigma_{W_t})] \,,$$

where

$$\begin{aligned} W_t &= \begin{cases} (1+\phi_1)Y_1, & t=1 \\ Y_t - \phi_1 Y_{t-1}, & t>1 \end{cases} \\ m_{W_t} &= \begin{cases} (1+\phi_1)\mu_0, & t=1 \\ (1-\phi_1)\mu_0, & t>1 \end{cases} \\ k_t &= \begin{cases} (1+\phi_1)k, & t=1 \\ (1-\phi_1)k, & t>1 \end{cases} \\ \sigma_{W_t} &= \begin{cases} \sqrt{\frac{1+\phi_1}{1-\phi_1}}\sigma_a, & t=1 \\ \sigma_a, & t>1 \end{cases} \,, \end{aligned}$$

and $k = (\mu_0 + \mu_A)/2$ is the reference value for detection of a shift towards μ_A for an i.i.d. chart. This means that at each time point $t > 1$ the previous value is modified by adding the amount the forecast error is larger than a certain percentile of its steady-state distribution if the resulting value is larger than zero, and by restarting from zero otherwise. This percentile is chosen according to the size of a shift which should be detected very soon with high probability. Additionally, an initialization U_0 has to be chosen. The filter for detecting downward shifts is defined accordingly. By its construction, a CUSUM-chart is not robust against artifacts since an extreme observation has large influence on the filters. Hence, artifacts can either result in false alarms or hinder the fast detection of a level shift.

As pointed out by Lin and Adams [48], a forecast based individuals-chart has high probability of detecting a level shift immediately, but it has low probability of shift detection after the first observation because of the recovery of the forecasts. On the other hand, EWMA charts have low probability of fast signals, but higher probability of detection at subsequent observations. The performance of the individuals chart applied to forecast errors is better than the performance of EWMA-charts in case of large level shifts, while the performance of the individuals chart is inferior to the performance of EWMA- and CUSUM-charts in case of small or moderate shifts [48], [49], [51]. In a comparison of several Shewhart and EWMA-charts, Lu and Reynolds [42] recommended a Shewhart-chart of the observations for the detection of large shifts in case of positive autocorrelations, an EWMA-chart of the residuals for medium shifts and an EWMA-chart of the observations for small shifts. Zhang [49] considered his modified EWMA-chart to be preferable to several other individuals- and EWMA-based charts in case of small and moderate autocorrelations and medium-sized level shifts. In the case of very strong positive autocorrelations and a medium-sized or large level shift he supposed a residuals-chart to be superior. However, Kramer and Schmid [43] concluded that Shewhart-charts based on the original observations perform better than Shewhart-charts of the residuals in case of positive autocorrelations. Run rules can be misleading for autocorrelated data [42]. Adams and Tseng [45] found individuals charts to be more robust against misspecification of the model parameters than EWMA- and CUSUM-charts. Simulations in Bauer et al. [29] provide evidence that the individuals chart for forecast errors is powerful against single outliers, but that an approach described in the next section is better for the detection of multiple outliers and level shifts. To summarize this short discussion, there is no single best control chart, but the decision which chart should be used strongly depends on the particular application and the most appropriate performance criterion. In intensive care the probability of the fast detection of a clinically relevant change, say within a few observations after the change, is much more important than the ARL, which is strongly influenced by the tail behavior of the run length distribution.

In Figure 5, several control charts are applied to example 1. Individuals-charts for forecast residuals and several EWMA-charts with $\lambda = 0.1$ are considered. CUSUM-charts have similar properties as EWMA-charts but are less elegant than methods based on exponential weighting. Forecast residuals are calculated from fitting an AR(2) model to an initial sequence since experience from earlier studies [28], [39], [52] shows that physiological time series can typically be described adequately by low order AR(p) models during the time interval in between extraordinary events. The choice $p = 2$ seems to be sufficient in most cases. Here, the

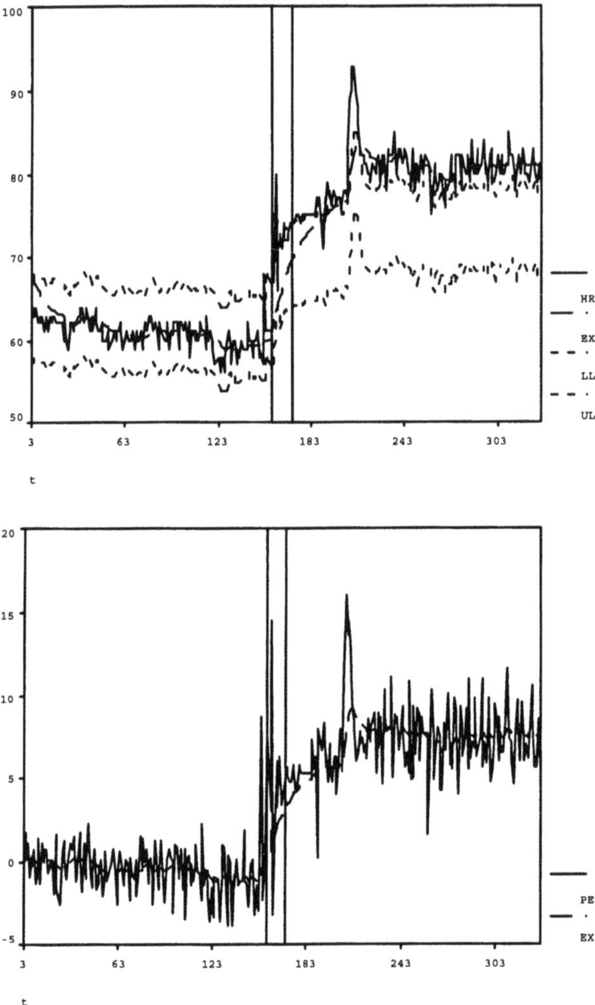

Figure 5. Several control charts applied to example 1. Top: Observed time series (HR), EWMA-chart for the observations with $\lambda = 0.1$ (EX) and confidence limits (UL, LL) for the predictions based on an AR(2) model fitted to a start sequence. An increase of the underlying signal is detected at t=158 if the predictions are simply compared to these confidence limits, which corresponds to using an individuals chart for the prediction errors, and at t=171 if a run rule is used to overcome possible problems caused by measurement artifacts. Bottom: Prediction errors (PE) for the AR(2) model and EWMA chart (EX) for these residuals with $\lambda = 0.1$. A disturbance is detected at t=158 by an individuals chart for the prediction errors and at t=169 by this EWMA-chart, cf. the vertical reference lines.

model estimated from the first 60 observations is

$$HR_t - 61.94 = 0.318(HR_{t-1} - 61.94) + 0.282(HR_{t-2} - 61.94) + a_t .$$

Obviously, in this example the individuals-chart for the forecast residuals detects the increase of the heart rate more quickly than the EWMA-charts, but the risk of false alarms because of, e.g., artifacts is much higher for an individuals-chart, too. While monitoring forecast residuals needs a starting sequence since a time series model has to be fitted to the data, applying a control chart to the measured values needs a starting sequence for estimation of the current level since usually there is no single target value available. In the EWMA-charts for the observations the target value was set to 70. It takes some observations until the EWMA-chart stabilizes at the true mean of the series before the shift.

Schack and Grieszbach [53] suggested an adaptive trend estimator based on exponential weighting. However, they applied it for visualization only and did not construct a significance test based on their approach. Montgomery and Mastrangelo [50] constructed a cumulative tracking signal using the sum of forecast errors from an EWMA chart, which is standardized by the mean absolute deviation. They specified critical values for this tracking signal from their experience. However, a cumulative sum is very sensitive against artifacts because of its long memory. The tracking signal developed earlier by Trigg [54], which uses exponential weighting of the forecast errors, has some advantages in this respect. Simulation studies [55] showed that Trigg's tracking signal TTS performs better than the cumulative tracking signal for small values of λ w.r.t. the average run length in case of independent measurements. However, TTS has difficulties to detect level shifts occurring in several steps [22] and does not distinguish between trends and level shifts [56]. Stoodley and Mirnia [57], who also use a cumulative sum of forecast errors provided by an EWMA chart, suggested rules based on the number of subsequent forecast errors having the same sign or being larger (smaller) than a certain value δ to recognize whether a trend, a level shift or an artifact has occurred.

The greatest problem of the usage of statistical control charts for online monitoring in intensive care are the implicit assumptions of (weak) stationarity of the data generating mechanism and the existence of a target value for the data. However, in the clinical application of monitoring vital signs it is impossible to determine more than a region of acceptable values for a physiological variable. Moreover, physiological processes are not stationary, but change according to the clinical state and the biorhythm of the patient. The mean value of a physiological variable does not remain fixed all of the time [58]. Rather, the aim in intensive care is to detect and limit extreme variation. Moreover, parameter estimation from past data, which is often suggested for industrial processes, is difficult for vital signs because of fundamental differences between the individuals.

7 Online Pattern Recognition Based on Statistical Time Series Analysis

The statistical control charts presented in the previous section are particularly designed for the detection of step changes and assume the existence of a target value the observations should be compared with. Statistical time series analysis, however, allows to estimate a time-varying mean value and to detect unusual changes in this mean. Intervention analysis does not work online since we need future observations of the variable to calculate the test statistics. Generally, pattern recognition will be more reliable the more observations we can use in the analysis. In the following we discuss several strategies for online pattern detection based on statistical time series analysis which can be found in the literature.

7.1 Dynamic Linear Models

In an early attempt to apply statistical time series analysis to online monitoring data, Smith and West [13] used a multiprocess dynamic linear model to monitor patients after renal transplantation. In dynamic linear models (DLMs) [59] the observation X_t at time t is regarded as a linear transform of an unobservable state parameter. This state is assumed to change dynamically in time according to a simple regression model. Particularly, the linear growth model

$$\begin{aligned} X_t &= \mu_t + \epsilon_t \\ \mu_t &= \mu_{t-1} + \beta_{t-1} + \delta_{t,1} \end{aligned}$$

$$\beta_t = \beta_{t-1} + \delta_{t,2}$$

is very appealing in the context of physiological variables since its state at time t consists of a level parameter μ_t and a slope parameter β_t which are easily interpretable. In the multiprocess version used by Smith and West different variances of the random observation error ϵ_t and the random systematic changes $\delta_{t,j}$, $j = 1, 2$, at time t are used for describing the steady state, artifacts, level changes and trends. For pattern classification they calculated the posterior probabilities of these states in a Bayesian framework using a multiprocess Kalman filter. Routine application of this model has not been practiced yet because of its very strong sensitivity against misspecification of the hyperparameters, particularly of the error variances, and its insensitivity against moderate level shifts.

The number of the hyperparameters and the necessary computational effort can be significantly reduced by using a single-process model. Wassermann and Sudjianto [60] constructed a control chart by comparing the current mean estimate to a specified target value. An alarm is triggered in their approach if the absolute difference is larger than a fixed multiple of the estimated standard deviation of the mean. However, this rule does not allow early detection of trends.

All patterns can be detected by assessing the influence of recent observations on the parameter estimates via suitably chosen influence statistics [61], which compare estimates of the state parameters calculated with these recent observations to estimates calculated without them. For a level change and a trend the recent observations should have a large influence on the level estimate and the slope parameter respectively, while an outlier should be far from the level estimate.

This approach based on influence statistics was successfully applied for retrospective pattern detection [61], [62]. For online application, however, we have to estimate the hyperparameters from a rather short estimation period. Practical experience shows that difficulties arise if the variability during the estimation period is low, if level changes occur in several little steps and if patterns of outliers occur in short time lags. Little variability during the estimation period causes the detection of outliers and level changes to be too sensitive subsequently. Level changes in several steps are hard to detect since the smoothed level parameter adjusts in each step, so that possibly the influence statistics are not significant at any time. Several outliers in short time lags may either mask each other or may be mistaken for a level change. Nevertheless, all kind of patterns in hemodynamic time series could correctly be identified in most of the cases [63].

7.2 ARMA Modeling

Alternatively, for the online recognition of patterns of change in a dynamical system we can model the underlying process during the equilibrium or steady state from past data and use a measure to detect deviations from this steady state. An intuitive rule for the detection of an outlier is to compare the incoming observation to the one-step ahead prediction in a suitable model, e.g., an AR(2) model fitted to past measurements. This approach is equivalent to using a forecast based Shewhart chart, which has already been mentioned in the previous section.

Bauer et al. [29] used an alternative approach to develop an automatic procedure for the online detection of outliers and level shifts in time series. They modeled the marginal distribution of m-dimensional vectors of subsequent observations $\mathbf{y}_t = (y_t, y_{t-1}, \ldots, y_{t-m+1})'$. In this way the dynamical information of the univariate time series is transformed into a spatial information within an m-dimensional space and rules for outlier identification in multivariate data can thus be transferred into the time series context. The embedding dimension m should be chosen according to the dependence structure of the underlying process. Since most physiological time series can be described by AR(2) models, $m = 3$ is an obvious choice.

Under the assumption of joint normality of the random variables during the steady-state, the time-delay vectors $\mathbf{y}_m, \ldots, \mathbf{y}_N$ should form an m-dimensional elliptical cloud. A control ellipsoid can be constructed using the Mahalanobis distance

$$MDTS(t) = \sqrt{(\mathbf{Y}_t - \bar{Y}_{N-m+1})'\mathbf{S}_{Y,N-m+1}^{-1}(\mathbf{Y}_t - \bar{Y}_{N-m+1})},$$

$t = m, \ldots, N$. Here $\bar{Y}_{N-m+1} = \frac{1}{N-m+1}\sum_{t=m}^{N} \mathbf{Y}_t$ is the arithmetic mean of the time-delay vectors and $\mathbf{S}_{X,N-m+1}$ is the sample covariance matrix

$$\mathbf{S}_{Y,N-m+1} = \begin{pmatrix} \hat{\gamma}_N(0) & \hat{\gamma}_N(1) & \cdots & \hat{\gamma}_N(m-1) \\ \hat{\gamma}_N(1) & \hat{\gamma}_N(0) & & \vdots \\ \vdots & \ddots & & \vdots \\ \hat{\gamma}_N(m-1) & \cdots & \cdots & \hat{\gamma}_N(0) \end{pmatrix},$$

with $\hat{\gamma}_N(h) = \frac{1}{N}\sum_{t=1}^{N-h}(Y_t - \bar{Y}_N)(Y_{t+h} - \bar{Y}_N)$, $h = 0, \ldots, m-1$, where $\bar{Y}_N = \frac{1}{N}\sum_{t=1}^{N} Y_t$.

Bauer et al. [29] compared this approach in a simulation study to forecast based detection for ARMA(p, q)-models as mentioned above. They find the forecast based detection to perform better for single outliers, while the approach based on the multivariate Mahalanobis distance is preferable for patterns such as level shifts which affect several subsequent observations. This is according to the forecast recovery of one-step ahead prediction. For a patch of outliers or a level shift the approach based on forecasts fails with high probability if the first outlier is not detected and not replaced by a prediction. This deficiency is even more serious in case of biological systems like the health state of a patient, which often shows a step-wise reaction to disturbances and interventions. On the other hand, using the marginal distribution means to judge m subsequent observations simultaneously. The power of the rule based on the Mahalanobis distance should be increasing with the number of subsequent outliers since they move the time-delay vector further out of the control ellipsoid than a single outlier does. Therefore, the rule based on the Mahalanobis distance will be better than an approach based on one-step ahead prediction for patchy outliers and level shifts at the slight expense of lower power against single outliers, which are clinically a much less relevant phenomenon. Robust estimators of the autocovariances and the mean can be used to overcome problems such as swamping and masking effects of outliers [64].

Applying the approach based on the marginal distribution to example 1 using the embedding dimension $m = 3$ and control limits which adjust automatically to the current variability in the series [29] results in the detection of upward shifts at $t = 154$ and $t = 208$, and of a downward shift at $t = 232$, cf. Figure 6.

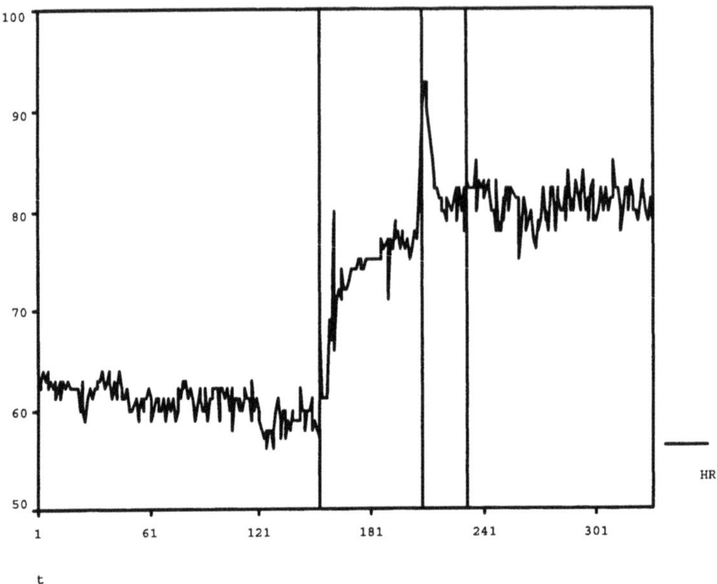

Figure 6. Modeling the marginal distribution of $m = 3$ subsequent observations during the steady state results in the detection of clinically relevant changes at the time points marked by the vertical reference lines.

7.3 Trend Detection

The approaches based on forecasts and the Mahalanobis distance are suitable for the detection of outliers and level shifts, but not for the detection of slow trends. Early detection of such monotone systematic behavior in a physiological time series is important since in intensive care many changes of actual interest have a duration ranging from some minutes to several hours [1].

For retrospective detection of slow trends in time series data regression based models can be used for instance [23]. This approach is a special case of curve fitting methods, which have generally been described in Section 2. Typically it is assumed that the observation Y_t at time point t is a measurement of the current process level $\mu_t = \alpha + \beta t$, which is disturbed by autocorrelated noise E_t. For the noise, often an AR(p)

model is assumed. Thus, the model reads

$$\begin{aligned} Y_t &= \alpha + \beta t + E_t\,, \\ E_t &= \phi_1 E_{t-1} + \ldots + \phi_p E_{t-p} + a_t\,. \end{aligned}$$

We either can use simple least squares estimators for α and β and adjust their variances for the autocorrelations within the noise process estimated from the regression residuals, or we can use maximum likelihood or robust techniques to estimate all model parameters jointly. Neglecting the autocorrelations increases the probability of erroneous conclusions [23]. Such a regression based approach can be modified for online monitoring by using a moving time window. The test statistic can be calculated for the current time window including the last n observations. An inherent problem is the appropriate choice of the window width n since for trends with different length different window widths should be optimal.

Another problem is that linearity of the trend is assumed since trends which are non-linear might not be detected this way. In [65] a rule for retrospective detection of any monotone trend which was proposed by Abelson and Tukey [66] for independent data and by Brillinger [67] for time series was adapted to the online monitoring situation. Let μ_t be the time-varying level of the process, which is disturbed by autocorrelated noise E_t, such that

$$Y_t = \mu_t + E_t, \qquad t \in \mathbb{Z}.$$

Then a weighted sum $\sum_{t=1}^{N} w_t Y_t$ of the observations of the current time window is used to test for any form of monotone change of μ_t during the time interval $t = 1, \ldots, n$, i.e., $\mu_1 \leq \mu_2 \leq \ldots \leq \mu_n$ with $\mu_t < \mu_{t+1}$, or $\mu_1 \geq \mu_2 \geq \ldots \geq \mu_n$ with $\mu_t > \mu_{t+1}$ for at least one $t \in \{1, \ldots, n-1\}$. Since the weights $w_1, \ldots, w_n$ are restricted to have arithmetic mean $\overline{w} = 0$, the weighted sum has mean zero if μ_t is constant over time. The weights are then determined to solve

$$\max_{w} \min_{\mu} \frac{|\sum(w_t - \overline{w})(\mu_t - \overline{\mu})|^2}{\sum(w_t - \overline{w})^2 \sum(\mu_t - \overline{\mu})^2}$$

with $\overline{\mu} = n^{-1} \sum \mu_t$. This means, they are chosen to have a worst case discriminatory power for an extremely unfavorable trend which is as high as possible. This results in

$$w_t = \left[(t-1)\left(1-\frac{t-1}{n}\right)\right]^{1/2} - \left[t\left(1-\frac{t}{n}\right)\right]^{1/2}$$

and the corresponding worst case is a single step change. Comparing the mean of time delayed moving windows with length m, which is a standard approach to detect systematic differences [68], has lower worst case discriminatory power since it corresponds to a weighted sum with weights $w_1 = -1/m, \ldots, -1/m, 0, \ldots, 0, 1/m, \ldots, 1/m$. The hypothesis of a constant mean should be rejected in favor of a monotone increasing (decreasing) mean if $\sum_{t=1}^{n} w_t Y_t$ is large (small) in comparison to its variance. During the steady state this variance is equal to

$$Var\left(\sum_{t=1}^{n} w_t Y_t\right) = \sum_{t=1}^{n}\sum_{s=1}^{n} w_t w_s \gamma(t-s)\,, \tag{3}$$

where $\gamma(h), h = 0, 1, \ldots,$ are the autocovariances of the noise process. Hence, parameter estimation can be accomplished easily if we have reliable estimates of these autocovariances. Since a trend has a serious impact on the usual sample autocovariances, we can try to eliminate a (local) linear trend $\alpha + \beta t$ by regression methods first and estimate the autocovariances from the residuals [65]. In a simulation study, suitable critical values for the standardized weighted sum were determined. This procedure seems reliable for the detection of both linear and non-linear trends which are not very slow. Nevertheless, very large positive autocorrelations may cause some problems. Since very large positive autocorrelations result in monotone sequences just like deterministic trends, these mechanisms are hard to distinguish within short time series anyway [23].

Figure 7 shows the standardized values of the weighted sums calculated from a moving time window of length $n = 60$ for example 2. If we compare this test statistic to the critical value $c = 5.0$ found by simulations in [65] we detect a systematic monotone decrease of the heart rate at time point $t = 66$.

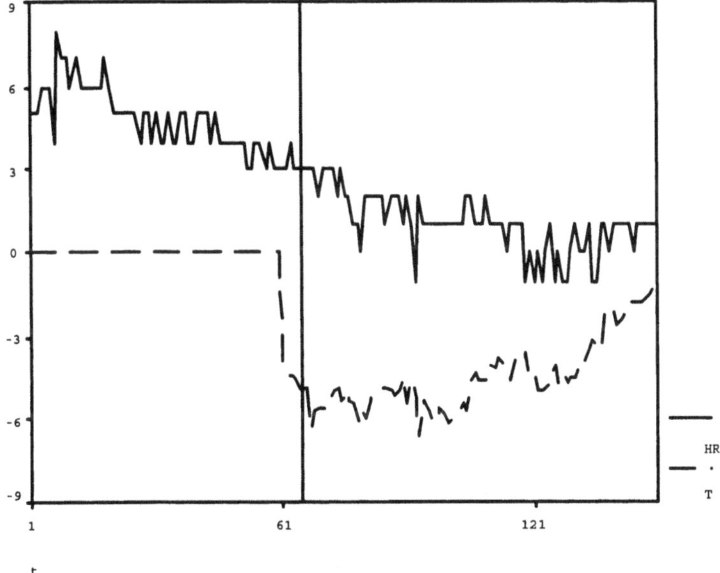

Figure 7. When applying the weighted sum chart with window width $n = 60$ to example 2 a systematic monotone decrease (negative trend) is detected at time point $t = 66$. We subtracted 70 from the heart rate measurements for the reason of illustration.

8 Conclusion

There are a couple of approaches to pattern detection within physiological time series. Each of them has its own strengths and weaknesses. In intensive care, very complex combinations of several patterns may be observed within rather short time intervals. This complicates the online application of curve fitting methods, for instance, since they require predefined functional forms. Fixed target values can hardly be specified in advance because of fundamental differences between individuals and because of the uncertain temporal development of the patient's state. Therefore, usual control charts and methods developed within other clinical contexts such as mechanical ventilation can hardly be used. Autocorrelations within subsequent measurements should be considered in the analysis since neglecting them may lead to false conclusions. Another problem is that methods which detect large changes with high probability and with a short time delay as is needed for life-threatening complications may be insensitive against small or moderate shifts. Reliable detection of the latter is important for assessing intervention effects and as an input for knowledge-based bedside decision support [69]. A particularly difficult problem is the fast and correct detection of a slow trend. Mäkivirta [1], e.g., stated that the trend detectors developed at that time had little practical use. Moreover, a useful system should not only detect a trend, but it should also be able to quantify it. In view of all these difficulties certainly the best approach for online pattern recognition within physiological time series is to search for a proper combination of several methods [14]. The individual methods should be further refined and improved using ideas from other approaches. For instance, the performance of the method for online detection of outliers [29] was improved for real time series by replacing the usual fixed significance levels by significance levels which adapt to the time-varying variability of the process. Such an adaptive significance level corresponds to a control limit, i.e., a certain allowable deviation from the process level which is fixed by the physician.

For multivariate monitoring, we either can combine the information contained in several variables using logical rules, or we can try to calculate joint control regions. The latter approach suffers from the "curse of dimensionality" [70], i.e., from the large number of unknown parameters which have to be estimated from the data. To fill a high-dimensional sample space we need huge sample sizes, which are rarely available in practice. Physicians usually select one variable out of a group of closely related variables and base their decisions on the patterns found in this variable only. Statistical methods for dimension reduction like factor analysis can be used to compress the relevant information into a few important variables [71]. Graphical models [72] provide additional information to guarantee that the results obtained from dynamic factor analysis are interpretable by the physician. In a case-study it was shown that latent factors can be more adequate for pattern detection in the observed variables than each single variable [14].

In conclusion, methods for automatic online analysis of physiological variables offer an opportunity for a more reliable evaluation of the individual treatment and lead to intelligent alarm systems. A future task is the construction of intelligent bedside decision support systems. Such a system can be based on techniques for data abstraction as we have outlined here. These techniques could be combined with methods of artificial intelligence which use the patterns found in the statistical analysis to assess the current state of the patient. By classifying these patterns according to existing knowledge gained from physicians and former data analysis [69] the physician in charge might then be given options of how to respond properly.

Acknowledgments

The financial support of the Deutsche Forschungsgemeinschaft (SFB 475, "Reduction of complexity in multivariate data structures") is gratefully acknowledged.

References

[1] Mäkivirta, A. (1989), "Towards Reliable and Intelligent Alarms by Using Median Filters," Research Report 660, Technical Research Centre of Finland.

[2] Morris, A. and Gardner, R. (1992), "Computer applications," in: Hall, J., Schmidt, G., and Wood, L. (Eds.), *Principles of Critical Care*, McGraw Hill, New York, pp. 500-514.

[3] Miller, G. (1956), "The marginal number seven, plus or minus two: some limits to our capacity for processing information," *Psychol. Rev.*, vol. 63, pp. 81-97.

[4] Jennings, D., Amabile, T., and Ross, L. (1982), "Informal covariation asessments: data-based versus theory-based judgements," in: Kahnemann, D., Slovic, P., and Tversky, A. (eds.), *Judgment under Uncertainty: Heuristics and Biases*, Cambridge University Press, Cambridge, pp. 211-230.

[5] Guyatt, G., Drummond, M., Feeny, D., Tugwell, P., Stoddart, G., Haynes, R., Bennett, K., and LaBelle, R. (1986), "Guidelines for the clinical and economic evaluation of health care technologies," *Soc. Sci. Med.*, vol. 22, pp. 393-408.

[6] Morris, A. (1998), "Algorithm-based decision making," in: Tobin, J.A. (Ed.), *Principles and practice of intensive care monitoring*, McGraw Hill, New York, pp. 1355-1381.

[7] Lawless, S.T. (1994), "Crying wolf: false alarms in a pediatric intensive care unit," *Critical Care Medicine*, vol. 22, pp. 981-985.

[8] Wiklund, L., Hök, B., Stähl, K., and Jordeby-Jönsson, A. (1994), "Postanaesthesia monitoring revisited: frequency of true and false alarms from different monitoring devices," *J. Clin. Anesth.*, vol. 6, pp. 182-188.

[9] O'Carrol, T. (1986), "Survey of alarms in an intensive therapy unit," *Anesthesia*, vol. 41, pp. 742-744.

[10] Haimowitz, I.J., Le, P.P., and Kohane, I.S. (1995), "Clinical monitoring using regression-based trend templates," *Art. Intel. Med.*, vol. 7, pp. 473-496.

[11] Haimowitz, I.J. and Kohane, I.S. (1996), "Managing temporal worlds for medical trend diagnosis," *Art. Intel. Med.*, vol. 8, pp. 299-321.

[12] Miksch, S., Horn, W., Popow, C., and Paky, F. (1996), "Utilizing temporal data abstraction for data validation and therapy planning for artificially ventilated newborn infants," *Art. Intel. Med.*, vol. 8, pp. 543-576.

[13] Smith, A.F.M. and West, M. (1983), "Monitoring renal transplants: an application of the multiprocess Kalman filter," *Biometrics*, vol. 39, pp. 867-878.

[14] Gather, U., Fried, R., Lanius, V., and Imhoff, M. (2001), "Online monitoring of high-dimensional physiological time series – a case-study," *Estadistica*. (To appear.)

[15] Shahar, Y. and Musen, M.A. (1993), "RESUME: a temporal-abstraction system for patient monitoring," *Computers and Biomedical Research*, vol. 26, pp. 255-273.

[16] Shahar, Y. and Musen, M.A. (1996), "Knowledge-based temporal abstraction in clinical domains," *Art. Intel. Med.*, vol. 8, pp. 267-298.

[17] Mäkivirta, A., Koski, E., Kari, A., and Sukuvaara, T. (1991), "The median filter as a preprocessor for a patient monitor limit alarm system in intensive care," *Computer Methods and Programs in Medicine*, vol. 34, pp. 134-149.

[18] Justusson, B.I. (1978), "Noise reduction by median filtering," *Pro. 4th Int. Joint Conf. Pattern recognition*, Kyoto, Japan, Nov. 1978, pp. 502-504.

[19] Justusson, B.I. (1981), "Median filtering: statistical properties," in: Huang, T.S. (Ed.), *Topics in Applied Physics, Two-Dimensional Signal Processing II*, Springer-Verlag, Berlin.

[20] Nieminen, A., Neuvo, Y., and Mitra, U. (1988), "Algorithms for real-time trend detection," *ICASSP International Conference on Acoust., Speech and Signal Proc. 1988*, IEEE, New York, pp. 1530-1532.

[21] Nieminen, A., Neuvo, Y., and Mitra, U. (1989), "Algorithms for real-time trend detection," *Signal processing*, vol. 18, pp. 1-15.

[22] Endresen, J. and Hill, D.W. (1977), "The present state of trend detection and prediction in patient monitoring," *Intensive Care Medicine*, vol. 3, pp. 15-26.

[23] Woodward, W.A. and Gray, H.L. (1993), "Global warming and the problem of testing for trend in time series data," *Journal of Climate*, vol. 6, pp. 953-962.

[24] Hill, D.W. and Endresen, J. (1978), "Trend recording and forecasting in intensive care therapy," *British Journal of Clinical Equipment*, January, pp. 5-14.

[25] Imhoff, M. and Bauer, M. (1996), "Time series analysis in critical care monitoring," *New Horizons*, vol. 4, pp. 519-531.

[26] Gordon, K. and Smith, A.S.M. (1990), "Modeling and monitoring biomedical time series," *J. Americ. Statist. Assoc.*, vol. 85, pp. 328-337.

[27] Hepworth, J.T., Handrickson, S.G., and Lopez, J. (1994), "Time series analysis of physiological response during ICU visitation," *West J. Nurs. Res.*, vol. 16, pp. 704-717.

[28] Imhoff, M., Bauer, M., Gather, U., and Löhlein, D. (1998), "Statistical pattern detection in univariate time series of intensive care on-line monitoring data," *Intensive Care Medicine*, vol. 24, pp. 1305-1314.

[29] Bauer, M., Gather, U., and Imhoff, M. (1999), "The Identification of Multiple Outliers in Online Monitoring Data," Technical Report 29/1999, SFB 475, Department of Statistics, University of Dortmund, Germany.

[30] Brockwell, P.J. and Davis, R.A. (1987), *Time Series: Theory and Methods*, 2nd ed., Springer, New York.

[31] Box, G.E.P., Jenkins, G.M., and Reinsel, G.C. (1994), *Time Series Analysis. Forecasting and Control*, 3rd ed., Prentice-Hall, Englewood Cliffs.

[32] Fox, A.J. (1972), "Outliers in time series," *J. Roy. Statist. Soc. Ser. B*, vol. 34, pp. 350-363.

[33] Chang, I., Tiao, G.C., and Chen, C. (1988), "Estimation of time series parameters in the presence of outliers," *Technometrics*, vol. 30, pp. 193-204.

[34] Tsay, R.S. (1988), "Outliers, level shifts and variance changes in time series," *J. Forecasting*, vol. 7, pp. 1-20.

[35] Chen, C., and Liu, L. (1993), "Joint estimation of model parameters and outlier effects in time series," *J. Am. Stat. Assoc.*, vol. 88, pp. 284-297.

[36] Peña, D., Tiao, G.C., and Tsay, R.S. (2000), *A Course in Time Series Analysis*, Wiley, New York.

[37] Muirhead, C.R. (1986), "Distinguishing outlier types in time series," *J. R. Statist. Soc. Ser. B*, vol. 48, pp. 39-47.

[38] Ljung, G.M. (1993), "On outlier detection in time series," *J. Roy. Stat. Soc. B*, vol. 55, pp. 559-567.

[39] Imhoff, M., Bauer, M., Gather, U., and Löhlein, D. (1997), "Time series analysis in intensive care medicine," *Applied Cardiopulmonary Pathophysiology*, vol. 6, pp. 263-281.

[40] Johnson, R.A. and Bagshaw, M. (1974), "The effects of serial correlation on the performance of the CUSUM tests," *Technometrics*, vol. 16, pp. 103-122.

[41] Padgett, C.S., Thombs, L.A., and Padgett, W.J. (1992), "On the α-risks for Shewhart control charts," *Communications in Statistics - Simulation and Computation*, vol. 21, pp. 1125-1147.

[42] Lu, C.-W. and Reynolds, M.R. Jr. (1999), "EWMA control charts for monitoring the mean of autocorrelated processes," *Journal of Quality Technology*, vol. 31, pp. 166-187.

[43] Kramer, H. and Schmid, W. (2000), "The influence of parameter estimation on the ARL of Shewhart type charts for time series," *Statistical Papers*, vol. 41, pp. 173-196.

[44] Tseng, S. and Adams, B.M. (1994), "Monitoring autocorrelated processes with an exponentially weighted moving average forecast," *J. Statist. Comp. Simul.*, vol. 50, pp. 187-195.

[45] Adams, B.M. and Tseng, I.-T. (1998), "Robustness of forecast-based monitoring schemes," *Journal of Quality Technology*, vol. 30, pp. 328-339.

[46] Runger, G.C. and Willeman, T.R. (1995), "Model-based and model-free control of autocorrelated processes," *Journal of Quality Technology*, vol. 27, pp. 283-292.

[47] Hembree, G.B. (1994), "Recursive estimation of the weighting factor for EWMA control charts from autocorrelated data," *ASAProQlPr*, 39-43.

[48] Lin, W.S.W. and Adams, B.M. (1996), "Combined control charts for forecast-based monitoring schemes," *Journal of Quality Technology*, vol. 28, pp. 289-301.

[49] Zhang, N.F. (1998), "A statistical control chart for stationary process data," *Technometrics*, vol. 40, pp. 24-38.

[50] Montgomery, D.C. and Mastrangelo, C.M. (1991), "Some statistical process control methods for autocorrelated data," *Journal of Quality Technology*, vol. 23, pp. 179-193.

[51] Timmer, D.H., Pignatiello, J. Jr., and Longnecker, M. (1998), "The development and evaluation of CUSUM-based control charts for an AR(1) process," *IEEE Transactions*, vol. 30, pp. 525-534.

[52] Lambert, C.R., Raymenants, E., and Pepine, C.J. (1995), "Time-series analysis of long-term ambulatory myocardial ischemia: effects of beta-adrenergic and calcium channel blockade," *Am. Heart J.*, vol. 129, pp. 677-684.

[53] Schack, B. and Grieszbach, G. (1994), "Adaptive methods of trend detection and their applications in analysing biosignals," *Biometrical Journal*, vol. 36, pp. 429-452.

[54] Trigg, D.W. (1964), "Monitoring a forecasting system," *Operational Research Quarterly*, vol. 15, pp. 271-274.

[55] Gardner, E.S. (1983), "Automatic monitoring of forecast errors," *Journal of Forecasting*, vol. 2, pp. 1-21.

[56] Kennedy, R.R. (1995), "A modified Trigg's tracking variable as an 'advisory alarm' during anaesthesia," *International Journal of Clinical Monitoring and Computing*, vol. 12, pp. 197-204.

[57] Stoodley, K.D.C. and Mirnia, M. (1979), "The automatic detection of transients, step changes and slope changes in the monitoring of medical time series," *The Statistician*, vol. 28, pp. 163-170.

[58] Högel, J. (2000), "Applications of statistical process control techniques in medical fields," *Allg. Stat. Archiv*, vol. 84, pp. 337-359.

[59] West, M. and Harrison, J. (1989), *Bayesian Forecasting and Dynamic Models*, Springer, New York.

[60] Wasserman, G.S. and Sudjianto, A. (1993), "Short run SPC based upon the second order dynamic linear model for trend detection," *Communications in Statistics - Computation and Simulation*, vol. 22, 1011-1036.

[61] Peña, D. (1990), "Influential observations in time series," *J. Business & Economic Statistics*, vol. 8, pp. 235-241.

[62] De Jong, P., and Penzer, J. (1998), "Diagnosing shocks in time series," *J. Americ. Statist. Assoc.*, vol. 93, pp. 796-806.

[63] Gather, U., Fried, R., and Imhoff, M. (2000), "Online classification of states in intensive care," in: Gaul, W., Opitz, O., and Schader, M. (Eds.), *Festschrift in Honor to Hans-Hermann Bock's 60th Birthday, Data Analysis, Classification, and Applications*, Springer, Berlin, pp. 413-428.

[64] Becker, C. and Gather, U. (2000), "The masking breakdown point of multivariate outlier identification rules," *J. Americ. Statist. Assoc.*, vol. 94, pp. 947-955.

[65] Fried, R., Gather, U., and Imhoff, M. (2001), "The online detection of a monotone trend in a time series," Preprint, Department of Statistics, University of Dortmund, Germany.

[66] Abelson, R.P. and Tukey, J.W. (1963), "Efficient utilization of non-numerical information in quantitative analysis: general theory and the case of simple order," *Ann. Math. Statist.*, vol. 34, pp. 1347-1369.

[67] Brillinger, D.R. (1989), "Consistent detection of a monotonic trend superposed by a stationary time series," *Biometrika*, vol. 76, pp. 23-30.

[68] Härdle, W. (1990), *Applied Nonparametric Regression*, Cambridge University Press, Cambridge.

[69] Morik, K., Imhoff, M., Brockhausen, P., Joachims, T., and Gather, U. (2000), "Knowledge discovery and knowledge validation in intensive care," *Art. Int. Med.*, vol. 19, pp. 225-249.

[70] Friedmann, J.H. (1994), "An overview of predictive learning and function approximation," in: Cherkassky, V., Friedmann, J.H., and Wechsler, H. (Eds.), *From Statistics to Neural Networks*, Springer, Berlin et al., pp. 1-61.

[71] Fried, R., Gather, U., and Imhoff, M. (2000), "Some statistical methods in intensive care online monitoring – a review," in: Brause, R.W. and Hanisch, E. (Eds.), *Medical Data Analysis*, Springer, Berlin, pp. 67-77.

[72] Gather, U., Imhoff, M., and Fried, R. (2000), "Graphical Models for Multivariate Time Series from Intensive Care Monitoring," Technical Report 33/2000, SFB 475, University of Dortmund, 44221 Dortmund, Germany.

Chapter 7

Artificial Neural Network Models for Timely Assessment of Trauma Complication Risk

R.P. Marble and J.C. Healy

This chapter espouses the deployment of neural network-based diagnostic aids for evaluation of morbidity risks in the prehospital, acute care, and rehabilitation circumstances evinced by traumatic injury. The potential effectiveness of such systems is addressed from several points of view. First, the ability of the underlying connectionist models to identify complex, highly nonlinear, and sometimes even counterintuitive patterns in trauma data is discussed. Prior work in the area is reviewed and the approach is illustrated with an application that succeeds in identifying coagulopathy outcomes in victims of blunt injury trauma. Second, the feasibility of the universal applicability of neural models in actual trauma situations is argued. Their ability to use standardized, widely available data and their capacity for reflecting local differences and changing conditions is exposed. Finally, the potential enhancements for such models are explored in the contexts of clinical decision support systems.

1 Artificial Neural Network Models

1.1 Background

The fields of psychology and medicine have known many attempts to model the processing of the human brain. McCulloch and Pitts [1] and Hebb [2] produced initial approaches to characterizing this activity by using networks of interconnected processing elements called neurons. The motivation for this approach was centered on a desire to address the phenomena of recognition and learning. With the advent of computer technology, and hence a workable means of simulating this activity, came a number of developments in the computational theory underlying these neural networks. In 1960, Widrow and Hoff [3] presented a feedback-oriented algorithm, for facilitating the "training" of neural networks. They, as well as Rosenblatt [4], were able to formalize a general framework for this theory and to report favorable results of computer experiments with it.

The theory defined sensory neurons as input variables whose collective values are used to stimulate response (or output) neurons. The stimulation takes place via unidirectional connections from the sensory to the response layer and the significance of individual inputs is modulated by weights that reflect the varied strengths of the individual connections. The processing elements of the response layer evaluate the weighted sums of their inputs. An activation function then determines the output that is issued by each response neuron, usually according to the dictates of a constant stimulation threshold. In the training of such a model, the outputs resulting from a given configuration of the input values are compared with desired (or target) output values. The Widrow-Hoff learning algorithm is then used to adjust the connection strengths in an iterative fashion, to reduce the error produced by the differences between actual and target responses.

In 1969, Minsky and Pappert [5] brought mathematical rigor to the scene and proved that this theory was incapable of solving an important class of problems in the domain of pattern recognition. To surmount the problem, the use of multi-layered adaptive linear elements (MADALINES) was developed and has now become the widely accepted means of defining the structure of multi-layered neural networks. In addition to input and output layers, these architectures include one or more hidden (or associative) layers of neurons, as illustrated in Figure 1. The input nodes correspond to independent variables whose values can be numeric or categorical. The input values are passed along via weighted connections to the hidden layer neurons. In the hidden layer processing elements, differentiable activation functions of the weighted sums of inputs then produce stimulation for the output layer, which in turn issues outputs as described above. As is well described in the volumes of Rumelhart and McClelland [6], supervised learning takes place with the use of target output values that are known to be associated with the input configurations of individual training cases. The Widrow-Hoff learning rule is supplemented here by an algorithm for propagating error back through the additional layer(s) of neurons and adapting the connection weights accordingly. Figure 1 depicts the input variable values as x_i, the hidden layer activations as g_j, and the outputs of the network as O_k. The weights representing connection strengths are shown with matrix subscripts that specify their destination and source endpoints, respectively.

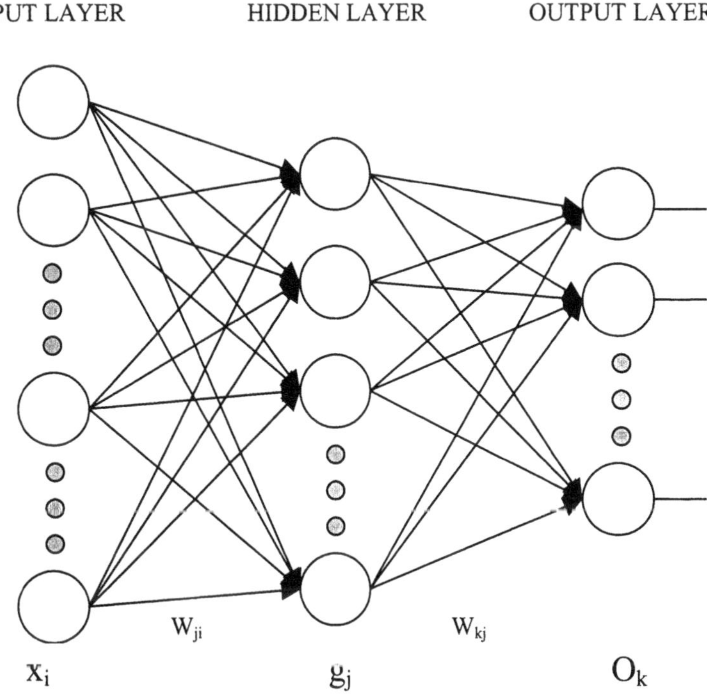

Figure 1. A multi-layer neural network.

The last 15 years or so have witnessed the publication of numerous successful applications of neural models. The problems they solve usually involve situations in which a relationship is assumed to exist between the variables of a data set, but the exact nature of the relationship is not well understood. A neural network is designed, which assigns certain variables to the role of input and others to that of output. The network is trained with a collection of cases from the data set, with the output variables' values used as targets in the training algorithm. Another collection of cases is withheld from training, to be used for testing the trained network. A trained network is defined by the final values of its connection weights, iteratively modified to minimize the differences between output and target for all cases in the training set. The trained network's effectiveness is evaluated by giving it input values from the withheld cases of the data set. The resulting output values are then compared with the corresponding actual values from the testing cases. The training cases can be viewed as examples of the relationship postulated to exist between the variables. A good performance with the testing cases indicates that the neural network has generalized the pattern in the training cases, to recognize it in cases it has never seen before.

The computational details of the algorithms mentioned here are beyond the scope of this exposition. Excellent coverage of the foundations of neural computing can be found in the literature, however. A comprehensive description of the field was produced by Hecht-Nielsen [7], for instance, and an elucidating guide to the construction of neural networks has been provided by Hagan and colleagues [8]. The reader is also encouraged to peruse the fascinating account of applications of neural models, contained in the paper of Widrow, Rumelhart, and Lehr [9]. It presents an interesting and diverse survey of successful deployments of neural models in such areas as financial analysis, manufacturing process control, optical pattern recognition, and marketing.

1.2 Neural Networks and Statistical Analysis

By far, the most respected tools for the analysis of relationships in quantitative data are statistical ones. With a long history of development and theory-building, the statistical sciences have established many rigorous and widely-used methods for drawing credible conclusions about populations using data drawn from samples of those populations. Indeed, the probabilistic statements enabled by statistical analysis can even specify the degree of credibility that these conclusions should enjoy. Neural networks, on the other hand, are notorious for their lack of ability to explain the conclusions drawn when they have succeeded in recognizing and characterizing patterns in data sets. While recent work has evinced improvements in this situation (as discussed later), the learning and generalizing that neural networks do has always been viewed as being somewhat mysterious.

A couple of key circumstances seem to contribute to that view. First, the training of neural models is a non-deterministic process. This is because the starting values for the weights of a neural architecture are random. Thus, unlike the estimation process of statistical regression, for instance, a neural network training episode can never be replicated with the same data to achieve the same values for its trained weights. (A replication of training *can* achieve comparable performance, but the final configurations of weight values will not be equivalent.) In addition, the extreme complexity of the multi-layer neural model usually prevents the mathematical derivation of any conclusions regarding the significance of individual input variables in the results they have helped to achieve. This is caused by the nature of the activation functions that must be employed for the learning algorithm to work properly. These transfer functions are decidedly non-linear and the general network structure nests them together in a way that defies isolation of individual input variables.

A number of comprehensive articles have appeared in recent years, which address the comparison of statistical and neural computing methods. The works of Ripley [10] and Cheng and Titterington [11] are particularly extensive examples of the scrutiny that neural network modeling has received in the statistical community. Indeed, the responses to the latter (published in the same issue of *Statistical Science*) offer a telling indication of the attention statisticians have given to neural computing concepts. Some researchers have developed ways of deriving statistical inferences from the results of neural modeling and of incorporating neural computing into statistical methods. The work of Warner and Misra [12], Wang [13], and Hwang and Ding [14] offer recent examples of developments in this direction. Further, a plethora of articles exists (in publications of many diverse academic fields, in fact), each of which compares the actual performance of neural network models in a particular problem area with that of some statistical technique used to address the same problem. See, for example, [15], [16], and [17].

In the latter body of research, one often finds results that show trained neural models to be better at predicting the output values of test cases than statistical estimation tools. While statistical tools often provide concise explanatory inferences, neural networks make fewer mistakes. (The following sections cite numerous results of this nature from the medical literature.) Indeed, some researchers have found that neural networks can find relationships in data sets, which statistical methods fail to discern at all. (See, for instance, Marble and Maier [18].) Evidently, the failure of statistical methods could belie the existence of significant relationships in the data, which neural networks might find. As pointed out by Denton [19], the statistical assumptions of noncorrelation between independent variables, independence of residuals for individual observations, and normality of these residuals (with a zero mean and constant variance) may be so strong as to render regression or logistic regression results suspect, for data whose character may not endorse these assumptions. Additionally, the linear functional form that regression analyses impose on the data may itself be too restrictive for the actual relationships embodied in the data.

In the face of this type of situation, Warner and Misra [12] suggest that the data themselves be allowed to define the functional form. They point out that the backpropagation training method for artificial neural networks is equivalent to maximum likelihood estimation and thus provides the power to do just that. This renders the multilayered feedforward neural network a powerful modeling tool. Furthermore, as shown by Cybenko [20], a two-layer feedforward neural network with a sufficient number of hidden units can approximate any continuous function to any degree of accuracy. This insures the coverage by such models of most of the useful functional relationships that might exist between the variables. Indeed, neural networks can be viewed as nonparametric regression methods [12] and should not be overlooked in areas where standard parametric assumptions may be doubted.

1.3 Neural Networks in Medicine

Recent years have witnessed an increasing attention in the literature to the analysis of medical data using neural network models. The increased awareness of neural computing in general has certainly contributed to this attention. The successes that neural networks have demonstrated in other fields have indeed been conspicuous. We suppose, however, that the nature of medical problems and the complexity of relationships in medical data are major drivers of this interest. Numerous medical researchers are embracing this method for modeling complex, non-linear functions. Baxt [21] recently discussed the shortcomings of linear estimation techniques in relating clinical signs and symptoms to specific disease states. He espoused the use of non-linear techniques. The difficulties inherent, however, in even postulating the precise nature of the non-linear relationships are great. He cited the propensity of artificial neural network models for reliably characterizing the chaotic functional dependencies evident in clinical data.

In 1991, Baxt [22] demonstrated the predictive reliability of artificial neural network models in medical diagnosis. He constructed a neural model for diagnosing myocardial infarction. It used input variables selected from the presenting symptoms, the past history findings, and the physical and laboratory findings of adult patients presenting to an emergency department with anterior chest pain. This model improved on the diagnostic accuracy of attending physicians markedly. With improvements in both the sensitivity and the specificity, when compared with clinicians' judgements, the model provided an early benchmark for studies of this type.

In the intervening years, a myriad of results have appeared in the application of neural computing to prediction and diagnosis in various fields of medicine. In many reports, these applications are compared with the results of applying statistical or other techniques to the same problems. Interesting examples can be found in oncology ([23], [24], and [25]), in radiology ([26] and [27]), and cardiology ([28], [29], and [30].) Further examples can be found addressing such conditions as auditory brainstem response [31], sleep classification in infants [32], glaucoma [33], and even interhospital transport mode [34].

In 1993, McGonigal and colleagues pioneered the application of neural network models in the area of trauma scoring [35]. They employed such a model to estimate the probability of survival for patients with penetrating trauma. The results show a significant improvement in sensitivity over the TRISS [36] and ASCOT [37] methods of survival prediction. The model was carefully constructed to utilize only data elements that are routinely available and included in the estimations of the other two techniques. McGonigal noted the increased ability of neural models to characterize the nonlinear behavior of biologic systems. Others have followed with various neural network models for trauma outcomes prediction. See, for instance, [38], [39], [40], and [41]. In 1995, Rutledge employed neural networks to assess injury severity by using them to predict the probability of survival in trauma cases [42]. This study also showed predictive improvement over standard methods and was able to accomplish this without needing data the lack of which in many state and regional trauma centers often hinders or prevents application of the standard methods.

Until recently, all the work on neural computing in trauma care has been in the area of survival scoring and its motivation has apparently been in the area of extramural quality control assessment. In fact, the patient outcomes of survival/death at discharge from the acute care hospital and length of stay in the intensive care unit and in the hospital are the target variables of the Major Trauma Outcome Study of Champion et al. [43]. To expand this orientation in neural network analysis of trauma circumstances, we have addressed the analysis of specific complication outcomes. In [44], we presented a neural model for diagnosing sepsis in victims of blunt injury trauma. With very good performance in sensitivity and specificity, the model uses data elements that are routinely recorded in regional trauma center TRACS [45] databases.

The purpose of this chapter is to advocate and facilitate the application of neural computing to improved diagnosis of all such morbidity outcomes in trauma care. The Committee on Trauma of the American College of Surgeons has itself advocated the expansion of trauma outcomes research beyond its previously limited scope of straight survival [45]. Citing the importance of research in morbidity outcomes, they have adopted a standardized and well-defined classification of trauma complications, which was proposed by Hoyt and colleagues in 1992 [46]. It is thanks to Hoyt's work that we have an organized framework for pursuit of our present aims. The following section introduces a new study we have completed on the complication outcome of coagulopathy.

2 A Neural Network Model for Predicting the Incidence of Coagulopathy in Victims of Blunt Injury Trauma

Many trauma patients will develop some degree of coagulopathy. Extensive tissue destruction, hypothermia and shock are clear indicators of future coagulation abnormalities. Massive transfusion is associated with the development of coagulopathy, however, this relationship is less predictable. The relationship of trauma index, Glascow Coma Score and other pre-hospital variables to coagulopathy complications in trauma victims is even less intuitive. The ability to identify these patients early in their hospital course may allow the clinical team to intervene early and potentially prevent and or treat these problems. The implications of early intervention are clear: shortened hospital stays and decreased blood product costs are just two of the many areas where significant health care resources can be saved.

2.1 Model Description

A neural network model was constructed to recognize any patterns that may exist in prehospital and emergency department patient data, which are consistent with the presence or absence of coagulopathy complications. The data elements were drawn from those available in the NATIONAL TRACS design [45], which includes elements on demographics (13 elements), injury (11), prehospital findings (22), referring hospital findings (15), emergency department admission (19), emergency department treatment (13), hospital diagnoses (22), operations (36), quality assurance indicators (17), complications (28), and outcomes (9). This structure, as mentioned earlier, has been endorsed by the Committee on Trauma of the American College of Surgeons, and is now a standard for trauma registries. It is widely used by regional trauma registries and includes procedures for reporting data to state and central trauma registries [30]. Our intention was to provide a neural model that can be applied everywhere this data structure is adopted.

The variables selected for input roles in the network architecture were patient age; prehospital pulse rate, respiratory rate, systolic blood pressure, Glasgow Coma Score, and Revised Trauma Score; emergency department temperature, pulse rate, respiratory rate, systolic blood pressure, Glasgow Coma Score, and Revised Trauma Score; Hematocrit value (recorded from blood testing in the emergency department); injury severity score, and number of units of blood given to the patient in the first 24 hours after injury. The indicator variable showing the presence or absence of coagulopathy complications was chosen as the single output of the neural network, with the value 0 representing absence of disease and 1 indicating its presence. Its values in the database were used as targets in the training algorithm, for patients whose records were used for training, and as benchmarks for evaluating the success of training, for the patient records reserved for testing.

The hidden layer of the neural model was assigned only one neuron, in the hopes that a parsimonious architecture might be enough to capture the patterns in the data. (As it turned out, this was in fact sufficient. In normal experimentation with neural models, it is not uncommon for a successful design to necessitate a bit more trial and error.) Extra constant-valued neurons, called bias nodes, were added to each of the input and hidden layers, to insure non-zero stimulation of their succeeding layers.

The data were selected from blunt injury trauma cases recorded between July, 1994 and April, 1995 in the TRACS database of the Creighton University Medical Center. Prehospital values were taken from accident scene data elements, unless a patient was admitted to the emergency department by referral from another hospital. In those cases, the referring hospital patient data were substituted. After culling records with missing or corrupt entries (of which we commonly find there to be many), we arrived at a set of 328 cases. Of these, 5 (1.5%) were coded as having involved coagulopathy complications. Descriptive statistics for the input variable values are given in Table 1. Half of the data set's cases were randomly selected for training the neural network and the remaining cases were reserved for testing.

2.2 Results

The model trained completely after only 809 iterations of presenting the training data and adjusting the weights. The average of squared errors resulting from training was .009. (Since the target values were coded as an indicator variable and the network output was normalized to fall within the unit interval, training and testing absolute error can never exceed 1 for any case.) With an individual absolute error tolerance level of .1, the training set witnessed 100% sensitivity, correctly adjusting to recognize the 3 cases of coagulopathy which were present in that randomly selected training set of 164 cases.

Additionally, the neural network was able to train to 100% specificity with the training cases. The input weights established in this training are given in Table 1.

With an absolute error tolerance of .2 for testing, the trained network performed correctly with 161 of the 164 testing cases. This included 100% sensitivity, with the 2 cases involving coagulopathy that were found in the testing data set. The network failed to recognize 3 patient cases in the testing set as being free of the complication, although the database indicated that the corresponding patients were free of it. This gives a specificity of 98.2%. It should be noted that the network output for these cases would not necessarily have led to a false diagnosis, since their network outputs were also far from the value of 1. (With error tolerance of .2, an absolute error exceeding .8 would have been necessary to incorrectly conclude that a value of 1 might be associated with the coagulopathy indicator.) The lack of certainty evinced by these testing errors simply indicate an inconclusive result for these cases.

Table 1. Input variable descriptives and weights.

i	x_i	Max	Min	Mean	StDev	w_i
1	AGE	93.6	1.2	41.45	24.76	-1.0092
2	PULSE	150.0	0	91.80	19.37	3.7892
3	RESP	44.0	0	19.77	5.57	-3.8602
4	BP	232.0	0	127.72	32.19	4.0356
5	GCS	15.0	3.0	13.61	3.08	-1.6976
6	TS	12.0	0	11.36	1.75	-0.4300
7	EDTEMP	100.8	90.5	97.53	1.48	1.8036
8	EDPULSE	170.0	42.0	95.77	19.96	-1.3366
9	EDRESP	70.0	0	21.79	6.92	4.1280
10	EDBP	240.0	44.0	141.76	28.48	-0.0762
11	EDGCS	15.0	3.0	14.09	2.65	-2.0802
12	EDTS	12.0	2.0	11.54	1.53	1.5864
13	HCT	52.6	11.3	39.14	5.70	-5.6254
14	ISS	41.0	0	9.10	8.05	0.3270
15	BLOOD	29.0	0	0.48	2.59	6.3572

As is the case with many medical conditions and data sets, prevalence of the coagulopathy complication here is quite low (1.5%). It is therefore very important that the results of this diagnostic model be reported for the separate populations of diseased and disease-free patients. Separate sensitivity and specificity results can help avoid the misleading indications of a prevalence-dependent overall assessment. As is well explained by Lett and colleagues [48], the skewed distribution of trauma data can also be addressed in evaluation by use of receiver operator characteristic curve analysis. Calculated using nonparametric assumptions, the area, A, under the ROC curve for our neural network diagnosis model was established as .995, with a standard error of .003. This can be interpreted as a further indication of the success of the modeling effort.

2.3 Remarks

The lean neural network architecture established in this experiment produced an advantage here that goes way beyond the normal result of reduced computational complexity. Although it is not a common eventuality in neural modeling, the simplicity of this particular network structure evinces some important opportunities to make inferences about the individual input variables' influences on the final result. This is because of the restriction to one processing neuron in the hidden layer. While the mathematical function that represents the final model is still non-linear, due to the activation functions of the hidden and output layers, the nesting of these activation functions is one dimensional. This results in an overall network characterization that is monotonically increasing.

The neural model can be expressed as

$$O_k = 1/(1 + \exp(4.1362 - 7.9994 * g(\mathbf{x}))),$$

where $\mathbf{x} = (x_1, \ldots, x_{15})$, representing the set of independent variable values for a given case, and

$$g(\mathbf{x}) = 1/(1 + \exp(2.1480 - \Sigma w_i * x_i)),$$

with the summation being over $i = 1, \ldots, 15$. The w_i 's are the trained weights for the hidden layer and the constants shown are those of the output layer and the bias elements. As can be readily seen, when its trained weight is positive, an increase in the value of an input variable leads directly to an increase in the overall network output. A negative weight estimate produces the opposite effect. This allows us to interpret the signs of the trained weights (shown in Table 1) as indicators of the effect their input variables have on the diagnoses of the model.

The data show a clear relationship between units of blood transfused and low hematocrit and the development of coagulopathy. The relationship of massive transfusion to dilutional coagulopathy is well recognized. Our data indicate that this relationship is much stronger than that of any of the other variables considered and that the effect of potential dilutional coagulapathy may be important. The model did not examine other variables such as hypothermia, so hematocrit and units transfused may just be serving as a proxy for other factors. The other variables examined show a somewhat paradoxical effect. On the one hand, elevated pulse, decreased respiratory rate, low Glascow Coma Score and low Revised Trauma Score in the field are related to the development of coagulopathy (as would be expected as these indicate the severity of injury and physiologic evidence of shock). These variables are also highly inter-related (pulse and respiratory rate are components of the Revised Trauma Score for example) and one would expect that they would each predict coagulopathy. Interestingly, elevated blood pressure in the field would not be expected to predict coagulopathy as this variable would be expected to be low if the patients were in shock. In the emergency department, an elevated respiratory rate, low blood pressure, low Glascow Coma Score, low hematocrit and high injury severity score are associated with the development of coagulopathy while an elevated trauma score is not.

We can postulate a variety of reasons for these discordant results. First, we have previously shown that data coding errors may have profound effects on a neural networks' evaluation of the data set [44]. Second, there are inherent inaccuracies in establishing data such as the Glascow Coma Score in the field. Additionally, we have a small number of cases and a limited number of variables, so the power of this analysis is limited. As previously noted, many of these data are inter-related so the strength of any individual datum may be limited.

We suggest several areas for further study. The paradoxical relationship between 'in the field' and emergency department relationships of trauma scores suggests errors in coding, evaluation or changes in patient response during transport and treatment. A real difference in physiologic parameters in the field and in the emergency department may indeed exist. A recent paper of Lang and colleagues [61] lends credibility to this possibility. They established significant differences in predicted outcomes of severe head injury, depending on whether the neural model used data collected 6 hours after trauma or 24 hours later. They stressed the implication that therapeutic decisions, such as cessation of therapy, should be based on the patient's status one day after injury and only rarely on admission status alone. However, in both of our studies, patients with the complication in question (sepsis or caogulopathy) was limited. Studies involving a larger number of patients are warranted.

3 Prospects for Refining and Utilizing Neural Models in Trauma Care Settings

3.1 Sensitivity Analysis, Pruning, and Rule Extraction

As noted previously, the simplicity of the neural model in the present study is rare. Only one additional neuron in the hidden layer would have completely prevented the above sensitivity analysis of input variables' individual effects on the model. The traditional approach to deriving knowledge from a trained neural network about its individual components has always utilized Hinton diagrams [49]. As illustrated by Figure 2, this diagram depicts the trained weights of a network layer as shades of gray, with the darker shades representing larger values. Here, the rows would show the neuron activation levels of 7 hidden neurons, while the columns would represent the levels for 8 inputs. A Hinton diagram can tell us at a glance which units are strongly active, which are "off", and which are indeterminate.

To evaluate the relative significance of individual network components, we use the Hinton diagram to visualize their connection strengths. Sensitivity analysis can then be conducted by selectively activating certain input nodes, while leaving the others silent. Often, we can identify the neurons which react most strongly to certain inputs. In fact, this kind of network testing can identify inputs which excite or inhibit different neurons. With a goal of determining which input variables are important to the successful training of a neural model, this and other approaches to sensitivity analysis appear to have promise. Developments have been isolated, however, and usually are somewhat computationally intensive. (A recent exception to this was just published by Hunter et al. [50].)

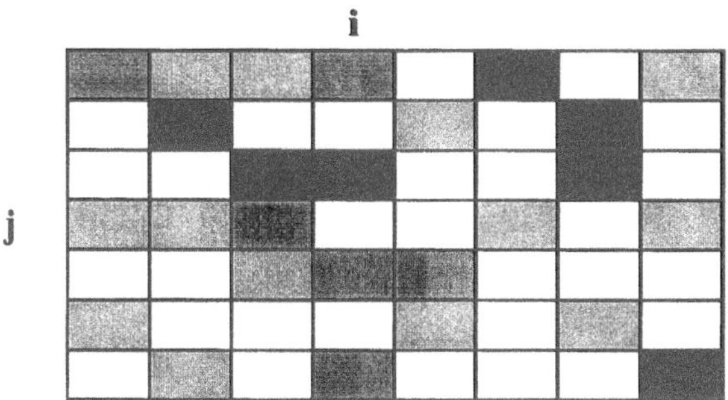

Figure 2. A Hinton diagram.

Another approach to simplifying the structures of successful neural models is called network pruning. It has sustained significant investigation in recent years and promises to help our efforts in sensitivity analysis. Oriented to removing neurons from a network, without sacrificing diagnostic power, this approach has evolved with great mathematical rigor in the theoretical journals. (See, for instance, [51]-[54].) Workable methods for its practice have not yet converged in the applied literature, but there does appear to be promise for attacking sensitivity analysis. The seemingly confounding interpretations evinced by the study exposed in this chapter could be greatly alleviated with fewer independent variables to consider. The heavy interaction noted earlier between groups of variables in the coagulopathy study would certainly lead us to suspect that some inputs can be eliminated without loss of the information that they embody.

One further step toward discovering particulars about the knowledge that trained neural networks embody has recently gained momentum in the literature. It involves the direct derivation of rules of behavior from the activity of neural networks. With the logical structure of expert system production rules, these kernels of knowledge can explain a network's responses in general terms and help guide further study in their problem domains. Interesting expositions of this pursuit have been forwarded by Towell and Shavlik [55] and Setiono et al. [56], for instance. In fact, Setiono has even applied rule extraction methods to breast cancer diagnosis models [57]. Also offering potential help for our program of refining neural models of trauma complications, his method depends heavily on a network pruning procedure to result in rules that are useful. For a relatively uncomplicated network architecture, however, extracted rules can offer interesting combinations of input cutoff values that, taken together, signal changes in a model's diagnosis output.

3.2 Trauma Systems Development

The practical implications of neural networks in trauma are several. First, development of systems that would identify patients at risk for life-threatening and costly complications may allow the clinical team to intervene early in the care of these patients. While there are no data, at present, to indicate the degree to which early intervention might decrease the morbidity and mortality associated with these complications, the potential certainly exists. For example, Barret, et. al. [58] suggest that transfusion with reconstituted whole blood may decrease the incidence of coagulopathy in severely burned patients. Current therapeutic guidelines, on the other hand, encourage the selective use of components (for example see Fakhry, et. al. [59]) because experimental models indicate that other factors present in the trauma patient (e.g. tissue destruction and hypothermia) have a much greater effect on coagulation. Our data, as well as Barret's, suggest that further study of prophylactic therapy may be warranted. Effective use of scare resources in this setting should decrease the costs associated with trauma care. Neural models such as ours might provide a means to identify these patients.

A second advantage of this approach is the potential for scrubbing data contained in trauma databases. We have previously shown the potential for these systems to identify miscoded cases [44]. As these databases are used in further research, clean data are essential. Problems and limitations with trauma registries often involve interpretive coding by trained medical personnel. Rutledge [42] notes that this coding involves a significant amount of time and money to perform coding on a case by case basis. Others [47] have noted that this data collection is occurring in the context of health care reform and cost cutting. Databases, such as TRACS, are ambitious undertakings that require extensive checking and analysis "to remove all inaccuracies and inconsistencies" [60]. As we have previously argued, use of regional (or large local) databases for research requires that significant resources be allocated to control for the quality of the data. To the extent that neural network models identify putative miscoding, significant resources should be saved.

One might envision a neural network trained with extensively reviewed data being used to examine a new record set. Cases that confound the network's learning (where non-convergent cycling, for example, occurs in the attempts to train) we might suspect an inconsistency in the way these cases' data elements were valued. These cases would then be flagged for further review and recoding. This could obviate the need for a rigorous review of all the cases in the set.

References

[1] McCulloch, W.S. and Pitts, W.H. (1943), "A logical calculus of the ideas immanent in nervous activity," *Bull. Math. Biophys.*, vol. 5, pp. 115-133.

[2] Hebb, D.O. (1949), *The Organization of Behavior*, John Wiley and Sons, New York.

[3] Widrow, B. and Hoff, M.E. (1960), "Adaptive switching circuits," *IRE Wescon Conv. Rec.*, pp. 96-104.

[4] Rosenblatt, F. (1961), *Principles of Neurodynamics: Perceptrons and the Theory of Brain Mechanisms,* Spartan Press, Washington, D.C.

[5] Minsky, M. and Papert, S. (1969), *Perceptrons: An Introduction to Computational Geometry,* MIT Press, Cambridge, Mass.

[6] Rumelhart, D.E. and McClelland, J.L. (1986), *Parallel Distributed Processing, Vol. I and II*, MIT Press, Cambridge, Mass.

[7] Hecht-Nielsen, R. (1990), *Neurocomputing*, Addison-Wesley, Reading, Pa.

[8] Hagan, M.T., Demuth, H.B. and Beale, M. (1996), *Neural Network Design*, PWS Publishing Company, Boston.

[9] Widrow, B., Rumelhart, D.E. and Lehr, M.A. (1994), "Neural networks: Applications in industry, business, and science," *CACM,* vol. 37(3), pp. 93-105.

[10] Ripley, B.D. (1993), "Statistical aspects of neural networks," *Networks and Chaos – Statistical and Probabilistic Aspects* (Barndorff-Nielsen, O.E., Jensen, J.L. and Kendall, W.S., eds.), Chapman & Hall, London, pp. 40-123.

[11] Cheng, B. and Titterington, D.M. (1994), "Neural networks: A review from a statistical perspective," *Statist. Sci.,* vol. 9(1), pp. 2-54.

[12] Warner, B. & Misra, M. (1996), "Understanding neural networks as statistical tools," *The American Statistician,* vol. 50, pp. 284-293.

[13] Wang, S. (1998), "An insight into the standard backpropagation neural network model for regression analysis," *Omega*, vol.26, pp. 133-140.

[14] Hwang, J.T. and Ding, A.A. (1997), "Prediction intervals for artificial neural networks," *J. Am. Statist. Assoc.,* vol. 92, pp. 748-757.

[15] Yoon, Y., Swales, G. and Margavio, T.M. (1993), "A comparison of discriminant analysis versus artificial neural networks," *J. Opl. Res. Soc.,* vol. 44, pp. 51-60.

[16] Markham, I.S. and Rakes, T.R. (1998), "The effect of sample size and variability of data on the comparative performance of artificial neural networks and regression," *Computers Ops. Res.,* vol. 25, pp. 251-263.

[17] Cooper, J.C.B. (1999), "Artificial neural networks versus multivariate statistics: An application from economics," *J. Appl. Statist.,* vol. 26, pp. 909-921.

[18] Marble, R.P. and Maier, F.H. (1999), "Distribution-independent confidence intervals for non-linear models: An application to manufacturing performance factors," *Proceedings of the 29th Annual Meeting of the Decision Sciences Institute,* pp. 1232-1234.

[19] Denton, J.W. (1995), "How good are neural networks for causal forecasting?" *J. Bus. Forecasting,* Summer, pp. 17-20.

[20] Cybenko, G. (1989), "Approximation by superpositions of a sigmoidal function," *Mathematics of Control, Signals, and Systems,* vol. 2, pp. 303-314.

[21] Baxt, W.G. (1994), "Complexity, chaos and human physiology: The justification for non-linear neural computational analysis," *Cancer Lett., vol.* 77, pp. 85-93.

[22] Baxt, W.G. (1991), "Use of an artificial neural network for the diagnosis of myocardial infarction," *Ann. Intern. Med.,* vol. 115, pp. 843-48.

[23] Downs, J., Harrison, R.F., Kennedy, R.L. and Cross,S.S. (1996), "Application of the fuzzy ARTMAP neural network model to medical pattern classification tasks," *Artificial Intelligence in Med.,* vol. 8, pp. 403-428.

[24] Mango, L.J. (1994), "Computer-assisted cervical cancer screening using neural networks," *Cancer Lett.,* vol. 77, pp. 155-162.

[25] Wilding, P., Morgan, M.A., Grygotis, A.E. et al. (1994), "Application of backpropagation neural networks to diagnosis of breast and ovarian cancer," *Cancer Lett.,* vol. 77, pp. 145-153.

[26] Lo, J.Y. et al. (1997), "Predicting breast cancer invasion with artificial neural networks on the basis of mammographic features," *Radiology,* vol. 203, pp. 159-63.

[27] Tourassi, G.D., Floyd, C.E., Sostman, H.D. and Coleman, R.E. (1993), "Acute pulmonary embolism: Artificial neural network approach for diagnosis," *Radiology,* vol. 189, pp. 555-558.

[28] Heden, B., Edenbrandt, L., Haisty, W.K. and Pahlm, O. (1994),

"Artificial neural networks for the electrocardiographic diagnosis of healed myocardial infarction," *Am. J. Cardio.,* vol. 74, pp. 5-8.

[29] Kennedy, R.L. et al. (1997), "An artificial neural network system for diagnosis of acute myocardial infarction (AMI) in the accident and emergency department: Evaluation and comparison with serum myoglobin measurements," *Comput. Methods Programs Biomed.,* vol. 52, pp. 93-103.

[30] Orr, R.K. (1997), "Use of a probabilistic neural network to estimate the risk of mortality after cardiac surgery," *Medical Decision Making,* vol. 17, pp. 178-185.

[31] Tian, J., Juhola, M. and Grönfors, T. (1997), "Latency estimation of auditory brainstem response by neural networks," *Artificial Intelligence in Medicine,* vol. 10, pp. 115-128.

[32] Koprinska, I., Pfurtscheller, G. and Flotzinger, D. (1996), "Sleep classification in infants by decision tree-based neural networks," *Artificial Intelligence in Medicine,* vol. 8, pp. 387-401.

[33] Henson, D.B., Spenceley, S.E. and Bull, D.R. (1997), "Artificial neural network analysis of noisy visual field data in glaucoma," *Artificial Intelligence in Medicine,* vol. 10, pp. 99-113.

[34] Hosseini-Nezhad, S.M. et al. (1995), "A neural network approach for the determination of interhospital transport mode," *Computers and Biomed. Res.,* vol. 28, pp. 319-334.

[35] McGonigal, M.D., Cole, J., Schwab, C.W. et al. (1993), "A new approach to probability of survival scoring for trauma quality assurance," *J. Trauma,* vol. 34, pp. 863-870.

[36] Boyd, C.R., Tolson, M.A. and Copes, W.S. (1987), "Evaluating trauma care: The TRISS method," *J. Trauma,* vol. 27, pp. 370-378.

[37] Champion, H.R., Copes, W.S., Sacco, W.J. et al. (1990), "A new characterization of injury severity," *J. Trauma,* vol. 30, pp. 539-546.

[38] Dombi, G.W. et al. (1995), "Prediction of rib fracture injury

outcome by an artificial neural network," *J. Trauma,* vol. 39, pp. 915-921.

[39] Izenberg, S.D., Williams, M.D. and Luterman, A. (1997), "Prediction of trauma mortality using a neural network," *Am. Surg.,* vol. 63, pp. 275-81.

[40] Lim, C.P., Harrison, R.F. and Kennedy, R.L. (1997), "Application of autonomous neural network systems to medical pattern classification tasks," *Artificial Intelligence in Med.,* vol. 11, pp. 215-239.

[41] Selker, H.P. et al. (1995), "A comparison of performance of mathematical predictive methods for medical diagnosis: Identifying acute cardiac ischemia among emergency department patients," *J. Investig. Med.,* vol. 43, pp. 468-476.

[42] Rutledge, R. (1995), "Injury severity and probability of survival assessment in trauma patients using a predictive hierarchical network model derived from ICD-9 codes," *J. Trauma,* vol. 38, pp. 590-601.

[43] Champion, H.R., Copes, W.S., Sacco, W.J. et al. (1990), "The major trauma outcome study: Establishing national norms for trauma care," *J. Trauma,* vol. 30, pp. 1356-1365.

[44] Marble, R.P. and Healy, J.C. (1999), "A neural network approach to the diagnosis of morbidity outcomes in trauma care," *Artificial Intelligence in Medicine,* vol. 15, pp. 299-307.

[45] Rice, C.L. and Rutledge, R. (1993), "Trauma registry, in: American College of Surgeons - Committee on Trauma," *Resources for Optimal Care of the Injured Patient,* pp. 97-101.

[46] Hoyt, D.B., Hollingsworth-Fridlund, P., Fortlage, D. et al. (1992), "An evaluation of provider-related and disease-related morbidity in a level I university trauma service: Directions for quality improvement," *J. Trauma,* vol. 33, pp. 586-601.

[47] Shapiro, M.J., Cole, K.E., Keegan, M., Prasad, C.N. and Thompson, R.J. (1994), "National survey of state trauma

registries- 1992," *J. Trauma,* vol.37, pp. 835-842.

[48] Lett, R.R., Hanley, J.A. and Smith, J.S. (1995), "The comparison of injury severity instrument performance using likelihood ratio and ROC curve analyses," *J. Trauma*, vol. 38, pp. 142-148.

[49] Skapura, D.M. (1995), *Building Neural Networks*, Addison-Wesley, Reading, Mass.

[50] Hunter, A., Kennedy, L. et al. (2000), "Application of neural networks and sensitivity analysis to improved prediction of trauma survival," *Comput. Methods Programs Biomed.*, vol. 62, pp. 11-19.

[51] Karnin, E.D. (1990), "A simple procedure for pruning back-propagation trained neural networks," *IEEE Transactions on Neural Networks,* vol. 1, pp. 239-242.

[52] Chung, F.L. and Lee, L. (1992), "A node pruning algorithm for backpropagation networks," *International Journal of Neural Systems,* vol. 3, pp. 301-314.

[53] Hagiwara, M. (1994), "A simple and effective method for removal of hidden units and weights," *Neurocomputing,* vol. 6, pp. 207-218.

[54] Setiono, R. (1997), "A penalty-function approach for pruning feedforward networks," *Neural Computation,* vol. 9, pp. 185-204.

[55] Towell, G.G. and Shavlik, J.W. (1993), "Extracting refined rules from knowledge-based neural networks," *Machine Learning,* vol. 13, pp. 71-101.

[56] Setiono, R., Thong, J.Y.L. and Yap, C. (1998), "Symbolic rule extraction from neural networks: An application to identifying organizations adopting IT," *Information & Management,* vol. 34, pp. 91-101.

[57] Setiono, R. (1996) Extracting rules from pruned neural networks for breast cancer diagnosis," *Artificial Intelligence in Medicine,* vol. 8, pp. 37-51.

[58] Barret, J.P., Desai, M.H. and Herndon, D.N. (1999), "Massive transfusion of reconstituted whole blood is well tolerated in pediatric burn surgery," *J. Trauma,* vol. 47, pp. 526-528.

[59] Fakhry, S.M., Messick, W.J. and Sheldon, G.F. (1996), "Metabolic effects of massive transfusion" in *Principles of Transfusion Medicine*, (Rossi, E.C., Simon, T.L., Moss, G.S. and Gould, S.A. eds.), pp. 615-25, Williams and Wilkins, Baltimore.

[60] Jones, J.M. (1995), "An approach to the analysis of trauma data having a response variable of death or survival," *J. Trauma,* vol. 38, pp. 123-128.

[61] Lang, E.W. et al. (1997), "Outcome after severe head injury: an analysis of prediction based upon comparison of neural network versus logistic regression analysis," *Neurol. Res.,* vol. 19, pp. 274-280.

Chapter 8

Artificial Neural Networks in Medical Diagnosis

Y. Fukuoka

The purpose of this chapter is to cover a broad range of topics relevant to artificial neural network techniques for biomedicine. The chapter consists of two parts: theoretical foundations of artificial neural networks and their applications to biomedicine. The first part deals with theoretical bases for understanding neural network models. The second part can be further divided into two subparts: the first half provides a general survey of applications of neural networks to biomedicine and the other half describes some examples from the first half in more detail.

1 Introduction

Artificial neural networks (ANNs) are computational techniques inspired by knowledge from neurophysiology. In the 1980s, various important ANN methods were developed [1], [2], [3]. Since the later part of the 1980s, their applications to biomedicine have been widely studied because of their ability to perform nonlinear data processing with a relatively simple algorithm [4], [5]. Now they are successfully applied to medical diagnosis [6], physiological system modeling [7], image and signal processing [8]. The first part of this chapter describes basic concepts of ANNs. In the later part, after categorizing ANN applications, we explore not only their diagnostic applications but also biomedical applications, in which ANNs may provide useful information for medical diagnosis, in a systematic manner. This chapter concludes with a view of future directions of ANNs in biomedicine.

2 Foundations of Artificial Neural Networks

2.1 Artificial Neuron

The basic element of ANNs is an artificial neuron, which is a simplified representation of the real neuron. As shown in Figure 1, an artificial neuron receives multiple inputs and calculates its activity level (its output), which corresponds to the impulse frequency of a real neuron. The total input to the artificial neuron, x is the weighted sum of the input values:

$$x = \sum_i w_i x_i \tag{1}$$

and its output, y is a function of x,

$$y = f(x) \tag{2}$$

where $f(\cdot)$ is called the activation function of the neuron.

Figure 2 illustrates examples of activation functions. The first artificial neuron proposed by McCullogh and Pitts in 1941 employed the binary activation function (Figure 2(a)) [9]. The linear activation function (Figure 2(b)) is used for an input neuron, which receives an input signal from the outside of an ANN. The sigmoidal activation function (Figure 2(c)) is similar to that of a real neuron and is widely employed.

In the literature, the terms “artificial neuron,” “unit” and “node” are used interchangeably. In what follows, we will use “unit” instead of “artificial neuron.”

2.2 Network Architectures

The arrangement of neural processing units and their interconnections can have a profound effect on the capabilities of ANNs. There are three major architectures of ANNs: feedforward network (Figure 3(a)), recurrent network (Figure 3(b)) and the mixture of both. There is no feedback information flow in a feedforward network, and this type of network is suitable for pattern classification tasks. In contrast, there exists some cir-

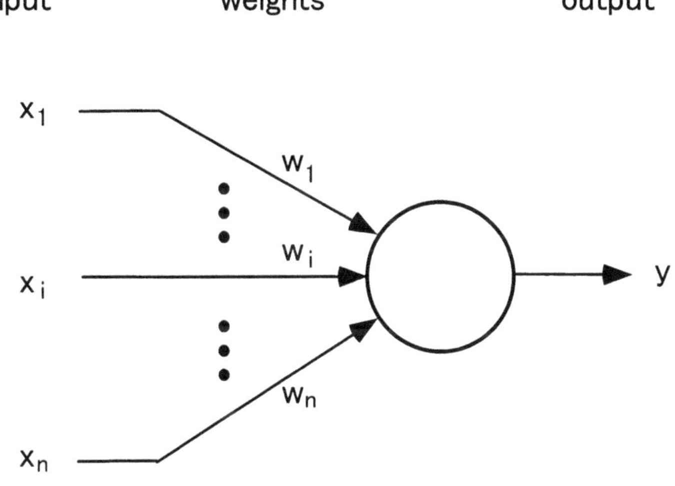

Figure 1. An artificial neuron.

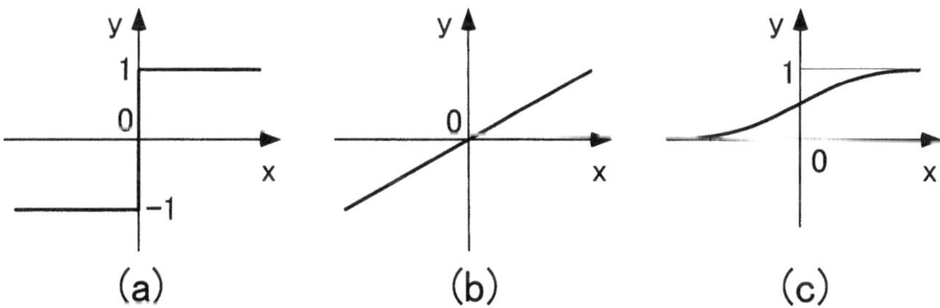

Figure 2. Examples of activation function.

cular flow in the other two types. They are suitable for time-series processing because they behave as dynamical systems.

2.3 Learning Algorithms

Learning algorithms can be divided into two classes: supervised learning and unsupervised learning. A supervised learning algorithm requires a data set of example inputs and their corresponding desired outputs. In contrast, no desired signal is needed for an unsupervised algorithm. The following two sections describe a typical learning algorithm in the two categories.

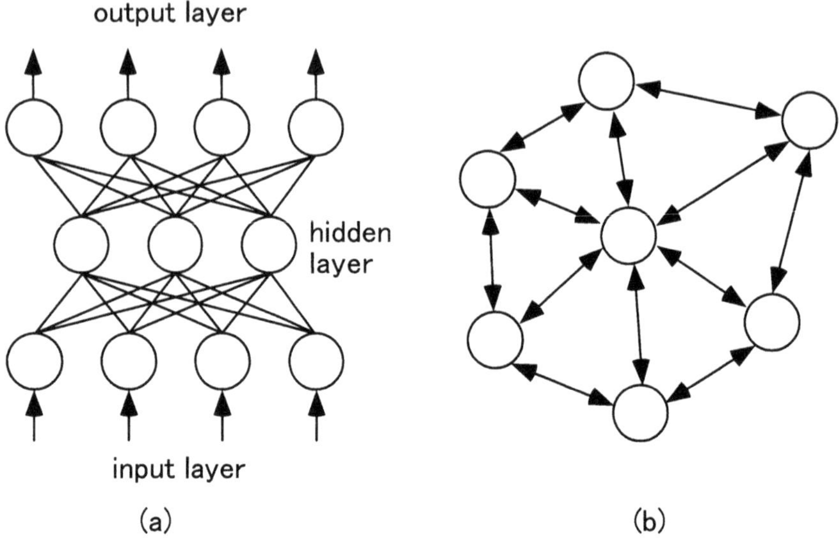

Figure 3. (a) A feedforward network and (b) a recurrent network.

2.3.1 Back-Propagation

The most popular learning algorithm is the back-propagation (BP) method [10], [11]. The algorithm has been widely used to train multilayer, feedforward neural networks. A feedforward network having three or more layers trained with this algorithm is often called "multilayer perceptron." This algorithm can evolve a set of weights to produce an arbitrary mapping from input to output by presenting pairs of input patterns and their corresponding output vectors. It is an iterative gradient algorithm designed to minimize a measure of the difference between the actual output and the desired output. Because many applications employ this learning method, we describe it in some depth here.

The total input to unit j in layer l, x_j^l, is determined by

$$x_j^l = \sum_i w_{i\,j}^{l-1\,l} y_i^{l-1}, \tag{3}$$

where $w_{i\,j}^{l-1\,l}$ represents the connecting weight between unit j in layer l and unit i in the next lower layer. The activity level of unit j in the hidden and output layers is a nonlinear function of its total input,

$$y_j^l = f(x_j^l) = \frac{1}{1 + \exp^{-ax_j^l}}, \tag{4}$$

where a is a slant parameter. It is usually set to 1 and often omitted. On the other hand, the activation function of the input units is linear.

Learning is carried out by iteratively updating the connecting weights so as to minimize the quadratic error function E which is defined as

$$E = \frac{1}{2} \sum_{c} \sum_{i} (y_{i,c}^{3} - d_{i,c})^{2}, \tag{5}$$

where c is an index over input-output vectors, $y_{i,c}^{3}$ is the actual activity level of output unit i and $d_{i,c}$ is its desired value. We will suppress index c in what follows. A weight change Δw is calculated from

$$\Delta w_{i\,j}^{l-1\,l} = -\varepsilon \frac{\partial E}{\partial w_{i\,j}^{l-1\,l}}, \tag{6}$$

where ε is the learning rate. The term on the right-hand side is proportional to the partial derivative of E with respect to that weight. Instead of Equation (6), Equation (6') is often used to adapt the step size as a function of the local curvature of the error surface:

$$\Delta w_{i\,j}^{l-1\,l}(t) = -\varepsilon \frac{\partial E}{\partial w_{i\,j}^{l-1\,l}} + \alpha \Delta w_{i\,j}^{l-1\,l}(t-1), \tag{6'}$$

where α is the momentum factor and t is the iteration number. The second term makes the current search direction an exponentially weighted average of past directions. The partial derivative $\partial E / \partial w$, which is denoted as δ, is calculated as

$$\delta_{i\,j}^{l-1\,l} = \frac{\partial E}{\partial w_{i\,j}^{l-1\,l}} = \frac{\partial E}{\partial x_{j}^{l}} \frac{\partial x_{j}^{l}}{\partial w_{i\,j}^{l-1\,l}} = \frac{\partial E}{\partial x_{j}^{l}} y_{i}^{l-1}, \tag{7}$$

where $\partial E / \partial x_{j}^{l}$ is the error signal,

$$\frac{\partial E}{\partial x_{j}^{l}} = \begin{cases} (y_{j}^{3} - d_{j}) f'(x_{j}^{3}) & \text{(for an output unit)} \\ (\sum_{k} w_{jk}^{23} \frac{\partial E}{\partial x_{k}^{3}}) f'(x_{j}^{2}) & \text{(otherwise).} \end{cases} \tag{8}$$

Here $f'(\cdot)$ is the sigmoidal derivative which is calculated as

$$f'(x) = a(1 - f(x)) f(x). \tag{9}$$

A weight change in the n-th layer counting backwards from the output layer involves $(f'(\cdot))^n$. This causes weight changes corresponding to different layers to differ considerably in magnitude, because $0 \leq f'(\cdot) \leq \frac{1}{4}$ when a=1.

The weights in the network are initialized with small random values at the onset of a learning process. This starts the search at a relatively safe position [12]. Two different schemes of updating weights can be used: the on-line mode and the batch mode. In the on-line mode, weights are updated after every input-output case. On the other hand, in the batch mode, δ is accumulated over all the input-output cases before updating the weights; i.e., a learning step, which is called an epoch, consists of a presentation of the whole set of patterns and a weight modification based on accumulated δ. The learning process continues until the sum of squared errors becomes less than a preset tolerance, e_{th}. The tolerance is usually set according to the nature of the desired output. For example, if the desired output has a continuous value, e_{th} should be set to a value small enough for the network to map the given function accurately. On the other hand, if the desired output is represented by a binary value, e_{th} need not to be set to a very small value because it is satisfactory as long as the difference between the actual and desired outputs is less than 0.5.

In an actual application of the BP algorithm, the user should set some parameters such as a learning rate. Unfortunately, there is no effective method to determine those parameters although the BP procedure is sensitive to different learning rates and initial weights. Here are some benchmark examples from [13], illustrating the influence of those parameters.

The symmetry problem, which was investigated previously by Rumelhart et al. [10], [11], was posed for a network consisting of a layer of six input units, a layer of two hidden units and one output unit. This pattern classification problem was solved using both the batch and the on-line modes. According to each mode, the connecting weights were updated to detect whether or not the binary output levels of a one-dimensional array of input units were symmetric about the center point. In what follows, the iteration number t is expressed by the unit of epoch (weight modification after presenting the whole input vectors) for the batch mode, but by the unit of iteration for the on-line mode. Initial weights were drawn from a uniform distribution between w_{init} and $-w_{init}$. Learning rates of 0.01, 0.02, 0.05, 0.07, 0.1, and 0.2 and w_{init} of 0.05, 0.1, and 0.3 were used. For each condition, a rate of successful convergence within a given iteration limit (rate of success) was estimated over 100 trials in which the momentum factor α and the tolerance e_{th} were fixed at 0.9 and 0.25, respectively. The iteration limit was 1600 epochs for the batch mode and 102,400 iterations for the on-line mode.

Table 1 summarizes the experimental results. In general, a small w_{init} as well as a small learning rate cause slow learning progress. For this problem, the batch mode provides better results than the on-line mode. Table 1(a) indicates that the rate of success obtained with the batch mode depends strongly on the learning rate ($\varepsilon = 0.05$ is optimal and 0.07 is near-optimal). There are two reasons why a high rate of success is not achieved using the other values: one is a small learning rate ($\varepsilon < 0.05$ in this case) that causes failure to converge within the iteration limit and the other is a false local minimum or premature saturation which frequently occurs with a relatively large learning rate ($\varepsilon > 0.07$ for this problem) [13].

Various improvements of the original BP algorithm have been proposed not only for feedforward networks [13], [14] but also for recurrent networks [15].

2.3.2 Self-Organizing Map

The self-organizing map (SOM), sometimes also referred as self-organizing feature map, was proposed by Kohonen [16], [17]. The SOM is a clustering method similar to *k*-means clustering. However, unlike the *k*-means method, the SOM facilitates easy visualization and interpretation. Kohonen has demonstrated that an SOM can be implemented as a feedforward ANN trained with an unsupervised learning algorithm. An ANN for a SOM is composed of input and competitive output layers. The input layer has n units that represent n-dimensional input vector while the output layer consists of c units that represent c decision regions. The output units are arranged with a simple topology such as a two-dimensional grid. The two layers are fully connected, i.e., every input unit is connected to every output unit.

Table 1. Experimental results. RS: rate of success, LI: learning epochs/iterations required for convergence. The data include only successful trials.

	$w_{init} = 0.05$		$w_{init} = 0.1$		$w_{init} = 0.3$	
ε	RS	LI	RS	LI	RS	LI
(a) Batch mode						
0.01	0	—	0	—	0	—
0.02	0	—	0	—	33	1398±81
0.05	100	938±62	100	893±91	93	872±221
0.07	100	1157±131	92	1095±211	68	845±284
0.1	0	—	9	1484±91	38	899±309
0.2	0	—	0	—	0	—
(b) On-line mode (×64 iterations)						
0.01	0	—	0	—	0	—
0.02	0	—	0	—	0	—
0.05	0	—	3	1556±31	24	1169±178
0.07	1	1333	8	1237±165	32	855±138
0.1	0	—	4	783±43	31	607±102
0.2	0	—	0	—	19	303±38

The connecting weights are modified to cluster input vectors into c classes. At the onset of a training process, the weights are initialized to small random values. On subsequent iterations, an input vector $\mathbf{P}_i$ is presented to the network in a random order and then the weights are adjusted on a winner-take-all basis. The winning unit is typically the unit whose weight vector $\mathbf{W}_j$ is the closest to $\mathbf{P}_i$. A weight vector $\mathbf{W}_j$ consists of the connecting weights from all input units to output unit j.

$$\mathrm{W}_j = [W_{1j}, W_{2j}, \ldots, W_{nj}] \tag{10}$$

where W_{1j} denotes the weight from the first input unit to the j-th out-

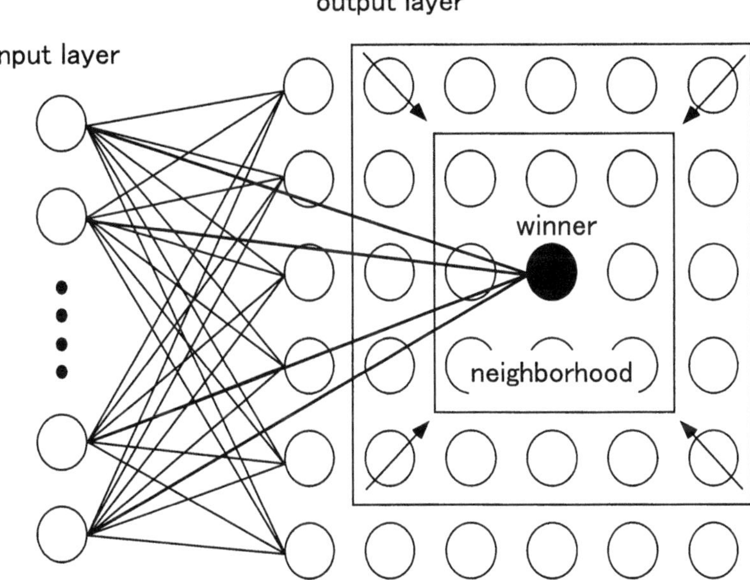

Figure 4. A neural network for an SOM.

put unit. To achieve a topological mapping, not only $\mathbf{W}_j$, but also the weights of the adjacent output units in the neighborhood of the winner are adjusted. The weights are moved in the direction of $\mathbf{P}_i$ according to Equation (11).

$$W_k(t+1) = \begin{cases} W_k(t) + \alpha(t)(P_i - W_k(t)) & \mathit{if\ unitk} \in \mathit{neighborhood} \\ W_k(t) & \mathit{otherwise} \end{cases} \tag{11}$$

where $\alpha(t)$ is the learning rate at t-th iteration. The learning rate and the size of the neighborhood decrease with iteration number t.

3 Applications to Biomedicine

Applications of ANNs to biomedicine can be categorized based on a learning algorithm and a task pose to the network (e.g., pattern classification or system modeling). The most widely employed scheme is pattern classification with the BP algorithm. The SOM is also employed for a pattern classification problem. In some applications, the BP algorithm is used for data compression and physiological system modeling (including time-series prediction). To illustrate points, we will summarize various applications of ANNs to biomedicine in **3.1** to **3.4**, followed by thorough reviews of two examples in **3.5**.

3.1 Pattern Classification with BP

Medical diagnoses can be regarded as pattern classification problems. Various studies have proven feedforward networks trained with the BP algorithm to be powerful tools for pattern classification problems, especially in the case of having a lot of examples without enough knowledge about classification rules. However, as Rogers et al. [18] have pointed out, those networks are not magical solutions with mystical abilities that work without good engineering. With good understanding of their capabilities and limitations, they can be applied productively to problems in biomedicine.

In the scheme of pattern classification with the BP algorithm, a feedforward network is trained to classify input patterns into c categories. Usually, a network having c output units, each of which represents one category, is employed. For example, if $c = 4$, the desired output signal of (1 0 0 0) is assigned for an input pattern belonging to category A. This signal requires the first output unit to fire and the other output units not to fire. Similarly, when an input pattern from category B is fed to the network, the desired signal of (0 1 0 0) is used.

The next step is the selection of input variables. Because it is inefficient to feed the original signal to the network, feature (e.g. average and standard deviation) extraction is carried out to choose variables more likely to influence the performance of the network and to reduce the number of input variables. If n features are extracted, n input units are employed. In general, this step involves some trial and error process using different sets of input variables.

The number of hidden units, h, can be determined arbitrarily. However, too few hidden units might cause a slow learning progress and a poor mapping capability. On the other hand, too many hidden units degrade the generalization ability, which is a measure of the trained network's performance on data not presented in the training. In most cases, h is set to a value between n and c. Since a three-layer network is capable of forming arbitrary close approximations to any continuous nonlinear mapping [19], [20], only one hidden layer is used in most applications. Hereafter, we will use a notation "n-h-c" for a network composed of n input units, h hidden units and c output units. In the case of having two or more hidden layers, n-h_1-h_2-$\cdots$-c will be used.

3.1.1 Clinical Data

Patil et al. [21] have applied ANNs to assess the likelihood of acute pulmonary embolism (PE) based on 50 variables including the history, physical examination, electrocardiogram, and chest radiograph. They also examined the combination of these variables and ventilation/perfusion scan results. In the latter case, the input of ANNs consisted of 54 variables. These variables were coded as either binary values (present or absent) or continuous values. The output was the likelihood of PE. That is, patients with/without PE were assigned a score of 1(100%)/0, respectively. The study involved 1213 patients. Network configurations of 50-4-1 and 54-4-1 were employed (in what follows, we use 50/54-4-1 instead of listing all configurations separately). Their results showed that ANNs were able to predict the likelihood of PW with an accuracy comparable to experienced clinicians.

In the paper by Wilding et al. [22], ANNs of 10/6/3-5/6/10-2 have been trained to improve the diagnosis of breast and ovarian cancer using laboratory tests (including albumin, cholesterol and tumor marker) and the age of the patient. The breast cancer study involved 104 patients (45 malignant and 59 benign subjects) while the 98 subjects (35 malignant, 36 benign and 27 control subjects) participated in the ovarian study. The best network in the breast cancer study provided little improvement compared to the use of the tumor marker CA 125 as a prediction parameter. The results obtained in the ovarian cancer study were better than those in the breast case. However, they concluded that more refinement was required using larger data sets before a clinical application.

Baxt and colleagues [23], [24]have used a network of 19-10-10-1 to identify the presence of acute myocardial infarction (MI). As an input vector of the network, they employed 19 variables including electrocardiographic findings (e.g., ST elevation and T wave inversion), the history of MI, hypertension and angina of the patient, location of pain, sex. Data from 706 patients who presented to the emergency department with anterior chest pain were used (356 for training and the other 350 for testing). They reported that the network achieved an accuracy of 97%.

3.1.2 Bioelectric Signals

There are a number of papers on ANN applications to bioelectric signals: electroencephalogram (EEG), electrocardiogram (ECG) and electromyogram (ECG). Here we explore such applications.

To develop an automated analysis system of evoked potential (EP), Holdaway et al. [25] have applied three-layer networks to classification of somatosensory evoked potentials collected from patients with severe head injuries. They examined three network configurations of 14-4/8/12-3. They divided the EP response interval into 14 latency bins. The input of the networks was a 14-point analog vector. Each component of this vector was the sum of the absolute values in amplitude of all EP peaks occurring the corresponding latency bin. Each output units represented the severity of head injury. They reported that the ANNs provided the accurate classification rate of 77 %, which was comparable with that of 77.3 % achieved by a human expert.

Hiraiwa et al. [26] have applied ANNs to recognition of 12-channel EEG topography prior to voluntary movements. They used a 24-10-5 network for five-syllable recognition and a 24-10-4 net for recognition of joystick movement in one of four directions. Two snapshots in time were selected as the input: one had a latency of -0.66 s and the other had -0.33 s. Accordingly, n was 24 and $c = 5$ in the syllable recognition while $c = 4$ in the joystick case. After 1000 training cycles, 16 out of 30 and 23 out of 24 testing patterns were correctly recognized in the syllable and joystick cases, respectively.

Jansen [27] has reported results of a preliminary study to see if ANNs can be used to detect K-complexes, which are relatively large waves in EEG often seen during sleep. Two approaches were employed, but the second approach, which used a 500-180-48-5-2 network, was unsuccessful. A too large network without sufficient training cycles might be responsible for the failure. In the first approach, ANNs of 40-5/10-2 and 40-

5/10-5/10-2 were examined. Ten second intervals of EEG signals were selected and classified by an expert neurophysiologist. Each 10 s signal was divided into 20 segments of 0.5-s duration. Two bandpass filters were used to capture the basic frequency of the K-complex (0.5 to 2.3 Hz) and the main frequency component of the sleep spindle (10 to 17 Hz). After filtering, the output of the filters was integrated over 0.5 s segments. A set of 40 integrated values was an input vector of the ANNs. The desired output of (1 0) indicated a K-complex and (0 1) a negative example. Jansen reported that all configurations could achieve 95 % or better accuracy for the training data, but their ability to classify testing data was poor (42 to 67 %). There were some factors that might explain the poor performance: i) classification was performed on a single channel basis (human experts probably detect K-complexes based on inter- and intra-channel comparison) and ii) a K-complex could occur at any location within the 10 s intervals. This paper illustrates the importance of data preprocessing and feature extraction prior to ANN training.

In the paper by Ouyang [28], a 40-12-2 network was trained to distinguish patients with and without anterior wall myocardial infarction (AI) based on ECG. The input vector of the network consisted of voltages of Q-, R-, S-, T-waves and ST deviation in eight leads of the standard 12-lead ECG. The two output units indicated the presence or absence of AI. The network correctly identified 90.2 % of the patients with AI and 93.3 % of the non-AI patients. It should be emphasized that all ECGs used in this study were diagnosed as AI by a commercially available computer-assisted ECG interpretation system.

Fukuoka and Ishida [29] have developed a 48-8-1 network for evaluating chronic stress of rats based on ECG. We will review this work more thoroughly in the later section.

Kelly et al. [30] have applied two types of ANNs to EMG analysis to explore reliable methods for control of arm prostheses. One ANN was a recurrent network called the Hopfield type [31], [32], which is not described in this chapter, for the calculation of the time series parameters for a moving average EMG model. The second ANN was a 2-4-4 network trained with the BP to distinguish four separate arm functions. The input was a set of the first time-series parameter and the EMG signal power level. Each of the output units represented each of the four arm functions. They reported that obtained results were positive.

Schizas et al. [33] have examined 26 ANN configurations to classify motor neuron diseases, Becker muscular dystrophy and spinal muscular atrophy based on macro motor unit potential (MMUP). Twenty MMUPs were recorded from biceps brachii muscle of each patient. The features extracted from each MMUP were amplitude, area, average power and duration. They employed two different input vectors: 8-input vector and 80-input vector. In the 8-input case, the average and the standard deviation of each of the above parameters were used while in the 80-input models, the values of individual parameters formed the input. Because one additional cluster (normal) was used, the output layer had four units. Their results showed that the 8-input ANNs required more training epochs than the 80-input networks and that the ANNs provided recognition rates between 50 and 60 % for testing data.

3.1.3 Image Analysis

The recent acceleration of PCs' processing speed enables us to use larger networks. Now ANNs have been widely applied to medical image analyses. This section describes some applications of ANNs to medical images.

Cios et al. [34] have reported results of a preliminary work on ANN detection of cardiac diseases from echocardiographic images. Echocardiographic images were obtained from two separate regions: the left ventricular (LV) septum and the LV posterior wall. For the first region, six normal subjects, five patients with myocardial infarctions (MI) of the anterior ventricular septum, and seven with hypertrophic cardiomyopathy were examined. For the second region, five normals and six patients with MI of the posterior wall were investigated. ANNs of 100-36-3/2 were employed for the first and second regions, respectively. Matrices of 10-by-10 pixels were used as the input of the ANNs. Two connecting schemes were examined: fully connected and locally connected. In the former, every unit was connected to all units in the next layer, whereas in the latter, units within a predetermined neighborhood were connected to reduce the number of connections and thus computational time. The experimental results indicated that the first scheme provided slightly better results and that generalization performance of both scheme was not sufficient at this preliminary stage.

In the paper by Buller et al. [35], ANNs have been employed to localize and classify malignant and benign tumors in ultrasonic images of the female breast. They used two 48-49-9-1 networks, one for each category. As the input of each ANN, they selected grey scale values of 48 pixels around the pixel being analyzed based on a spider web topology. The key idea was to choose many points from a localized neighborhood and fewer points outside it. The correct localization and classification rates for the malignant and benign cases were 69 % and 66 %, respectively.

A similar study has been conducted by Chen et al. [36]. A 25-10-1 network was trained to classify benign and malignant breast tumors in ultrasound images. As the input of the ANN, they employed a predetermined threshold and two-dimensional normalized autocorrelation coefficients between a pixel and 24 pixels within the 5×5 neighborhood centered at the pixel. The trained network achieved 95 % accuracy for classifying malignancy.

3.2 Pattern Classification with SOM

In contrast to the BP algorithm, SOM networks organize themselves according to the natural structure of the data. The desired features are not known beforehand and hence, this approach falls within the family of clustering methods. In this section we review some applications of the SOM.

Frankel et al. [37] have compared an ANN trained with the SOM to those with the BP algorithm. Their purpose was to analyze marine plankton population using flow cytometry. Although the data used in their study were from oceanographic research, the methodology can be applicable to flow cytometry data of any sort. Their SOM network was fed five input variables: fluorescence in the red (660-700 nm) and orange-red (540-630 nm) excited by each of the 488 nm and 514 nm lines of an argon ion laser in a flow cytometer. The network was trained to classify prochlorophytes, synechococcus, large phytoplankton, and two types of calibration beads using a data set containing 530 cells. The trained network was able to classify synechococcus and the two types of beads accurately. However, about 20 % of the prochlorophytes were misclassified as large phytoplankton. On the other hand, networks of 6-3/4/8-7 and 6-3/4/8-3/4/8-7 were trained with the BP algorithm using a larger data set containing 4800 cells. In the BP case, one additional variable was added to both the input and output used in the SOM case: depth of sea where the sample was obtained to the input and the class "noise" to the output. It should be noted that one output unit was unused (the reason was not clear). The BP networks except 6-3-7 could achieve an accuracy of nearly 100 %. These results suggested that an ANN trained with the BP algorithm might provide a better result compared with the SOM network for the same problem. However, one of the advantages of the SOM is easy interpretation of the results, which cannot be achieved by the BP algorithm.

In the paper by Lagerholm et al. [38], a method for unsupervised characterization of ECG signals has been developed. Their approach involved Hermite function representation of the QRS complexes in combination with the SOM. Forty-eight ECG recordings from MIT-BIH arrhythmia database were used in their study. The QRS complexes were expanded onto a linear combination of N Hermite basis. In addition to N coefficients for the N Hermite polynomials, the width parameter, which approximated the half-power duration, was employed. An input vector consisted of 2N+4 variables: N+1 for each of two ECG leads and two parameters that contained local rhythm information. The output layer had 5×5 units. They examined three network configurations: N=4, 5 and 6. They reported that all three configurations were found to exhibit a very low degree of misclassification (1.5 %).

As the first step to classify a great variety of tumor profiles induced chemical carcinogens, Benigni and Pino [39] have applied the SOM to the analysis of tumors induced 536 carcinogens in rodents. Each carcinogen was associated with the information on the induction of 44 tumor types (target organs), for four experimental groups (rat and mouse, male and female). Thus, the number of input units was 176 while the output layer had 10×10 units. The input values had either 1 (tumor induction) or 0 (no induction). Their results pointed the efficiency in highlighting the associations among chemicals.

Chen et al. [40] have applied the SOM to the classification of benign and malignant sonographic breast lesions. In [36], they conducted a similar research using an ANN with the BP algorithm (see Section 3.1.3 for the details). The input variables were two-dimensional normalized autocorrelation coefficients between a pixel and 24 pixels within the 5×5 neighborhood centered at the pixel and accordingly, the input layer composed of 24 units. The output layer had 6×5 units. The results showed that the accuracy was 85.6 %, which was lower than that achieved by the ANN with the BP algorithm [36]. Again the comparison between the BP algorithm and the SOM for the same problem suggested the superiority of the BP algorithm.

Researchers at Whitehead Institute/Massachusetts Institute of Technology [41], [42] have applied the SOM to molecular classification of cancer based on gene expression data obtained from DNA microarrays [43], [44], [45]. Since this approach is applicable to other diseases and may become more common to analyze gene expression data in the near future, we will review it more thoroughly in the later section.

3.3 Data Compression with BP

In this scheme, a network is trained to establish identity mapping, i.e., the same values are used as the input and target signals of the network. The purpose of this approach is to compress the data using fewer units in the middle layer than the input and output layers [46]. A five-layer network of n-h-c-h-n, where $c<n$, is usually employed for this purpose (see the network shown in the upper, left corner of Figure 5) because Funahashi [47] has proved that the accuracy of identity mapping realization by three-layer networks is less than that of the principal component analysis.

Iwata et al. [48] have employed a 70-2-70 network to compress ECG data recorded with a digital Holter monitor. They used 70 points of the digitized ECG waveform as the input and target signal and developed a dual ANN system. One network, ANN1, was used for data compression without learning while the other, ANN2, was always trained using newly acquired data. Instead of the original ECG waveforms, the activation levels of the two hidden units and the weights in ANN1, which were enough to reproduce the original waveforms, were stored in the memory. While ANN1 worked well, only the activation levels were stored. When ANN1 could not work well because of body motion, etc., the network was replaced with ANN2 and the weights were stored in the memory. The system achieved a data compression rate of one-fifteenth to one-hundredth. It should be noted that this work had been carried out before Funahashi published his results [47].

Fukuoka and Ishida [29] have employed a 2-10-1-10-2 network, which was trained to establish identity mapping, for nonlinear feature extraction. They tried to derive a new index from two-dimensional input vectors. We will review this work more thoroughly in the later section.

3.4 System Modeling with BP

In this scheme, input and desired signals are continuous. Accordingly, the error tolerance is set to a smaller value than that in classification tasks. Some applications in this paradigm are reviewed in what follows.

Chon and Cohen [49] have developed a method to identify linear and nonlinear autoregressive moving average (ARMA/NARMA) parameters using an ANN incorporating polynomial activation functions in the hidden layer. Although the activation functions of the hidden units were different from the conventional sigmoidal function, the network was trained with the BP algorithm. They have demonstrated that the parameters of ARMA/NARMA models can be obtained from analysis of ANN models. A network of 25-4-1 was applied to measurements of heart rate and instantaneous lung volume fluctuations. The method provided slightly better results than did the conventional least squares method for ARMA parameter estimation.

In the paper by Fukuoka et al. [15], ANNs have been used to identify nonlinearity in the P_{CO_2} system in humans. An ARMA model and ANNs were employed to model the system, and three types of ANNs (Jordan, Elman and fully interconnected) were compared. All of these networks had recurrent information flow, but all of them could be trained with slightly modified BP algorithms. The input and output of the ANNs were end-tidal P_{CO_2} and minute ventilation, respectively. By comparing mean squared errors obtained with the ARMA model and the ANNs, the degree of nonlinearity was evaluated. Their results indicated that the fully interconnected network provided the best results and that the linear assumption of the P_{CO_2} system was invalid for three out of the seven normal subjects investigated.

Prank et al. [50] have used ANNs to predict blood glucose levels from concentrations of counterregulatory hormones (glucagon, epinephrine, norepinephrine, growth hormone, cortisol) and insulin. To determine blood levels of these hormones and glucose, blood samples were drawn every two minutes (before and after hypoglycemia, small concentration of blood glucose, induced by an insulin injection) via a central venous catheter from six healthy young male subjects. Glucose levels were predicted from the actual as well as m succeeding and m preceding measurements of hormone concentrations. They examined various network sizes ranged from 18 to 54 input units and from zero to 15 hidden units. The ANNs were trained with resilient propagation, a modified BP algorithm, using pruning techniques. Among the architectures investigated, a 30-10-1 network (m=2) provided the best result. The input vector of this network composed of five adjacent samples in time series for each of the six hormones and the target signal was the blood glucose level. They found that the predictive impact of glucagon, epinephrine and growth hormone secretion were dominant in restoring normal levels of blood glucose following hypoglycemia.

Robinson et al. [51] have applied an ANN of 7-3-1 to predict peak wavelengths, λ_{max}, in the absorption spectra of human red and green visual pigments. A training data set was created based on a listing of two wild type and 34 mutant red-green pigments with their λ_{max}. These mutations consisted of an amino acid change from the red to green or vice versa in one or more of the seven possible positions affecting wavelength modulation. These data, which were already in binary form, were used as input vectors and the target signal was λ_{max} scaled between 0 and 0.93. Testing the trained network using a total of 12 mutation patterns revealed that the ANN was able to predict λ_{max} correctly.

3.5 More Detailed Reviews

3.5.1 Chronic Stress Evaluation Using ANNs

Here we review the work by Fukuoka and Ishida [29] more thoroughly. As shown in Figure 5, their work employed two types of ANNs: one belonged to the class of pattern classification with the BP algorithm and the other was used to perform nonlinear feature extraction. Although the former network was trained using a continuous target value, we regard it as a pattern classifier.

The word "stress" is widely used in daily conversation. The omnipresent signs of stress are adrenal enlargement, thymicolymphatic shrinkage and gastrointestinal ulcers. These changes could be used as an objective index of stress. However, they cannot be measured by a noninvasive method. Thus it is not practical to use them as an index of stress in clinical cases. Techniques for stress assessment are divided into two categories: subjective measures and physiological or behavioral measures. The former is widely used in clinical cases. A doctor assesses a patient's stress based on an interview and a questionnaire. As many researchers have reported, the omnipresent signs are accompanied by various changes; for instance, behavioral and hormonal changes. It is useful to evaluate stress based on such changes that are noninvasively measurable. Fukuoka and Ishida [29] have developed a method to assess chronic stress of rats based on ECG using ANNs. As a first approach, they focused on whether an objective index derived from ECG reflected the previously mentioned internal changes.

Thirty-one male rats were repeatedly exposed to daily six- or two-hour restraint and water-immersion stress from Monday to Friday for four weeks. Another 16 rats, which were not exposed to any stress, were used as control. In the fourth week, an eight-minute ECG measurement was carried out under non-anesthetic conditions. During the measurement, the rat was stimulated using a one-minute tail pinch to examine its reaction to another type of stress. Then the adrenals and the thymus were weighed soon after sacrifice.

Training of an ANN using the BP algorithm requires pairs of input data and target values. Presently, there is no reliable numeric index of stress;

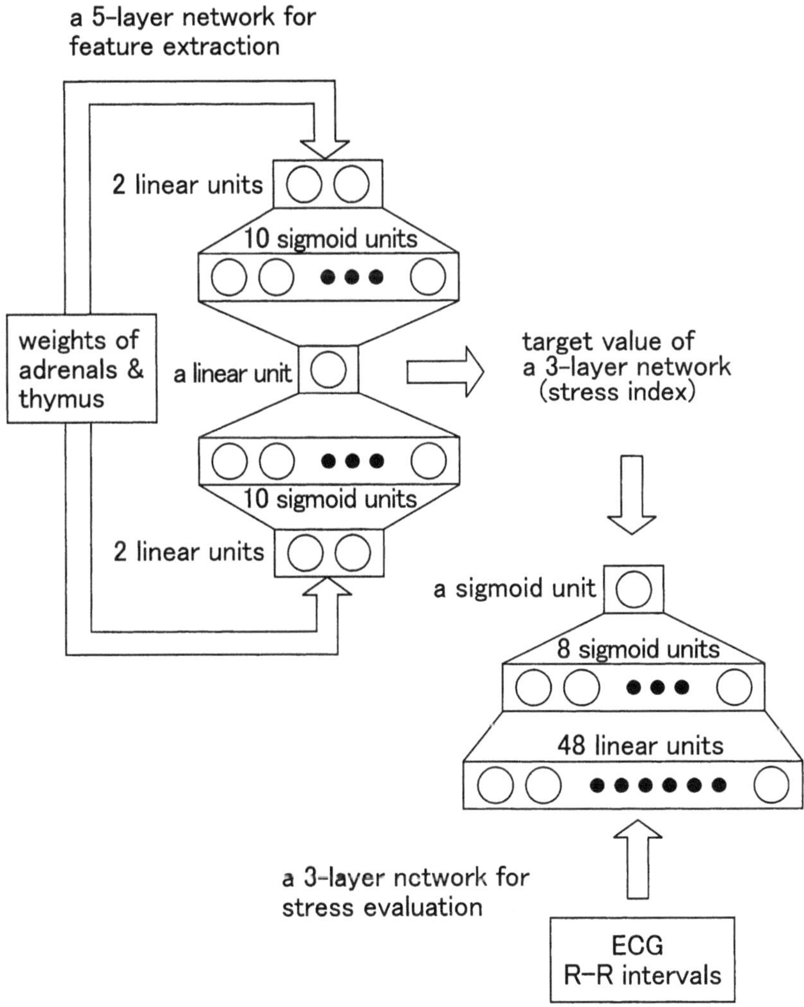

Figure 5. Artificial neural networks used in [29].

that is, no target value is available for neural estimation of stress. Accordingly, in [29] the authors should define the desired signal of an ANN for stress estimation. The most reliable signs of stress are the adrenal enlargement and the thymicolymphatic shrinkage. They, therefore, employed the weights of the adrenals and thymus to calculate an objective index of stress. The weights were normalized by the body weight

Table 2. Stress indices estimated by the three-layer network.

	stress	control
Training data	0.195±0.055	0.285±0.065
Testing data	0.197±0.037	0.261±0.040

of the rat and then put into a five-layer network of 2-10-1-10-2, which was trained to establish identity mapping, for nonlinear feature extraction (see Figure 5). The units in the input, third, and output layers had a linear activation function. The activation function of the other units was sigmoidal. Because the network was designed to extract the most important feature at the unit in the third layer, its output value was employed as the objective index. All weight data of the adrenals and thymus were used to train the network. The training was carried out with $W_{init} = 0.3$ and ε =0.05, and the process was terminated when the sum of squared errors became less than 0.02.

A three-layer network of 48-8-1, shown in Figure 5, was trained to estimate the index based on ECG. R waves were detected only from those ECG signals that were not contaminated by noise. The average and standard deviation of the R-R intervals were calculated for every 20-second data segment in the eight-minute ECG recording. These 48 values were used as the input vector of the three-layer network whose desired signal was the stress index derived by the five-layer network. ECG data of 22 stressed and ten control rats were employed for training while the data from the other nine stressed and five control rats were used for testing the trained network. The training of the network was carried out with $W_{init} = 0.3$ and ε =0.035, and the process was terminated when the sum of squared errors became less than 0.002.

As shown in Table 2, the trained three-layer network could estimate the index accurately not only for the training data but also for testing data. These results indicated that the index estimated from ECG reflected the internal changes and demonstrated that the usefulness of stress assessment by their approach.

3.5.2 Gene Expression Data Analysis with SOM

Here we review a new application area of the SOM, analysis of gene expression data. A gene is expressed to produce the protein encoded by the gene. Investigating gene expression, i.e. how much of each gene product is made, provides important insight into pathological as well as biological processes. As a consequence of large scale DNA sequencing activities, the number of DNA sequences in public databases is increasing rapidly. These data are used to create probes for gene expression analysis. DNA microarray technology [43], [44], [45] makes it possible to simultaneously monitor expression patterns of thousand of genes. For example, high-density oligonucleotide microarrays produced by Affymetrix in 1999 contained probes for 6,817 human genes. (The number of probes is increasing year by year.) DNA microarrays have been used to explore molecular markers of cancer [42], [52]. One of the key goals here is to extract the fundamental patterns inherent in expression data.

Gene expression data have been analyzed mainly with hierarchical clustering [53] and the SOM [41], [42]. In the hierarchical clustering method, the closest pair of data points is grouped and replaced by a single point representing their average, and the next closest pair is processed in the same way. This process is repeated until all data points form a phylogenic tree whose branch lengths represent the degree of similarity between the pairs. Although the value of this method has been demonstrated in [52] and [53], the method has shortcomings for gene expression analysis: lack of robustness, nonuniqueness, its deterministic nature and so on. On the other hand, the SOM is well suited to exploratory data analysis and thus, clustering and analysis of gene expression patterns. Other features of the SOM are its scalability to a large data set and easy visualization and interpretation. Researchers at Whitehead Institute/Massachusetts Institute of Technology [41], [42] have applied the SOM to molecular classification of cancer based on gene expression data.

Their method is as follows. First, a variation filter is used to eliminate genes that do not change significantly across samples. Then expression levels are normalized to have mean=0 and variance=1. The normalized data are put into a SOM network, which has a set of output units arranged in a two-dimensional grid (see Figure 4). The connecting weights are initialized with small random values at the onset of training. On the subsequent training, an expression pattern of a gene is presented in a random order and the weights are updated according to Equation (11).

Before they applied the SOM to expression data obtained from leukemia patients, they had tested their method using data sets of yeast cell cycle and hematopoietic differentiation [41]. In the yeast case, 828 genes that passed the variation filter were used as the input of the SOM, which had 6×5 output units. Their results indicated that multiple clusters exhibited periodic behavior and that adjacent clusters had similar behavior. In the hematopoietic case, normalized expression levels of 567 genes were fed into the SOM having 4×3 output units. They reported that the clusters corresponded to patterns of clear biological relevance although they were generated without preconceptions. These results demonstrated their method's ability to assist in interpretation of gene expression.

In [42], the authors have applied the method to expression data from 38 acute leukemia patients. The data set consisted of gene expression data obtained from 27 acute lymphoblastic leukemia (ALL) patients and 11 acute myeloid leukemia (AML) patients. First, the expression patterns of all 6,817 genes (i.e. no variation filter was used) were employed to train a two-cluster SOM. The trained SOM assigned 25 patients (24 ALL, 1 AML) in one cluster and the other 13 samples (10 AML, 3 ALL) in the other class. This result indicated that the SOM was effective at automatically discovering the two types of leukemia. Next, the authors selected informative genes whose expression pattern was strongly correlated with the ALL-AML class distinction. They employed a wide range of different number of informative genes (10 to 200 genes) for SOM training. All SOM networks performed well; for instance, a 20-gene SOM classified 34 patients correctly and there were one error and three uncertain cases. These results indicated that the SOM could distinguish ALL from AML. They also examined a four-cluster SOM and reported that it was able to divide ALL-patients into two subclasses: T-lineage ALL and B-lineage ALL. The success of the SOM methodology suggests that genome-wide profiling is likely to provide valuable insights into biological process that are not yet understood at the molecular level. This approach is applicable to other diseases and may become more common to analyze gene expression data in the near future.

4 Conclusion

This chapter described basic concepts of ANNs and explored various applications of ANNs to biomedicine. Based on a learning algorithm and a task, we categorized applications of ANNs into the four paradigms: i) pattern classification with the BP algorithm, ii) pattern classification with the SOM, iii) data compression with the BP algorithm, and iv) system modeling with the BP algorithm. Although this chapter tried to summarize as many applications as possible, there are a vast number of publications that we could not review here. One of the areas we have not reviewed is combinations of ANNs with other computing techniques like fuzzy [54], [55] and the genetic algorithm [56]. Since the mid 1990s, these combined methods have been becoming powerful tools for medical diagnosis. Another such area is molecular biology. ANNs have been applied to secondary structure prediction of protein [57] and peptides [58], analysis of sequences [59], [60], [61].

This chapter concludes with a view of future directions of ANNs in biomedicine. As we have seen so far, ANNs are capable of classifying patterns based on relationships in data that are not apparent to human analysis. However, at the moment, it is generally difficult to identify how ANNs derive their output. Therefore one direction is "rule discovery" in clinical data by ANNs. A number of researchers have been studying approaches for the goal [24], [62]. Another direction is applications to data mining [63], in which a huge amount of data is processed. Data mining applications require methodology to handle high dimension, unknown-structure data. As we reviewed previously, one such application area is gene expression data analysis. ANNs, especially the SOM, will gain more and more popularity in this field.

References

[1] Haykin, S. (1994), *Neural Networks: a Comprehensive Foundation*, Macmillan College Publishing Company, New York.

[2] Bishop, C.M. (1995), *Neural Networks for Pattern Recognition*, Oxford University Press, New York.

[3] Anderson, L. (1995), *Introduction to Neural Networks*, MIT Press, Cambridge.

[4] Hudson, D.L. and Cohen, M.E. (2000), *Neural Networks and Artificial Intelligence for Biomedical Engineering*, IEEE Press, Piscataway.

[5] Lisboa, P.J.G., Ifeachor, E.C. and Szczepaniak, P. (Eds.) (1999), *Artificial Neural Networks in Biomedicine (Perspectives in Neural Computing)*, Springer, London, Berlin, Heidelberg, New York.

[6] Penny, W. and Frost, D. (1996), "Neural networks in clinical medicine," *Medical Decision Making*, vol. 16, pp. 386-398.

[7] Marmarelis, V.Z. (Ed.) (1994), *Advanced Methods of Physiological System Modeling Vol. III*, Plenum Press, New York.

[8] Miller, A.S., Blott, B.H. and Hames, T.K. (1992), "Review of neural network applications in medical imaging and signal processing," *Medical & Biological Engineering & Computing*, vol. 30, pp. 449-464.

[9] McCullogh, W.W. and Pitts, W. (1941), "A logical calculus of the ideas immanent in nervous activity," *Bulletin of Mathematical Biophysics*, vol. 5, pp. 115-133.

[10] Rumelhart, D.E., Hinton, G.E., and Williams, R.J. (1986), "Learning representations by back-propagating errors," *Nature*, vol. 323, pp. 533-536.

[11] Rumelhart, D.E., McClelland, J.L. and the PDP Research Group (1986), *Parallel Distribute Processing*, MIT Press, Cambridge.

[12] Hush, D.R., Horne, B., and Salas, J.M. (1992), "Error surfaces for multilayer perceptrons," *IEEE Transactions on Systems, Man, and Cybernetics*, vol. 22, pp. 1152-1161.

[13] Fukuoka, Y., Matsuki, H., Ishida, A. and Minamitani, H. (1998), "A modified back-propagation method to avoid false local minima," *Neural Networks*, vol. 11, pp. 1059-1072.

[14] Xu, L., Klasa, S. and Yuille, A. (1992), "Recent advances on techniques of static feedforward networks with supervised learning," *Int. J. Neural Syst.*, vol. 3, pp. 253-290.

[15] Fukuoka, Y., Noshiro, M., Shindo, H., Minamitani, H. and Ishikawa, M. (1997), "Nonlinearity identified by neural network models in P_{CO_2} system in humans," *Med. & Biol. Comput. & Eng.*, vol. 35, pp. 33-39.

[16] Kohonen, T. (1989), *Self-organization and associative memory*, Springer-Verlag, New York, Berlin, Heidelberg.

[17] Kohonen, T. (1990), "Self-organizazing map," *Proc. of the IEEE*, vol. 78, pp. 1464-1480.

[18] Rogers, S.K, Ruck, D.W. and Kabrisky, M. (1994), "Artificial neural networks for early detection and diagnosis of cancer," *Cancer Lett.*, vol. 77, pp. 79-83.

[19] Cybenko, G. (1989), "Approximation by superpositions of a sigmoid function," *Mathematics of Control, Signals, and Systems*, vol. 2, pp. 303-314.

[20] Funahashi, K. (1989), "On the approximate realization of continuous mapping by neural networks," *Neural Networks*, vol. 2, pp. 183-192.

[21] Patil, S., Henry, J.W., Rubenfire, M. and Stein, P.D. (1993), "Neural network in the clinical diagnosis of acute pulmonary embolism," *Chest*, vol. 104, pp. 1685-1689.

[22] Wilding, P., Morgan, M.A., Grygotis, A.E., Shoffner, M.A. and Rosato, E.F. (1994), "Application of backpropagation neural networks to diagnosis of breast and ovarian cancer," *Cancer Lett.*, vol. 77, pp. 145-153.

[23] Baxt, W.G. (1991), "Use of an artificial neural network for the diagnosis of myocardial infarction," *Ann. Intern. Med.*, vol. 115, pp. 843-848.

[24] Baxt, W.G. and White, H. (1995), "Bootstrapping confidence intervals for clinical input variable effects in a network trained to identify the presence of myocardial infarction," *Neural Computation*, vol. 7 pp. 624-638.

[25] Holdaway, R.M., White, M.W. and Marmarou, A. (1990), "Classification of somatosensory-evoke potentials recorded from patients with severe head injuries," *IEEE Eng. in Med. & Biol. Mag.*, vol. 9, pp. 43-49.

[26] Hiraiwa, A., Shimohara, K. and Tokunaga, Y. (1990), "EEG topography recognition by neural networks," *IEEE Eng. in Med. & Biol. Mag.*, vol. 9, pp. 39-42.

[27] Jansen, B.H. (1990), "Artificial neural nets for K-complex detection," *IEEE Eng. in Med. & Biol. Mag.*, vol. 9, pp. 50-52.

[28] Ouyang, N., Ikeda, M. and Yamauchi, K. (1997), "Use of an artificial neural network to analyse an ECG with QS complex in V_{1-2} leads," *Med. & Biol. Eng. & Comput.*, vol. 35, pp. 556-560.

[29] Fukuoka, Y. and Ishida, A. (2000), "Chronic stress evaluation using neural networks," *IEEE Eng. in Med. & Biol. Mag.*, vol. 19, pp. 34-38.

[30] Kelly, M.F., PA Parker, P.A. and Scott, R.N. (1990), "The application of neural networks to myoelectric signal analysis: a preliminary study," *IEEE Trans. Biomed. Eng.*, vol. 37, pp. 221-230.

[31] Hopfield, J.J. and Tank, D.W. (1986), "Computing with neural circuits: A model," *Science*, vol. 223, 625-633.

[32] Tank, D.W. and Hopfield, J.J. (1986), "Simple 'neural' optimization networks: An A/D converter, signal decision circuitry, and a linear programming circuit," *IEEE Trans. Circuits Syst.*, vol. 33, pp. 533-541.

[33] Schizas, C.N., Pattichis, C.S., Schofield, I.S., Fawcett, P.R. and Middleton, L.T. (1990), "Artificial neural nets in computer-aided macro motor unit potential classification," *IEEE Eng. in Med. & Biol. Mag.*, vol. 9, pp. 31-38.

[34] Cios, K.J., Chen, K. and Langenderfer, R.A. (1990), "Use of neural networks in detecting cardiac diseases from echocardiographic images," *IEEE Eng. in Med. & Biol. Mag.*, vol. 9, pp. 58-60.

[35] Buller, D., Buller, A., Innocent, P.R. and Pawlak, W. (1996), "Determining and classifying the region of interest in ultrasonic images of the breast using neural networks," *Artif. Intell. Med.*, vol. 8, pp. 53-66.

[36] Chen, D.R., Chang, R.F. and Huang, Y.L. (1999), "Computer-aided diagnosis applied to US of solid breast nodules by using neural networks," *Radiology*, vol. 213, pp. 407-412.

[37] Frankcl, D.S., Olson, R.J., Frankel, S.L. and Chisholm, S.W. (1989), "Use of a neural net computer system for analysis of flow cytometric data of phytoplankton populations," *Cytometry*, vol. 10, pp. 540-550.

[38] Lagerholm, M., Peterson, C., Braccini, G., Edenbrandt, L. and Sörnmo, L. (2000), "Clustering ECG complexes using Hermite functions and self-organizing maps," *IEEE Trans. Biomed. Eng.*, vol. 47, pp. 838-848.

[39] Benigni, R. and Pino, A. (1998), "Profiles of chemically-induced tumors in rodents: quantitative relationships," *Mutation Res. Fundamental & Molecular Mechanism Mutagenesis*, vol. 421, pp. 93-107.

[40] Chen, D.R., Chang, R.F. and Huang, Y.L. (2000), "Breast cancer diagnosis using self-organizing map for sonography," *Ultrasound in Med. & Biol.*, vol. 26, pp. 405-411.

[41] Tamayo, P., Slonim, D., Mesirov, J., Zhu., Q., Kitareewan, S., Dmitrovsky, E. and Lander E.S., Gloub, T.R. (1999), "Interpreting patterns of gene expression with self-organizing maps: Methods and application to hematopoietic differentiation," *Proc. Natl. Acad. Sci. USA*, vol. 96, pp. 2907-2912.

[42] Gloub, T.R., Slonim, D.K., Tamayo, P., Huard, C., Gaasenbeek, M., Mesirov, J.P., Coller, M., Loh, M.L., Downing, J.R., Caligiuri, M.A., Bloomfield, C.D. and Lander E.S. (1999), "Molecular classification of cancer: Class discovery and class prediction by gene expression monitoring," *Science*, vol. 286, pp. 531-537.

[43] DeRisi, J., Penland, L., Brown, P.O., Bittner, M.L., Meltzer, P.S., Ray, M., Chen, Y., Su, Y.A. and Trent, J.M. (1996), "Use of a cDNA microarray to analyse gene expression patterns in human cancer," *Nature Genet.*, vol. 14, pp. 457-460.

[44] Wodicka, L., Dong, H., Mittmann, M., Ho, M.H. and Lockhart, D.J. (1997), "Genome-wide expression monitoring in *Saccharomyces cerevisiae*," *Nature Biotechnol.*, vol. 15, pp. 1359-1367.

[45] Lockhart, D.J., Dong, H., Byrne, M.C., Follettie, M.T., Gallo, M.V., Chee, M.S., Mittmann, M., Wang, C., Kobayashi, M., Horton, H. and Brown, E.L. (1996), "Expression monitoring by hybridization to high-density oligonucleotide arrays," *Nature Biotechnol.*, vol. 14, pp. 1675-1680.

[46] Cottrell, G.W. and Munro, P. (1988), "Principal component analysis of image via back propagation," *SPIE*, vol. 1001 Visual Communication and Image Processing '88, pp. 1070-1076.

[47] Funahashi, K. (1990), "On the approximation realization of identity mappings by three-layer neural networks," *IEICE Trans.*, vol. J73-A, pp. 139-145. (in Japanese).

[48] Iwata, A., Nagasaka, Y. and Suzumura, N. (1990), "Data compression of the ECG using neural network for digital Holter monitor," *IEEE Eng. in Med. & Biol. Mag.*, vol. 9, pp. 53-57.

[49] Chon, K.H. and Cohen, R.J. (1997), "Linear and nonlinear ARMA model parameter estimation using an artificial neural network," *IEEE Trans. Biomed. Eng.*, vol. 44, pp. 168-174.

[50] Prank, K., Jürgens, C., von zur Mühlen, A. and Brabant, G. (1998), "Predictive neural networks for learning the time course of blood glucose levels from the complex interaction of counterregulatory hormones," *Neural Computation*, vol. 10, pp. 941-953.

[51] Robinson, P.R., Griffith, K., Gross, J.M. and O'Neill, M.C. (1999), "A back-propagation neural network predicts absorption maxima of chimeric human red/green visual pigments," *Vision Res.*, vol. 39, pp. 1707-1712.

[52] Alizadeh, A.A., Eisen, M.B., Davis, R.E., Ma, C., Lossos, I.S., Rosenwald, A., Boldrick, J.C., Sabet, H., Tran, T., Yu., X., Powell, J.I., Yang, L., Marti, G.E., Moore, T., Hudson, J. Jr., Lu, L., Lewis, D.B., Tibshirani, R., Sherlock, G., Chan, W.C., Greiner, T.C., Weisenburger, D.D., Armitage, J.O., Warnke, R., Levy, R., Wilson, W., Grever, M.R., Byrd, J.C., Botstein, D., Brown, P.O. and Staudt, L.M. (2000), "Distinct types of diffuse large B-cell lymphoma identified by gene expression profiling," *Nature*, vol. 403, pp. 503-511.

[53] Eisen, M.B., Spellman, P.T., Brown, P.O. and Bosteon, D. (1998), "Cluster analysis and display of genome-wide expression patterns," *Proc. Natl. Acad. Sci. USA*, vol. 95, pp. 14863-14868.

[54] Pavlopoulos, S., Kyriacou E., Koutsouris, D., Blekas, K., Stafylopatis, A. and Zoumpoulis, P. (2000), "Fuzzy neural network-based texture analysis of ultrasonic images," *IEEE Eng. in Med. & Biol. Mag.*, vol. 19, pp. 39-47.

[55] Zahlmann, G., Scherf, M., Wegner, A., Obermainer, M. and Mertz, M. (2000), "Situation assessment of glaucoma using a hybrid fuzzy neural network," *IEEE Eng. in Med. & Biol. Mag.*, vol. 19, pp. 84-91.

[56] Dybowski, R., Weller, P., Chang, R. and Gant, V. (1996), "Prediction of outcome in critically ill patients using artificial neural net-

work synthesized by genetic algorithm," *Lancet*, vol. 347, pp. 1146-1150.

[57] Stolorz, P., Lapedes, A. and Xia, Y. (1992), "Predicting protein secondary structure using neural net and statistical methods," *J. Mol. Biol.*, vol. 225, pp. 363-377.

[58] Ruggiero, C., Sacile, R. and Rauch, G. (1993), "Peptides secondary structure prediction with neural networks: a criterion for building appropriate learning sets," *IEEE Trans. Biomed. Eng.*, vol. 40, pp. 1114-1121.

[59] Farber, R. and Lapedes, A. (1992), "Determination of eukaryotic protein coding regions using neural networks and information theory," *J. Mol. Biol.*, vol. 226, pp. 471-479.

[60] Frishman, D. and Argos, P. (1992), "Recognition of distantly related protein sequences using conserved motifs and neural networks," *J. Mol. Biol.*, vol. 228, pp. 951-962.

[61] Mahadevan, I. and Ghosh, I. (1994), "Analysis of E.coli promoter structures using neural networks," *Nucl. Acids Res.*, vol. 22, pp. 2158-2165.

[62] Cloete, I. and Zurada, J.M. (Eds.) (2000), *Knowledge-Based Neurocomputing*, MIT Press, Cambridge.

[63] Bigus, J.P. (1996), *Data Mining with Neural Networks*, McGraw-Hill, New York.

Chapter 9

The Application of Neural Networks in the Classification of the Electrocardiogram

C.D. Nugent, J.A. Lopez, N.D. Black, and J.A.C. Webb

The introduction of computers to assist with classification of the Electrocardiogram (ECG) is considered to be one of the earliest instances of computers in medicine. Over the past 4 decades, since its inception, research techniques in the given field have proliferated. Approaches adopted have included the use of rule based approaches such as Decision Trees, Fuzzy Logic and Expert Systems, to the use of Multivariate Statistical Analysis. The past decade has seen the approach of Neural Networks (NNs) being not only employed in the field, but a very popular and successful approach in comparison with the aforementioned techniques.

It is the aim of this chapter to initially introduce the area of Computerized Classification of the ECG both from a historical and biological perspective. A general outline of the basic theoretical issues of NNs is presented prior to explanations as to how such an approach can be used in the given scenario. Exemplar studies are mentioned and the chapter concludes with a comparison of NN and other computerized methodologies, along with an outlook for future trends.

1 Introduction to the Classification of the Electrocardiogram

Heart disease still remains one of the largest single causes of premature deaths. With careful clinical evaluation, some of the causes of heart disease can be foreseen and prevented. Consideration of cholesterol levels, blood pressure levels, family history, etc., all commonly referred to as risk factors [1], can contribute to the likelihood of developing a heart disease. Therefore, by changes in lifestyle the prevalence of heart disease can be reduced. In addition to the assessment of the aforementioned risk factors, various other clinical tests may be performed to evaluate the status of the heart. One of the most commonly employed, non-invasive techniques is that of the 12-lead ECG, which records the electrical activity of the heart. Attentive consideration of this electrical information provides sufficient detail to diagnose a number of cardiac abnormalities, e.g. Myocardial Infarction, Left Ventricular Hypertrophy, which may induce untimely deaths.

Over the past century, much attention has been given to the acquisition, preparation and the analysis of the ECG with the intention of assisting the fight against heart disease. Although modern technologies permit more accurate methods of acquisition and instrumentation of a portable nature, the basic conceptual ideas of Electrocardiography have changed little since its inception at the outset of the century. Perhaps the largest contributory element has been the introduction of computers to assist with the analysis of the acquired data. Indeed, the introduction of computers to the field of ECG classification is considered to be one of the earliest forms of computers in medicine [2]. Computers have the ability to consider larger amounts of data, in more complex permutations and with greater speeds than the human interpreter, whilst still maintaining comparable levels of classification. At present Computerized ECG Classification still remains an active and prolific field of research, with the common goal of enhancing current levels of automated classification performance.

1.1 Diagnostic Utilities of the ECG

The ECG has become a standard tool of modern clinical medicine. It is a non-invasive technique, painless to the patient, inexpensive, simple to use and the most popular practical means of recording the cardiac activity, in electrical terms [3]. It is possible to correlate the recorded electrical activity of the ECG with the fundamental behavior of the heart. Hence it is further possible to establish relationships between measured signals and electrophysiological cardiac events. The diagnostic capabilities of the ECG may be considered to be two fold. Firstly, analysis of the wave shapes attempts to describe the state of the working muscle masses. Secondly, analysis of the rate of the cardiac cycle provides rhythm statements which give additional diagnostic information. In the examination of a normal subject, the ECG is found to remain reasonably constant. However, under pathological conditions, several pertinent differences in comparison to normal, may be observed. These differences, coupled with the ability of the ECG to display them, largely account for its diagnostic utility. Although the ECG can provide invaluable diagnostic information it does have its limitations. Not all cardiac abnormalities are identifiable by the ECG. Nevertheless in conjunction with other clinical techniques, e.g., angiography and echocardiography, a more global and complete picture of the heart may be obtained for analytical purposes.

A range of information sources are presently available, describing the role of the ECG in clinical evaluation, indicating the relevant signs of deviation from the norm. However, no standards are currently available for diagnostic classification [4]-[7]. This largely impedes the acceptance of a specified set of diagnostic criteria and may be partly attributed to the variance of diagnosis between human observers.

1.2 Introduction to Computerized Classification

In an effort to enhance the process of ECG classification, in terms of accuracy, speed, precision and reliability, computerized techniques were introduced. 1957 saw the advent of Computerized ECG classification by Pipberger [8]. Since its inception, research techniques in the field have proliferated, with the driving factor being the common goal of enhancing the aforementioned criteria and hence improving the overall level of performance of classification.

From the outset it was envisaged that computerized systems would provide significantly more accurate classification of ECGs. Although recent findings indicate that computer based systems provide only comparable results with humans [9], they do offer higher levels of consistency and lower levels of observer variability [10]. It was never considered that the computer would totally replace the expert, however, devoid of their presence, via the computerized system an intermediate classification may be produced, prior to overreading.

The classification of the ECG, by computerized techniques, is essentially a pattern recognition task. This may be subsequently subdivided into two succinct steps. Firstly a pre-processing stage examines the digitally recorded data. Boundaries are determined and inter-wave intervals are measured [11]. This pre-processing stage gathers all the information required which is subsequently processed by the secondary stage of the classification algorithm. Based on the provided information, the ECG is allocated to one from a set of possible diagnostic categories.

Over the past 4 decades, much effort has been made in the enhancement of both areas of the pattern recognition problem. With regard to the pre-processing, techniques have begun to reach their limit, in pushing forward the accuracy of automated analysis [12]. On the other hand, the problem of the optimum classification algorithm still remains unsolved [9], [13]. From the outset, two different approaches have generally been adopted [14]. Firstly, there have been methods referred to as deterministic, based on a rule based approach. Secondly, multivariate statistical techniques have been applied, where large amounts of information have been considered in parallel, during classification. Many comparisons, between systems employing the two techniques, have been performed over the years. Comparisons between the fundamental algorithmic nature of each technique, indicate advantages and disadvantages for each. Rule based techniques are considered to be more flexible in that new rules may be added without disturbing the rest of the system. However, as the rules are defined by the human expert, this may be seen as a potential introduction of bias into the system. Conversely statistical techniques tend to be slightly biased towards the prior information or training population. Nevertheless, they avoid human bias in the design of the classification function. Additionally these techniques cannot provide an explanation as to their method of arriving at the final answer.

A number of major advances have taken place in the development of new pattern recognition approaches, especially in the past 2 decades. The most significant, in the field of ECG classification, has been the resurgence of interest in NNs. Developments in training algorithms in the late 1980's have lead to NN classification techniques which provide comparable levels of classification, to existing methods, with short developmental times [15], [16]. Indeed, recent literature supports researchers in the field of computerized ECG classification, in moving from traditional approaches to adopt neural ones [17].

As previously mentioned, an inherent problem in clinical diagnosis of the ECG is the lack of standards with regard to classification rules. This problem is subsequently propagated into the computerized field. With rule based systems, the rules are defined by the human, with no guarantee that the rule set is optimal. With statistical and NN techniques this problem of lack of standards is overcome to a certain extent by the development of adaptive algorithms.

The results from the Common Standards for Quantitative Electrocardiography (CSE) study indicated that human classification performed better in comparison with the majority of computerized approaches [13]. It was also suggested that the proposed algorithms were not close to the optimal solution, hence improvements were still achievable. A comparison between the multivariate and deterministic approaches indicated the former to provide, the optimal level of classification; with average classification levels of 76.55% and 67.68% respectively.

Though the techniques over the years have shown a steady progression in the enhancement of computerized classification of the ECG, it is still a well recognized fact that there is still considerable room for improvement.

2 Fundamentals of the 12-Lead ECG

The first recording of an ECG from a human subject dates back to 1887, performed by Waller [18]. 1903 saw the introduction of the string galvanometer to electrocardiography by Einthoven [19]. Not only did Einthoven largely contribute to the development of the first recognized electrocardiograph, but also to standard nomenclature and electrode placement locations. With regard to the lead configurations it was some 30 years following Einthoven's premises before Wilson introduced the six unipolar chest leads V_1-V_6 [20]. Together the 12 leads comprising Einthoven's 3 limb leads (I, II, III), Wilson's 6 unipolar chest leads and Goldberger's 3 augmented unipolar Limb leads (aV_L, aV_F, aV_R) [21] form the basis of today's present 12-lead ECG.

2.1 The 12-Lead ECG and Associated Nomenclature.

Electrocardiography is concerned with the measurement of the electrical activity of the atrial and ventricular masses, rather than individual cells. The basic waves of the ECG (Figure 1), as defined nearly a century ago by Einthoven [19], have become standard [22] and may be described as: P wave (atrial depolarization), QRS complex (ventricular depolarization), ST segment (ventricular repolarization), T wave (ventricular repolarization) and U wave (ventricular repolarization). The P wave is representative of atrial contraction, caused by the spread of stimulus from the sinoatrial node across both the left and right atrium. The QRS complex is representative of ventricular depolarization following the spread of stimulus from the AV junction, through the bundle of His and left and right bundle branches. The ST segment, T wave and U wave may be attributed to ventricular repolarization.

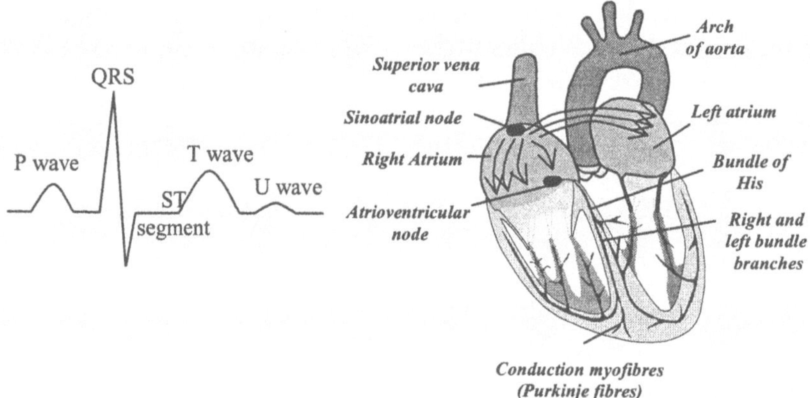

Figure 1. *Left*: Typical normal ECG for a single cardiac cycle indicating P wave, QRS complex, ST segment and T and U waves. *Right*: Conduction system within the heart, indicating the main components required to maintain the spread of stimulus.

The standard 12-lead ECG records differences in voltages between the body surface electrodes. The 12 leads may be subdivided into two groups of 6. The extremity/limb leads and the precordial/chest leads. The limb leads I, II, III, aV_R, aV_L and aV_F measure the cardiac activity in the frontal plane of the body and the chest leads V_1-V_6 measure the cardiac activity in the horizontal plane of the body. Figure 2 shows the typical representation of the 12-Lead ECG.

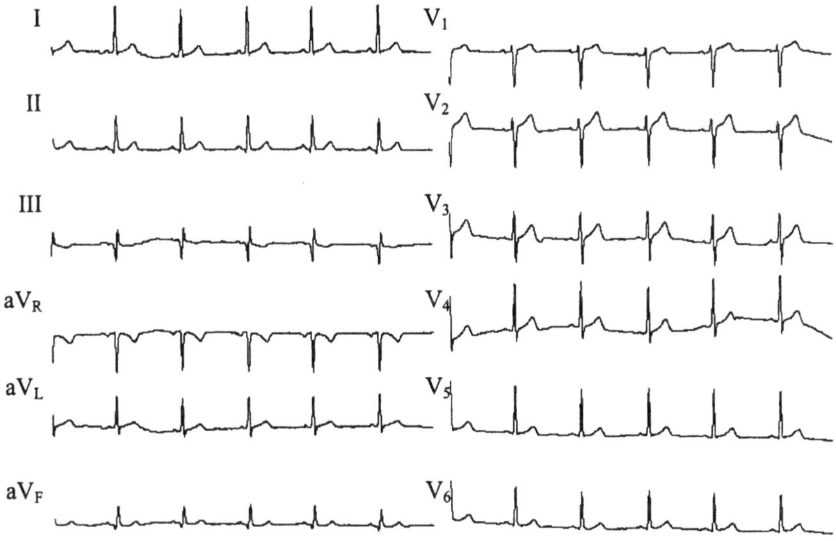

Figure 2. Typical format of the 12-lead ECG recording.

The 12-lead ECG allows the heart to be viewed from both the horizontal and vertical planes. The horizontal plane is represented by the chest leads and the vertical plane by the limb leads. When considering the 12-lead ECG as encompassing the electrical activity of the heart in both the horizontal and vertical planes, a three dimensional picture of the heart is obtained (Figure 3). It is also possible to record and analyze the electrical activity of the heart by other techniques. For example the Vector Cardiograph, Body Surface Mapping, single lead recordings and >12-lead ECG recordings. However, as previously stated, the 12-lead ECG is one of the most commonly employed and accepted techniques.

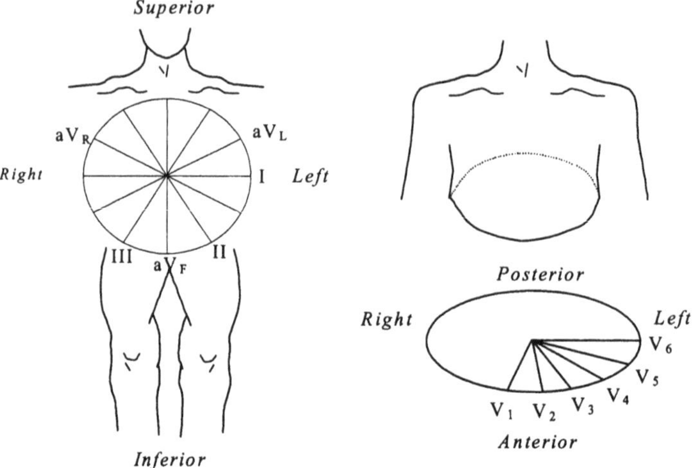

Figure 3. Horizontal and vertical planes as recorded by the limb and chest leads.

Each different diagnostic category, from a 12-lead perspective exhibits to a certain degree a unique change on the ECG. Each of these changes may be attributed to the presence of a certain pathology. It is possible to classify an unknown recording by understanding which pathology induces which changes on the ECG. The general ability to classify is based on the deviation from the norm and the ability and clinical knowledge to describe varying pathologies. It must be appreciated that there is a fine and unclear definition between some cases of abnormalities and Normal and indeed various types of abnormality may be unclearly segregated. Nevertheless, from a clinicians standpoint, experience, heuristics and clinical knowledge provide a suitable platform to distinguish between unknown recordings. Computerized techniques also have the ability to distinguish between normal and abnormal. As with computerized techniques in any field, the train of thought applied by the human is simulated to produce the same human process and end result. Computerized techniques were introduced in a general scenario to assist the clinician with patient evaluation and not aiming to replace them. A number of different techniques exist which support computerized analysis of the ECG, each with the common goal to provide a level of classification with minimal error.

3 Computerized Classification of the 12-Lead ECG

The aim of computerized classification of the ECG is to analyze the recorded activity of the heart and similar to the human, assign the patient to one from a possible set of diagnostic classes. The entire process can generally be subdivided into a number of disjoint processing modules; Beat Detection, Feature Extraction/Selection and Classification (Figure 4). Firstly the ECG is pre-processed by the stages of Beat Detection, Feature Extraction and possibly Feature Selection [23]. The initial pre-processing module of Beat Detection aims to locate each cardiac cycle in each of the recording leads and insert reference markers for the beginning and end of each inter-wave component. These reference markers are used by the module of Feature Extraction to generate measurements for inter-wave intervals, durations and amplitudes, hence producing a 'feature vector'. Feature Selection aims to reduce the dimensionality of the feature vector, selecting features to maximize the divergence between pattern classes, whilst still maintaining sufficient information to permit discrimination. The feature vector is subsequently processed by the final stage of the classification algorithm. Based on the information provided, the ECG is classified into a diagnostic category.

In general terms the overall problem facing Feature Extraction is the identification and extraction of parameters which will aid in classification. Following QRS detection and insertion of the relevant beat markers during the process of Beat Detection, the intervals, durations and amplitudes required are extracted during Feature Extraction and collectively form an input feature vector, x. Standards exist, as described by the American Heart Association [24], [25], defining the exact location from which measurements are to be made, along with respective nomenclature to be used. Nevertheless, due to the shear numbers of possible features, it is often difficult to choose the appropriate combination.

Currently, no standard set of features (or feature vector) has been agreed [6] and hence many different sets have been seen to be used. The formation of the feature vector during Feature Extraction is therefore dictated mainly by expert medical opinion, medical criteria and to a certain degree, trial and error. Although there are a certain amount of electrocardiographic changes which indicate various pathologies, e.g., R-R interval for the analysis of heart rate variability, Q-wave location for specific types of MI and large QRS complexes for the indication of ventricular hypertrophy, a standard definition of a feature vector, optimal for the diagnosis of a given pathology or discrimination between pathologies has not been made. As a consequence, various approaches of Feature Selection exist, usually in the form of a post-processing stage, whereby the recorded feature vector, produced during Feature Extraction, is reduced to an optimal sub-set as required.

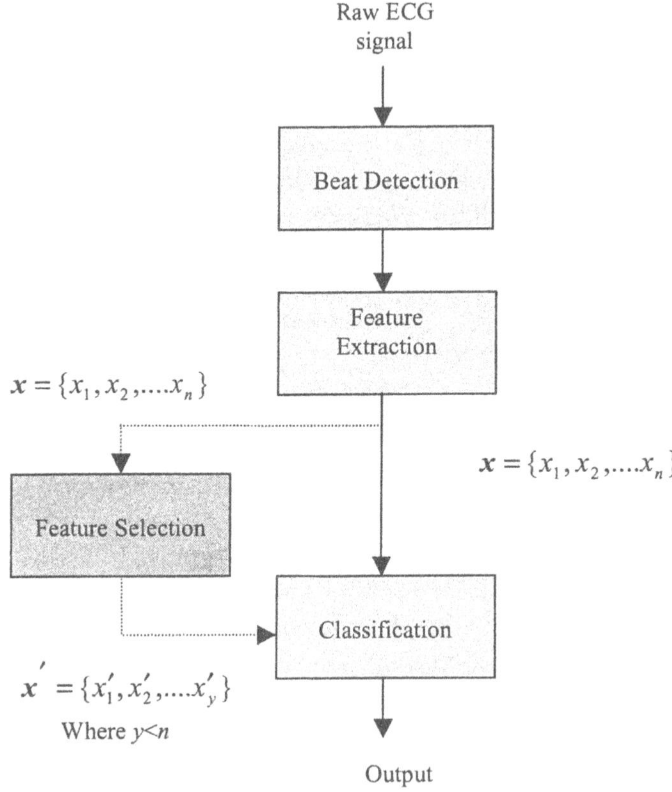

Figure 4. Overview of functional modules in a typical computerized electrocardiographic system.

A vast number of features can be measured from the ECG. Typically ECG classification systems make use of hundreds of ECG features spread over all recording leads. The selection of a good subset of features from the possible hundreds of features is one of the major problems in computerized diagnosis. It has been found that by increasing the number of measurements examined, the performance does increase up to a certain level but starts to deteriorate thereafter [26]. Thus, rather than increase the size of the feature vector to improve performance, it has been found that by employing an intermediate stage of Feature Selection prior to classification, the feature vector is reduced with the goal of enhancing performance. It is therefore the aim of Feature Selection to reduce the dimensionality of the feature vector, selecting features to maximize the divergence between pattern classes, whilst still maintaining sufficient information to permit discrimination without loss of classification performance. Hence, the essential problem is to find methods whereby 'optimal' features are identified and extracted with no loss in classification generalization.

The most optimal approach to Feature Selection would be to obtain a subset by exhaustive comparisons of all possible subset permutations. This, however, has been identified as an impracticable solution [27], [28]. Cornfield *et al.* [29] adopted an approach whereby following Beat Detection, every measurable ECG variable known, as described in electrocardiography literature, was extracted. This is considered to be a suitable starting point for Feature Selection. Various approaches have been applied of statistical [29], [30], stepwise [31], transformational [32], [33] and structural natures [34]-[36]. The processes indicate that it is possible to reduce the dimensionality of the feature vector without loss of classification accuracy.

3.1 Classification

In general terms the process of classification involves arranging a number of given objects into one of a possible number of groupings. Regardless of the nature of the object, its attributes depict to which grouping or class it belongs. The differing attributes of each class provide the relevant information necessary for the classification process to decide if objects belong to different classes and to which class they should be assigned. For example, given a collection of people based on the information regarding their dietary requirements, it is possible to group them into a number of different classes stating whether they are vegans, vegetarians or omnivorous. Thus it is possible, by considering the attributes (dietary requirements) of the objects (people) to generate a classification function with the capabilities to classify any given object following examination of its attributes, into one of the possible classes (vegans, vegetarians, omnivorous).

Classification of the recorded ECG signal is the final phase of a computerized ECG diagnostic system. The initial pre-processing stages present suitable information, in the form of the feature vector, to the classification module for analysis. A multitude of computerized algorithms have been adopted to accommodate the final process of classification of the ECG. Historically the techniques employed may be segregated into the following categories:

- Rule based or Heuristic Classifiers
- Multivariate Classifiers

Heuristic classifiers include Decision Trees [37], [38], Expert Systems [39], [40] and Fuzzy Logic [41], [42], all simulating the reasoning of the human expert in interpreting the ECG with multivariate classifiers including statistical [43] and NN techniques. Pipberger [43] referred to Heuristic Classifiers as First Generation Programs and multivariate techniques as Second Generation Programs. The labels assigned by Pipberger were not indicative of their time of occurrence but with regard to their respective complexity level. With the growth of computing power, programming techniques and research experience, 'Potential Third Generation' Programs, whereby a number of algorithms are considered to produce a combined diagnostic result, or a hybrid approach is employed [44], are now in vogue. Figure 5 represents, in diagrammatic form, segregation of computerized methods.

As with so much of what we have seen so far with automatic classification, there are no standardized classification rules [6]. This mainly effects rule based classifiers since they are dependent on the knowledge provided by the human expert, which can be considered to bias the classification function. Increasingly, Artificial Intelligence has become popular in which rules are not required to generate the classification function; the classifier adapts itself through a process of training.

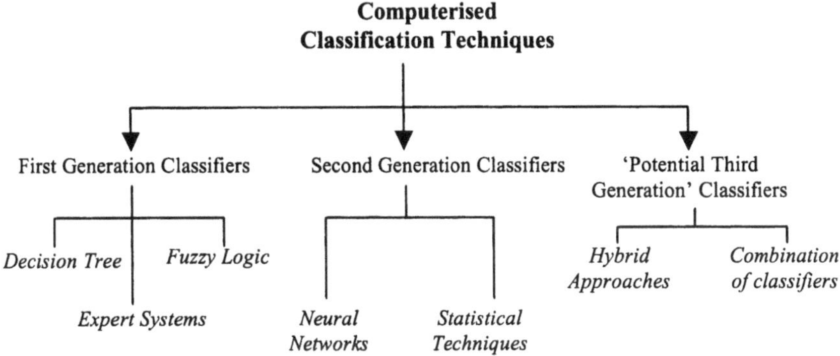

Figure 5. Hierarchical structure of computerized classification techniques.

4 Neural Networks in 12-Lead ECG Classification

Over the past decade, applications of NNs have proliferated in many areas of decision support. They have been applied to problems in areas of pattern recognition/classification, control and optimization and have been particularly useful in cases where the problem to be solved is ill defined and development of an algorithmic solution is difficult [45]. They have also been applied in many fields of medicine [46], [47] such as cancer prognosis, magnetic resonance imaging [45], diagnosis of low back disorders [48], plus many more. In the past decade, the research in the field of ECG classification employing NNs has been widespread and ever increasing. Their popularity to some extent may be attributed to their powerful parallel processing powers and their adaptability to prior expert knowledge for a given application.

4.1 The Artificial Neuron

NNs consist of dense interconnections of simple computational elements called neurons. The general structure of these artificial networks is greatly based on the understanding of the biological nervous system, attempting to simulate the functions of the brain.

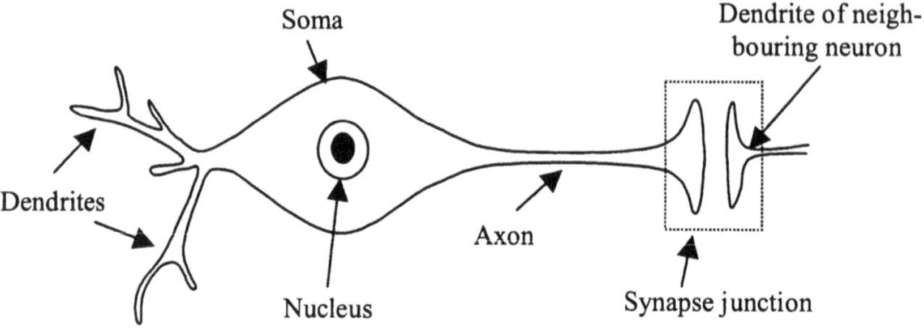

Figure 6. Representation of a biological neuron.

As the neuron is the building block of the brain, so too is it the building block of the NN. NNs, due to their large numbers of neurons and interconnections, utilize a parallel processing structure. Each individual neuron is connected to a number of neighboring neurons, thus establishing more interconnections in the network than neurons themselves. Figure 6 represents the structure of a biological neuron, the fundamental unit of the brain. The biological neuron itself is a simple processing unit, receiving and combining signals from other neurons through its input paths. The biological neuron may therefore be considered to have three complimentary parts; an input stage (dendrites), a processing unit (soma) and an output (axon and synapse junction). In a similar manner, an artificial neuron exhibits these basic parts, as indicated in Figure 7 and is therefore able to model the action of the biological neuron. The general structure of this artificial neuron consists of a set of n inputs x_j (dendrites) multiplied by a weight factor w_{jn} (synaptic junction) before reaching the main body or processing unit (soma) and finally an output (axon). The bias term (θ) is included to mimic the process of thresholding in the biological neuron.

The output O_i of the neuron may be characterized by

$$O_i = f\left(\sum_{i=1}^{n} w_{ji} x_i - \vartheta\right). \tag{1}$$

The role of the non-linearity function or activation function is to limit the output range of the neuron to some finite value. Usually the activation function is chosen for mathematical convenience and does not necessarily replicate the equivalent biological process. Figure 8 represents typical types of artificial activation functions. The nonlinearities are all bounded each having upper and lower limits.

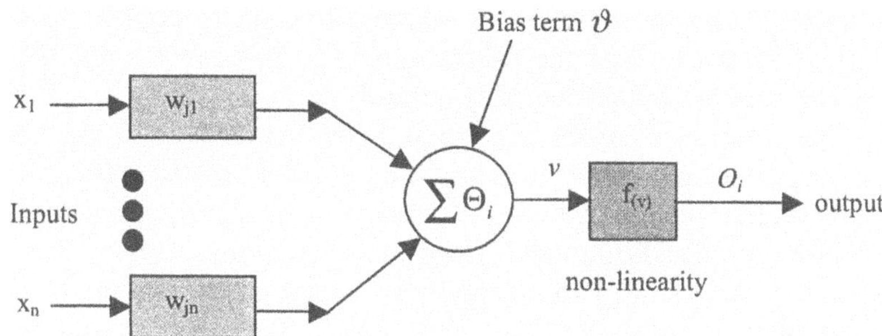

Figure 7. A basic artificial neuron, simulating a biological neuron.

Before an NN is presented in its desired environmental role its must perform a learning procedure to adapt itself. During learning, the NN adjusts its internal parameters (synaptic weights), whereby the adjustment of the weights forces the actual network to converge to the desired response. Learning of the NN may be very broadly categorized into: supervised (the presented input pattern to the network is accompanied with the desired network response) and un-supervised (the desired response is not known or it is not provided to the network along with the input).

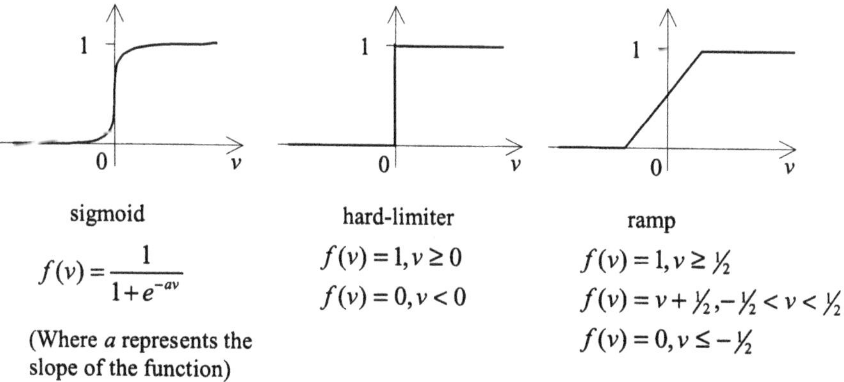

Figure 8. Commonly employed artificial activation functions: Sigmoid, Hard-limiter and Ramp.

There are a variety of ways in which NNs may be interconnected. The so called topology of the network is an important design parameter with regard to its efficiency. In the general case, the NN is a structured layer of neurons. The layer of neurons where the input is applied to the network is referred to as the input layer. The layer from which the output is produced is the output layer and the intermediate layers are referred to as hidden layers. Although a number of techniques exist which may be employed for classification of static patterns [49], it is well outside the bounds of this text to describe their intricacies. Given the problem of classifying the 12-lead ECG based on an input feature vector and the availability of an annotated training database, the Multi-Layered Perceptron (MLP), trained with supervised learning is considered an appropriate choice. Supporting literature (refer to next Section) additionally confirms that the MLP approach has been the most commonly employed.

4.2 The MLP and ECG Classification

If artificial neurons, such as those shown in Figure 7, are grouped together in layers, with weighted synapses interconnecting only neurons in successive layers, the NN structure is referred to as an MLP. An MLP consists of an input layer, an output layer, with the possibility of one or more layers in between, see Figure 9. The number of nodes and the number of hidden layers are not fixed and are highly application specific [50], [51]. An MLP may be considered as a multi-dimensional non-linear mapping function, from the input space to a hypercube. With sufficient numbers of synapse weights and number of internal layers, such a structure is capable of approximating any non-linear functional mapping with an arbitrary degree of accuracy.

The MLP is trained through a process of supervised learning. A commonly employed algorithm for the training of MLPs is that of Back Propagation (BP). It is beyond the scope of this chapter to explain the intricacies of this training, refer to [52] for further details. Following training, as determined by the training algorithm, the network is exposed to a set of unseen data in order to evaluate the performance of the network.

When employing an MLP as a classifier of an unknown ECG signal, the input to the network is the input feature vector.

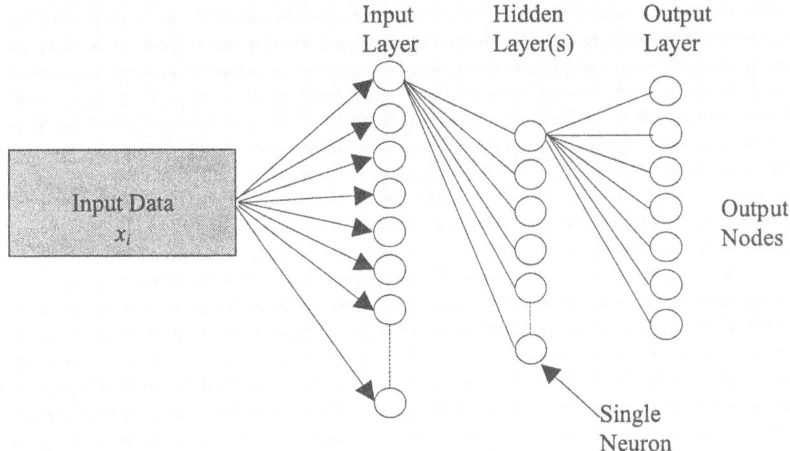

Figure 9. Multi-layer perceptron (MLP) architecture consisting of input layer, output layer and one hidden layer.

The feature vector itself describes the underlying structure of the ECG waveform across the given number of recording leads. It is possible that a form of Feature Selection is employed to ensure that only the necessary features are presented to the network with sufficient discriminatory abilities to segregate between the different diagnostic classes under investigation. Figure 10 summarizes the conventional approach to the classification of the ECG with MLP NNs.

Each feature in the input vector is presented to a dedicated node in the input layer of the MLP. The output of the MLP has a given number of neurons in the output layer, each one representative of a diagnostic class. An output of '1' from any neuron is indicative of the presence of a diagnostic class, a '0' its absence. To provide the NN with the desired classification abilities it must undergo suitable design and training procedures. This involves a given design for the NN and associated architecture being identified and the network presented with the training set until a suitable level of performance is established. Following training the network can then be exposed to unseen data for performance evaluation. Although conventional approaches to the design of such MLPs have been reported in the field to provide comparable results with existing algorithmic approaches in computerized classification of the ECG, careful consideration of design issues and architecture of the network can lead to enhanced performance. Typical factors of consideration are for example the number of nodes in the hidden layer and the number of hidden layers themselves. Depending on the variation of the training algorithm employed, factors such as the learning constant and momentum term

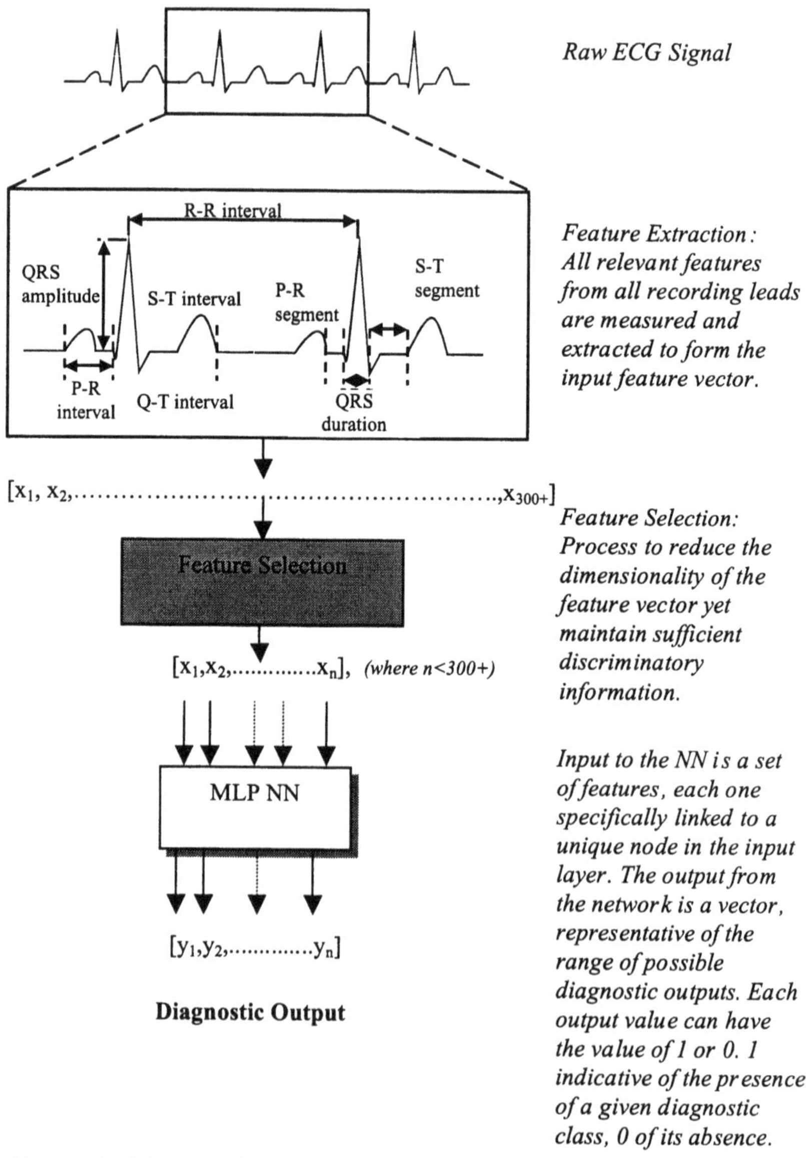

Figure 10. The use of an MLP NN to classify the feature vector of a recorded ECG signal.

must also be chosen correctly. Also of importance is the construction of the data sets used. It is a requirement to identify a training set, representative of the population of ECGs under consideration and their respective diagnostic classes. It is essential at this stage that a wide variation is included to ensure generalization of the final network will not be limited.

A secondary requirement is to identify a test set of data, which can be considered to be totally unseen to the network and hence can be used to give generalization figures. To avoid any instances of biasing the design of the network, both on the training and test data, a validation set should also be identified. This allows the network, during its design, to be evaluated in terms of both training capabilities and test capabilities. Once these figures have reached desired threshold levels, the network can then be considered to be fully trained, its free parameters are frozen and the network is presented with the test set to obtain the true indication of its generalization abilities. Indeed, the training and selection of the network is a heuristic procedure and many efforts have been described to produce the optimal classifier.

Bortolan *et al.* [53] described an MLP structure to discriminate a recorded ECG signal as belonging to one from the possible Category A classes as identified by the CSE [54], based on 39 features. A number of different internal hidden layer structures were tested along with a set of non-linear functional transforms of the input parameters (x), added as additional input nodes. The optimum results following testing were obtained by the network employing the additional non-linear functional transforms of $sin(\pi x)$, $sin(2\pi x)$, $cos(\pi x)$ and $cos(2\pi x)$. The network attained a correct classification level of 63% following testing in comparison with 67.2% for a Linear Discriminant analysis and 66.3% for Logistic Discriminant analysis. Although the results were lower than the conventional statistical approaches, the ease of use and development of the networks were suggested by the authors to warrant further investigation. The study was repeated in 1993 [55] whereby several further investigations were undertaken in order to enhance the classification performance and hence find the optimal NN solution. The following were investigated:-

i) Non-linear transformation of the input data.
ii) Dimensions of the NN were modified.
iii) Different activation functions were tested.

iv) Partitioning of the learning set with cluster analysis.
v) Combination of multiple NNs.

The authors noted very little change when altering the basic parameters of the network (i-iv above). However, they discovered that the combination of NNs employing majority rule enhanced the overall classification results. The best single NN attained a level of 66.3%, but when 7 different NNs were used and combined employing the majority rule, the performance was enhanced to 68.8%. In comparison with the findings of their previous study [53], the NN approach now provided superior results in comparison with conventional statistical techniques. Baxt [56] designed an NN to diagnose the presence of acute MI in patients presented to an emergency department. An MLP was employed with heuristic selection of the final architecture. The inputs to the network were selected from presenting symptoms, past history findings and physical and laboratory findings in relation to the patient. The output layer of the network contained only 1 neuron, indicating the presence or absence of acute MI. Following testing the network attained a correct classification of patients exhibiting acute MI of 97.2% in comparison to 77.7% attained by the physicians. Baxt concluded the study by suggesting one of the potential advantages of an NN was its ability to identify relationships between ECG features that may not have been identified by physicians. Harrison *et al.* [57] also developed an MLP to diagnose the presence of acute MI. The authors identified a major criticism of the employed structure as its inability to explain the final diagnosis. Therefore, a method was devised to monitor the effect of each individual input feature on the overall output with the goal of identifying the most contributing features. This was performed by monitoring the effect on the output caused by a unit change on the input. From a total of 38 features comprising ECG findings, patient history and clinical tests, the results presented by the authors indicated that the proposed method of selecting features which best contributed to the final output agreed with those as suggested by expert clinicians, namely the presence or absence of electrocardiographic changes. Edenbrandt *et al.* [58] developed an MLP to classify the recorded ECG into one from 7 possible different classes of ST-T segment diagnosis based on information from the ST-T segment. The input feature vector was varied during development, containing variable amounts of continuous sample information of the ST-T segment amplitude and specific feature measurements. Effects of additional hidden layers and hidden nodes were monitored, along with the effects of the size and composition of the training set. A number of significant results following testing were observed. Firstly it was found that increasing the amount of information in the feature vector did not substantially increase the performance. A certain number of nodes in the hidden layer was found, whereby increasing the number did not greatly improve the performance, however, decreasing the number, decreased the performance. The authors also found that by combining the outputs from 10 networks a higher level of classification, in comparison with the best network in isolation, was obtained; 95.2% and 94.4% respectively. Finally it was found that with more training examples used in the training set, higher levels of generalization were obtained. Marques de Sa *et al.* [59] investigated the ability to design the best NN based on the most discriminating feature set. Like others an MLP was employed with varying numbers of hidden nodes. Each feature was removed individually and its effect on the overall classification error recorded. Following examination of all features, those with the least significant effects were eliminated. By reducing the initial number of 27 features to 15 the authors were able to attain comparable levels of classification whilst reducing the computational requirements of the network.

Rather than attempt to select the most appropriate features for classification, it is possible when employing NNs to present a sampled signal pattern to the input of the network. This avoids the necessity of employing techniques of Feature Extraction and Selection, as all that is required of the pre-processing algorithms is temporal alignment of the signals in all recording leads. Figure 11 illustrates this approach.

During training of the NN, the data set used contains representative ECG complexes from each of the diagnostic categories under consideration. There are, however, problems when employing a technique like this with the difficulties imposed by time scale variations of the ECG waveform, as recognized by Xue *et al.* [60]. To overcome the problem, Xue *et al.*, rather than adjusting the network architecture, adjusted the training samples. Additional shifted signals, along with the original signals were used to form the database. Their proposed method gave comparable results with an architecture designed specifically to accommodate for time scale variations. Taur and Kung [61] also recognized the need to accommodate situations of temporal misalignment in continuously sampled input waveforms. The database for training included the original waveforms plus the waveforms shifted forwards or backwards by one or two time samples, thus expanding the original data set by five fold. Following testing an average of 87% correct classification was obtained for 4 NN classifiers in comparison to 96% correct classification for the same NN trained with the shifted training sets, hence indicating the improvement in performance by using the shifted training data. Edenbrandt *et al.* [62], however, found that networks trained with specific feature measurements, in comparison to a continuously sampled waveform, proved to be superior, producing sensitivity levels of 93% and 63% respectively.

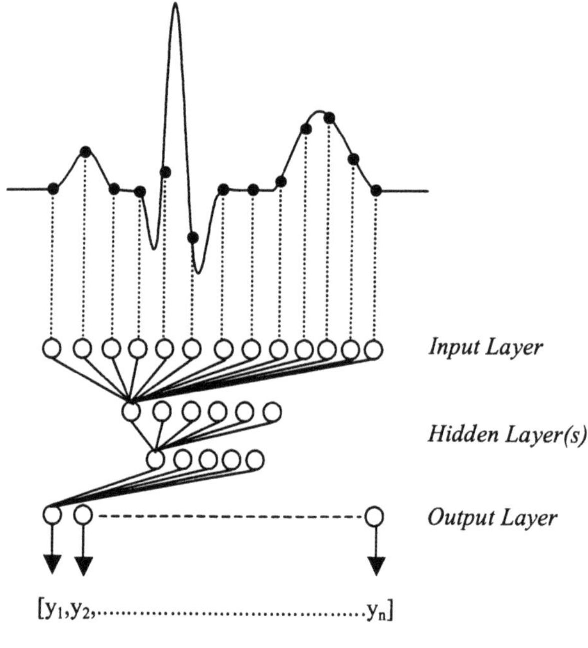

Figure 11. Classification of the ECG by applying a sampled version of the ECG itself as input to the MLP NN.

5 Summary

A number of different techniques have been applied to the field of computerized electrocardiography. It has been identified that one of the most common and successful approaches is that of NNs. Based on the presented literature and theoretical nature of each technique, Table 1 aims to summarize the relative advantages and disadvantages of each technique with respect to their specific application to the problem of ECG classification.

Table 1. Comparison of computerized ECG classification techniques.

Technique	Advantages	Disadvantages
DECISION TREE	The final result may be traced through the tree and reasoning can be given as to the final output. Easily modified to account for new situations without disturbing the rest of the system. The IF-THEN rule is close to human inference. The logic employed is easily followed and comprehended.	Requires arbitrary rule based knowledge from an expert. Due to the rigid structure and hard thresholds, it has been found that Decision Trees are unable to reproduce the same diagnostic output when quite similar ECG signals are considered.
EXPERT SYSTEM	The IF-THEN rule is close to human inference. Have the ability to produce a reasoning behind the final diagnosis. They are specifically designed to be user friendly. Well structured.	Requires rule-based knowledge from an expert.
FUZZY LOGIC	Common terms used by Cardiologists like *High*, *Low* etc. may be represented in a natural way. A linguistic strength to the final diagnosis may be given.	Ranking of fuzzy sets causes much dispute as no accepted method to date has been found. Requires rule based knowledge from an expert.

(continued on next page)

Table 1 (continued).

Technique	Advantages	Disadvantages
FUZZY LOGIC (continued)	The vagueness in any rules may be treated by the use of fuzzy sets, operating with the vagueness and in-exactness arising in the human thought process.	The final result may not be in a form understandable to a general user.
	No critical border lines must be drawn between two different classes / measurements.	The numerical transformation from linguistics is a very vague process.
MULTIVARIATE STATISTICAL ANALYSIS	The method has a solid basis in the theory of probability.	Most statistical techniques are limited to a diagnosis of single group disease only.
	It is independent of the bias of the expert.	No clear explanation as to why a result was reached may be given.
	The final result may be used to indicate a degree of presence of a disease.	A large data base is required for development and testing.
	Large amounts of information are used simultaneously as opposed to individual sequential analysis.	The effect of prior probabilities strongly influences the diagnostic output.
NEURAL NETWORKS	No stringent rules are required for classification.	A large database is required for developing and testing.
	It is independent of the bias of the expert.	No specific rules have been obtained as for optimum architectural design.
	They are easily implemented and evaluated.	The network has no way of explaining its method of reasoning.
	Can in some cases identify non-linear mappings not visually identified by the human. State of the art.	Temporal alignment problems are inherent if a continuous waveform is used.

It is apparent that each classification method has its own relative advantages and disadvantages. The most profound disadvantage a classifier can have is its reliability on the expert to detail the rule set for classification. This immediately highlights concerns with rule based classifiers. On the other hand, multivariate classifiers have the possible danger of specifically generalizing to the training set. Additionally, multivariate classifiers do not have the ability to explain the reasoning behind the diagnostic outcome and thus do not appeal to clinicians.

Many comparisons between the methods have been undertaken over the years. The CSE study [13] indicates that statistical methods have proved to yield better results than rule based approaches. NN research in the field has indicated that the techniques can provide superior results in comparison with statistical approaches [55]. This result clearly indicates the potential of employing NNs to the problem of ECG classification.

In many of the studies in the field it has been demonstrated that the neural approach is successful, however, there is an inherent danger that the methodologies are treated as black boxes. If treated in such a manner the powers and generalization abilities will not be exploited to their optimal levels. The literature reports on many studies where classification models based on NNs are further refined and developed to increase performance. Developments include the careful selection of features in the input feature vector, non-linear functional transformations of the network inputs, variations in the structure of the test and training data and architectural modifications of the network itself. An interesting result, in terms of classification performance enhancement is the combination of several NN models all addressing the same classification problem. As each NN generalization ability differs, the combination of many such models offers a greater consensus to the whole problem. Not only have NNs been combined with each other, but also instances of NNs being combined with other classification methods have been reported as being successful. Again this is an indication that the combination of classifiers with different generalization abilities is a successful approach. As the results from this approach demonstrate the potential for the combination or fusion of classifiers, research activities are also focusing on the best methodology to combine the classifiers themselves.

Clearly the neural approach has established itself as not only a viable option to the classification of the ECG, but also a successful one. As research in the field continues we can expect to see NN approaches becoming even more popular and successful.

References

[1] Tortora, J.G. and Grabowski, S.R. (1996), *Principles of Anatomy and Physiology*, 8th Edition, Harper Collins College Publishers, New York.

[2] Berbari, E.J. (1995), "Principles of electrocardiography," in: Bronzino, J.D. (Ed.), *The Biomedical Engineering Handbook*, CRC Press, Florida, pp. 181-190.

[3] Wellens, H.J.J. (1990), "Electrocardiography past, present and future," *Annals of the New York Academy of Sciences*, vol. 601 Electrocardiography, pp. 305-311.

[4] Willems, J.L. (1990), "Quantitative electrocardiography standardization and performance evaluation," *Annals of the New York Academy of Sciences*, vol. 601 Electrocardiography, pp. 329-342.

[5] Blake, T.M. (1994), *The Practice of Electrocardiography*, 5th Edition, Humana Press, Totowa, New Jersey.

[6] Kors, J.A. and van Bemmel, J.H. (1990), "Classification methods for computerized interpretation of the electrocardiogram," *Meth. Inform. Med.*, vol. 29, pp. 330-336.

[7] Van Bemmel, J.H., Zywietz, C., and Kors, J.A. (1990), "Signal analysis for ECG interpretation," *Meth. Inform. Med.*, vol. 29, pp. 317-329.

[8] Stallman, F.W. and Pipberger, H.V. (1961), "Automatic recognition of electrocardiographic waves by digital computer," *Circ. Res., vol.* 9, pp. 1138-1143.

[9] Macfarlane, P. (1992), "Recent developments in computer analysis of ECGs," *Clinical Physiology*, vol. 12, pp. 313-317.

[10] Willems, J.L., Abreu-Lima, C., Arnaud, P., van Bemmel, J.H., Brohet, C., Degani, R., Denis, B., Graham, I., van Herpen, G., Macfarlane, P.W., Michaelis, J., Moulopoulos, S., Poppl, S., and Zywietz, C. (1988), "Effect of combining electrocardiographic

interpretation results on diagnostic accuracy," *European Heart Journal*, vol. 9, pp. 1348-1355.

[11] Jenkins, J. (1981), "Computerized electrocardiography," *CRC Crit. Rev. Bioeng.*, vol. 6, pp. 307-350.

[12] Rowlandson, I. (1990), "Computerized electrocardiography a historical perspective," *Annals of the New York Academy of Sciences*, vol. 601 Electrocardiography, pp. 343-352.

[13] Willems, J.L., Arnaud, P., van Bemmel, J.H., Degani, R., MacFarlane, P.W., and Zywietz, C. (1992), "Comparison of diagnostic results of ECG computer programs and cardiologists," *Computers in Cardiology*, pp. 93-96.

[14] Macfarlane, P. (1990), "A brief history of computer assisted electrocardiography," *Meth. Inform. Med.*, vol. 29, pp. 272-281.

[15] Yang, T.F., Devine, B., and Macfarlane, P.W. (1993), "Deterministic logic versus software-based artificial neural networks in the diagnosis of atrial fibrillation," *J. Electrocardiology*, vol. 26, pp. 90-94.

[16] Bortolan, G. and Willems, J.L. (1993), "Diagnostic ECG classification based on neural networks," *Journal of Electrocardiology*, vol. 26, pp. 75-79.

[17] Bortolan, G., Brohet, C., and Fusaro, S. (1996), "Possibilities of using neural networks for ECG classification," *Journal of Electrocardiology*, vol. 29, pp. 10-16.

[18] Waller, A.D. (1887), "A demonstration on man of electromotive changes accompanying the heart's beat," *Journal Physiol.*, vol. 8, pp. 229-234.

[19] Mathewson, F.A.L. and Jackh, H. (1955), "The Telecardiogram," *Amer. Heart Jnl.*, vol. 49, pp. 77-82. (Translation of: Einthoven, W. (1906) "Le Telecardiogramme," *Arch. Int. Physiol*, vol. 4, pp. 132-161.)

[20] Wilson, F.N., Macleod, A.G., and Barker, P.S. (1931), "The potential variations produced by the heart beat at the apices of einthoven's triangle," *Amer. Heart Jnl.*, vol. 7, pp. 207-211.

[21] Goldberger, A.L. and Goldberger, E. (1994), *Clinical electrocardiography a simplified approach*, 5th Edition, Mosby, Missouri.

[22] "Report of Committee of the American Heart Association, The standardization of electrocardiographic nomenclature," (1943), *J.A.M.A*, vol. 121, pp. 1347-1351.

[23] Nugent, C.D., Webb, J.A.C., Wright, G.T.H., and Black, N.D. (1998), "Electrocardiogram 1: pre-processing prior to classification," *Automedica*, vol. 16, pp. 263-282.

[24] "Report of Committee on Electrocardiography, American Heart Association, Recommendations for standarization of electrcardiographic and vectorcardiographic leads," (1954) *Circulation*, vol. 10, pp. 564-573.

[25] "A report for health professionals by an ad hoc writing group of the Committee on Electrocardiography and Cardiac Electrophysiology of the Council on Clinical Cardiology, American Heart Association," (1990), *Circulation*, vol. 81, pp. 730-739.

[26] Chandrasekaran, B. and Jain, A.K. (1974), "Quantization complexity and independent measurements," *IEEE Trans. Comp.*, pp. 102-106.

[27] Mucciardi, A.N. and Gose, E.E. (1971), "A comparison of seven techniques for choosing subsets of pattern recognition properties," *IEEE Trans. Comp.*, vol. C-20, no. 9, pp. 1023-1031.

[28] Jain, U., Rautaharju, P.M., and Warren, J. (1981), "Selection of optimal features for classification of electrocardiograms," *J. Electrocardiography*, vol. 14, no. 3, pp. 239-248.

[29] Cornfield, J., Dunn, R.A., Batchlor, C.D., and Pipberger, H.V. (1973), "Multigroup diagnosis of electrocardiograms," *Computers and Biomedical Research*, vol. 6, pp. 97-120.

[30] Siddiqui, K.J., Greco, E.C., Kadri, N., Mohiuddin, S., and Sketch, M. (1993), "Best feature selection using successive elimination of poor performers," *Proc. Ann. Conf. IEEE Eng. in Med. and Biol.*, vol. 15, pp. 725-726.

[31] Jain, U. and Rautaharju, P.M. (1980), "Diagnostic accuracy of the conventional 12-lead and the orthogonal frank-lead electrocardiograms in detection of myocardial infarctions with classifiers using continuous and bernoulli features," *J. Electrocardiology*, vol. 13, no. 2, pp. 159-166.

[32] Chein, Y.T. and Fu, K.S. (1968), "Selection and ordering of feature observations in a pattern recognition system," *Information and Control*, vol. 12, pp. 395-414.

[33] Fukunaga, K. and Koontz, W.L.G. (1970), "Application of the Karhunen-Loeve expansion to feature selection and ordering," *IEEE Trans. Comp.*, vol. C-19, no. 4, pp. 311-318.

[34] Udupa, J.K. and Murthy, I.S.N. (1980), "Syntactic approach to ECG rhythm analysis," *IEEE Trans. Biomed. Eng.*, vol. BME-27, no. 7, pp. 370-375.

[35] Pietka, E. (1991), "Feature extraction in computerized approach to the ECG analysis," *Pattern Recognition*, vol. 24, pp. 139-146.

[36] Pavlidis, T. (1973), "Waveform segmentation through functional approximation," *IEEE Trans. Comp.*, vol. C-22, no. 7, pp. 689-697.

[37] Willems, J.L. (1977), "Introduction to multivariate and conventional computer ECG analysis: pro's and contra's," in: van Bemmel, J.H. and Willems, J.L. (Ed.), *Trends in Computer-Processed Electrocardiograms*, North-Holland Publishing Company, Holland, pp. 213-220.

[38] Okajima, M., Okamoto, N., Yokoi, M., Iwatsuka, T., and Ohsawa, N. (1990), "Methodology of ECG interpretation in the Nagoya program," *Meth. Inform. Med.*, vol. 29, pp. 341-345.

[39] Georgeson, S. and Warner, H. (1992), "Expert system diagnosis of wide complex tachycardia," *Computers in Cardiology*, pp. 671-674.

[40] Held, C. and Kurien, J. (1992), "FLAPECAN: a knowledge based system in electrocardiogram diagnosis," *Proc. 5th Annual Symposium on Computer-Based Medical Systems*, pp. 648-655.

[41] Degani, R. and Bortolan, G. (1987), "Fuzzy numbers in computerized electrocardiography," *Fuzzy Sets and Systems*, vol. 24, pp. 345-362.

[42] Degani, R. (1992), "Computerized electrocardiogram diagnosis: fuzzy approach," *Methods of Information in Medicine*, vol. 31, pp. 225-233.

[43] Pipberger, H.V. et al. (1975), "Clinical application of a second generation electrocardiographic computer program," *Amer. J. Cardiol.*, vol. 35, pp. 597-608.

[44] Nugent, C.D., Webb, J.A.C., and Black, N.D. (2000), "Feature and classifier fusion for 12-lead ECG classification," *Medical Informatics & the Internet in Medicine*, vol. 25, no. 3, pp. 225 - 235.

[45] Attikiouzel, Y. and deSilva, C.J.S. (1995), "Applications of neural networks in medicine," *Australasian Physical and Engineering Sciences in Medicine*, vol. 18, pp. 158-164.

[46] Lim, C.P., Harrison, R.F., and Kennedy, R.L. (1997), "Applications of autonomous neural network systems to medical pattern classification tasks," *Artificial Intelligence in Medicine*, vol. 11, pp. 215-239.

[47] Dorffner, G. and Porenta, G. (1994), "On using feedforward neural networks for clinical diagnostic tasks," *Artificial Intelligence in Medicine*, vol. 6, pp. 417-435.

[48] Bounds, D.G. and Lloyd, P.J. (1990), "A comparison of neural network and other pattern recognition approaches to the diagnosis of low back disorders," *Neural Networks*, vol. 3, pp. 583-591.

[49] Lippmann, R. (1987), "An introduction to computing with neural nets," *IEEE ASAP Magazine*, pp. 4-22.

[50] Zahner, D.A. and Micheli-Tzanakou, E. (1995), "Artificial neural networks: Definitions, methods, applications," in: Bronzino, J.D. (Ed.), *The Biomedical Engineering Handbook*, CRC Press, pp. 2699-2715.

[51] Kartalopoulos, S.V. (1996), *Understanding Neural Networks and Fuzzy Logic: Basic Concepts and Applications*, IEEE Press, New York.

[52] Haykin, S. (1994), *Neural Networks: a Comprehensive Foundation*, Macmillan College Publishing Company, New York.

[53] Bortolan, G., Degani, R., and Willems, J.L. (1990), "Design of neural networks for classification of electrocardiographic signals," *Proc. Ann. IEEE Eng. in Med. and Bio. Soc.*, vol. 12, no. 3, pp. 1467-1468.

[54] Willems, J.L., Arnaud, P., van Bemmel, J.H., Degani, R., MacFarlane, P.W., and Zwictz, C. (1990), "Common standards for quantitative electrocardiography: goals and main results," *Meth. Inform. Med.*, vol. 29, pp. 263-271.

[55] Bortolan, G. and Willems, J.L. (1993), "Diagnostic ECG classification based on neural networks," *Journal of Electrocardiology*, vol. 26, pp. 75-79.

[56] Baxt, W.G. (1991), "Use of an artificial neural network for the diagnosis of myocardial infarction," *Annals of Internal Medicine*, vol. 115, pp. 843-848.

[57] Harrison, R.F., Marshall, S.J., and Kennedy, R.L. (1991), "The early diagnosis of heart attacks: a neurocomputational approach," *Proc. Int. Joint Conf. on Neural Networks*, vol. 1, pp. 1-5.

[58] Edenbrandt, L., Devine, B., and MacFarlane, P.W. (1992), "Neural networks for classification of ECG st-t segments," *J. Electrocardiology*, vol. 25, pp. 167-173.

[59] Marques de Sa, J.P., Goncalves, A.P., Ferreira, F.O., and Abreu-Lima, C. (1994), "Comparison of artificial neural network based ECG classifiers using different features types," *Computers in Cardiology*, pp. 545-547.

[60] Xue, Q., Hu, Y.H., and Tompkins, W.J. (1990), "Training of ECG signals in neural network pattern recognition," *Proc. Ann. IEEE Eng. in Med. and Bio. Soc.*, pp. 1465-1466, vol. 12, no. 3.

[61] Taur, J.S. and Kung, S.Y. (1993), "Prediction-based networks with ECG application," *Proc. IEEE Int. Conf. Neural Networks*, vol. 3, pp. 1920-1925.

[62] Edenbrandt, L., Heden, B., and Pahlm, O. (1993), "Neural networks for analysis of ECG complexes," *J. Electrocardiology*, vol. 26, pp. 74.

Chapter 10

Neural Network Predictions of Significant Coronary Artery Stenosis in Women

B.A. Mobley, W.E. Moore, E. Schechter, J.E. Eichner, and P.A. McKee

Records from 1009 female coronary angiography patients were analyzed by artificial neural network. An outcome in each record, any coronary artery stenosis greater than 50% formed the dichotomous supervisory variable for the neural network. The network contained 19 and 30 elements in the input and middle layers respectively. A single output element corresponded to the supervisory variable. Patient records were ordered according to the date of the angiography and placed into four files. The first 409 records comprised the training file on which the network was trained by the method of backpropagation of errors with momentum. The next 400 records formed a cross-validation file on which the performance of the trained network was optimized. The next 100 records was a cutoff determination file, the file used to determine the output of the network, which distinguished significant stenosis. The cutoff was applied to the 100 records of the test file. ROC analysis revealed that a cutoff of 0.30 maximized specificity while maintaining perfect sensitivity in the cutoff determination file. The cutoff of 0.30 also maintained perfect sensitivity in the test file, while the trained network made output predictions less than 0.30 for 9 of 38 (24%) women who had no stenosis greater than 50%. Therefore the neural network system allowed the correct identification of 24% of the false positives from pre-angiographic diagnostic systems without making a single false negative prediction.

1 Introduction

1.1 Systems Enabling the Avoidance of Unnecessary Angiography

Some individuals, referred for coronary angiography because of angina, are found to be free of significant coronary artery stenosis; i.e., they are said to have "normal" coronary arteries. Some normal angiography findings would seem to be unavoidable as a practical matter because the non-invasive diagnostic systems utilized prior to angiography are not perfectly sensitive and specific, and therefore, in addition, the predictive values of negative and positive tests respectively of those same systems are not perfect either. The patient, medical, legal and business communities are reasonably more accepting of a few false positive pre-angiographic diagnoses than they are accepting of any false negative diagnoses from any diagnostic system. The neural network system presented in this chapter is a preliminary version of a system designed to reduce the number of false positives, i.e., individuals who, in retrospect, may have been unnecessarily subjected to the concerns, risks and expenses of the invasive procedure of coronary angiography. Other systems have also been proposed or utilized which reduce the number of false positives for coronary stenosis by means other than neural network analysis.

Among other systems developed to shield patients from what may become unnecessary coronary angiographies is one by Graboys et al. [1], who instituted a second-opinion trial among patients recommended for coronary angiography. The physicians granting the second opinions employed very strict guidelines for those second opinions, guidelines reflecting a skepticism concerning coronary angiography and the common subsequent procedures of angioplasty and coronary artery bypass grafting. One hundred thirty four of 168 patients were judged to not require angiography, immediately at least, therefore it is a virtual certainty that some false positives for coronary stenosis from the initial testing were spared angiography. Results of the trial were determined by cardiac events during a follow-up period and obviously not by angiography. Therefore it is impossible to say exactly how many false positives were spared angiography by this method.

Ouzan et al. [2] concluded that exercise radionuclide angiography (ERNA) might reduce the number of unnecessary coronary angiographies. They placed data from ERNA along with data from the clinical examination and exercise testing in a logistic regression system in order to optimize the diagnosis of significant coronary artery disease (defined as 70% stenosis or greater). Patients having significant stenosis were identified with 80% sensitivity and 77% specificity. Electron Beam Computed Tomography scanning can detect and gauge the extent of coronary heart disease more accurately than conventional risk factors according to Guerci et al. [3]. The electron beam scanner measures the accumulation of calcium in the coronary arteries. Obstructive coronary disease was defined as 50% or greater stenosis. Age, the ratio of total cholesterol to HDL and the coronary calcium score from the electron beam CT scan were significantly and independently associated both with the presence of any coronary disease and with obstructive coronary disease.

1.2 Women and Angiography

Sometimes, following an assessment of chest pain along with the risk factors for coronary artery disease and other tests such as a stress test, there nonetheless remains a concern about whether angiography is appropriate. Such situations of uncertainty probably occur more frequently in women patients than in men. The evidence-based neural network system in this chapter is proposed to address such concerns. The project was specifically devoted to identifying women who do not have significant coronary stenosis, with significant being defined as > 50% stenosis.

An earlier study by Mobley et al. [4] illustrated our rationale for devoting this project to the study of coronary stenosis in women. Data from 763 consecutive coronary angiography patients were collected and recorded during a period of 17 months. Demographic and medical data from this cross-sectional study were entered into a database. Patient records were ordered consecutively according to the date of the angiography and placed into three files (training file – 332 patients, cross validation file – 331 patients and test file – 100 patients) and analysed by an artificial neural network system. The neural network was developed to predict the presence or absence of any coronary artery stenosis. Data from the angiography patient records comprised 14 input variables to the neural network (age, sex, race, smoking, diabetes, hypertension, body-mass index, creatinine, triglycerides, cholesterol, HDL, cholesterol/HDL, fibrinogen, and lipoprotein a). The network was trained on the training file, optimized on the cross validation file and then tested on the records from the last 100 patients in the series, i.e. the test file. If 0.40 had been chosen as the output of the neural network distinguishing stenosis from no stenosis, all 81 patients in the test file who had stenosis would have been identified as such. Furthermore, 9 of 19 patients who did not have stenosis might have been advised against undergoing angiography. Importantly, all nine of the patients identified by the network to be without stenosis were women. For whatever reasons, the neural network system seemed to make better overall predictions distinguishing coronary artery stenosis in women patients than men. Mobley et al. [4] suggested that the explanation might lie in the fact that a greater fraction of women than men referred for coronary angiography are shown by the angiography to have normal coronary arteries. Therefore the neural network system had more frequent opportunities to "learn" the patterns for normal arteries in records from women than in men.

Women are referred for angiography less frequently than are men, and women hospitalized for coronary heart disease, undergo fewer major diagnostic procedures than men [5]. However, women referred for angiography are also more likely than are men to have normal coronary arteries. Jong et al. [6] examined the sex differences in the features of coronary artery disease for patients undergoing coronary angiography. Records from five hundred fifteen patients who underwent coronary angiography by one cardiologist between 1990 and 1995 were analyzed. Women were three times more likely than men to have normal coronary arteries, with normal being defined as less than 50% stenosis in the left main coronary or less than 70% stenosis in any one of the three major epicardial vessels.

1.3 Other Clinical Predictions by Neural Network

We are not aware of the development of neural network systems by others to predict a degree of coronary artery stenosis or its absence. However, neural networks have been developed for medical systems including cardiology and emergency medicine. Probably the medical neural network that is best known was used to predict acute myocardial infarction in prospective studies [7], [8]. The small prevalence of acute myocardial infarction for patients presenting with anterior chest pain caused the predictive power of a positive test by the neural network to be low even though the sensitivity, specificity and predictive power of a negative test were impressive. By contrast, as a result of pre-angiographic evaluation and testing, most of the patients referred for coronary angiography do indeed have significant coronary artery stenosis. Therefore it is the low prevalence of controls, i.e. individuals without significant coronary artery stenosis, that make the recognition and elimination of false positive cases from the pre-angiographic testing and the resulting effect on the predictive power of a negative test the major concern in this study. As noted above however, women referred for angiography present with a greater fraction of controls than men do. Therefore a neural network study devoted to women may obviate the problem of the prevalence of controls and be able to provide an important diagnostic service by shielding some women from unnecessary angiography.

2 Methods

2.1 Development of the Data Set from the SCA&I Database

A database from the Society for Cardiac Angiography and Interventions (SCA&I) provided the data on patients and their angiographies. The SCA&I database is a relational database of seven tables, to which records were contributed by 37 catheterization laboratories. Only four of the seven tables contained data pertinent to this project. The number of records originally contained in each of the four tables were, PATIENT table (n = 85,850), CATH table (n = 100,649), PROCED table (n = 306,492), and CAD table (n = 218,806). All four of the tables of interest were linked into a database, which was created within the database management program, Microsoft ACCESS (Redmond, WA). ACCESS was then used to produce a data set by joining tables and by allowing the selection of information from the tables and joined tables. Tables in the database were joined at the laboratory number and either a patient serial number or a catheterization serial number.

Only records that were complete in terms of all input and supervisory variables chosen for the neural network model were accepted for inclusion in the data set (Table 1). The individual tables and the joined tables of the database that were used in preparation of the data set were checked numerous times for duplicate records, which were deleted. 1 insistence on complete records; 2 elimination of duplicate records; 3 restriction of the records to those of women; 4 restriction of the records to those of first time elective angiography patients diagnosed with angina; and other reasons, to which we allude below, are responsible for the fact that the data set derived for this project had only 1,009 records. Nonetheless the records were all from women whose anginal symptoms leading to their angiographies were quite uniform. Thus it would seem that the neural network system was put to a rigorous test on these patient records.

The notations in the brackets following the names of input variables listed below indicate whether the variables chosen for the neural network model were continuous or dichotomous numbers. The PATIENT table contained the sex, age, race, ethnicity, height and weight of each patient. Initially records from both men and women were retained in the table so that data sets for both sexes could be prepared, while adhering to equivalent standards for the variables for both sexes. Because the number of Asian and Native American patients was relatively small, the original five categories of Race were compressed into three categories, Black [dichotomous], Caucasian [dichotomous] and Other [dichotomous], while the Ethnicity [dichotomous] categorization (Hispanic or not) was retained. As a consequence of some obviously erroneous entries for age, height and weight, those three variables were restricted in range. Age [continuous] was restricted to 30 – 89 years, height [continuous] was restricted to 54 – 78 inches, and weight [continuous] was restricted to 95 – 350 pounds. Body-mass index [continuous] was calculated from the height and weight of individuals (kg / m^2); however it too was restricted to a range, albeit a sizeable one (15 – 54). As a result of the restrictions described, the number of records in the PATIENT table was reduced to 44,533.

The CATH table and the PATIENT table contained the data used as input variables for the neural network model. The single output variable of the neural network in this project is significant coronary artery stenosis (CAS). The supervisory variable for the output of the network, which provided the correct answers from the results of the angiography and with which the network output could be compared, was whether or not any coronary artery was stenotic to a greater degree than 50% of its diameter. The supervisory variable was originally contained in the CATH table. The inclusive dates of the catheterizations were 7/1/1996 through 12/31/1998. Numerous categories were also included in the CATH table, which allowed us to be assured that suspicion of coronary artery disease was the primary reason for the catheterization. In addition we could also be assured that any record finally adopted for a given patient was a record of the patient at the time of his or her first catheterization. No prior catheterizations had occurred and no catheterization-based procedures such as angioplasty, etc., or no coronary artery bypass grafting had been performed on these patients.

We also used some data entries from the CATH table which allowed us to be assured that, although all of the patients were experiencing chest pain, which was diagnosed as either unstable angina or stable angina [dichotomous, 1/0 representing UA/SA respectively], the angiographies were elective and not deemed urgent or emergent. Unstable angina might or might not include rest pain [dichotomous]. Only if the patients were diagnosed with angina and not atypical chest pain/discomfort or no chest discomfort, could we be sure that the entries represented as Canadian Cardiovascular Society (CCS) classification were indeed appropriate for this project. The CCS classification of the chest pain was included in the CATH table, and it was restricted to the more serious classifications of 3 or 4 [dichotomous, 1/0 representing CCS classifications 4/3 respectively]. All records noting rest pain that were improperly classified CCS classification = 1, 2, or 3 were removed. All records with a diagnosis of unstable angina and an improper CCS classification of 1 or 2 were also removed. Although the diagnosis of chest pain was restricted to stable or unstable angina to assure proper CCS classifications, nevertheless the CCS classifications of 1 and 2 were applied improperly in some cases as indicated. Hence the records were further restricted to those with the CCS classifications of 3 or 4, as noted.

Table 1
SIG-CAS Model

Supervisory Variable

Variable	Definition	Proportion (%)	Range
1. SIG-CAS	Significant coronary artery stenosis (>50%) (Y/N, 1/0)	63.7	0,1

Dependent Output Variable

Variable	Definition	Range
1. SIG-CAS Prediction	Neural network prediction of significant coronary artery stenosis (>50%) (Y/N depending on the cutoff prediction chosen between 1 and 0)	0-1

Independent Input Variables

Variable	Units	Mean	Prop. (%)	S.D.	Median	Range
1. Age	years	64.4		11.3	66.0	33-88
2. Height	inches	63.5		2.6	63.0	54-74
3. Weight	pounds	168.3		36.0	162.0	95-329
4. Body-Mass Index	kg/m^2	29.3		6.0	28.5	16-53
5. Race: Black	Y/N, 1/0		6.8			0,1
6. Race: Caucasian	Y/N, 1/0		90.9			0,1
7. Race: Other	Y/N, 1/0		2.3			0,1
8. Ethnicity: Hispanic	Y/N, 1/0		7.7			0,1
9. Unstable/Stable Ang.	Y/N, 1/0		69.7			0,1
10. Rest Pain	Y/N, 1/0		12.1			0,1
11. CCSC4/CCSC3	Y/N, 1/0		29.9			0,1
12. Stress Test: Positive	Y/N, 1/0		85.1			0,1
13. Stress Test: Negative	Y/N, 1/0		9.1			0,1
14. Stress Test: Inconcl.	Y/N, 1/0		5.7			0,1
15. Diabetes	Y/N, 1/0		23.1			0,1
16. Hypertension	Y/N, 1/0		60.7			0,1
17. Peripheral Vasc. Dis.	Y/N, 1/0		5.9			0,1
18. High Creatinine	Y/N, 1/0		1.5			0,1
19. Dialysis	Y/N, 1/0		0.7			0,1

Positive [dichotomous], negative [dichotomous] or inconclusive [dichotomous] stress test results were included in the CATH table. Database rules were not restrictive as to the type of stress test allowed, and unfortunately the rules did not call for the identification of the type of stress test. Entries stating yes or no to diabetes, hypertension, peripheral vascular disease, high creatinine and dialysis were retained. Diabetes classification [dichotomous] meant diabetes mellitus requiring oral hypoglycemic agents or insulin. Hypertension classification [dichotomous] meant a history of hypertension or treatment for the same. Creatinine [dichotomous] meant a serum creatinine greater than 2.0 (mg%). Peripheral vascular disease [dichotomous] actually encompassed a number of vascular problems including stroke or stenosis of at least 50% in the carotid/cerebrovascular, aorto-iliac, or femoral/popliteal vessels. Dialysis [dichotomous] meant the patient was currently undergoing dialysis, either hemo- or peritoneal. The CATH table, modified as described, was reduced to 4,708 records.

The PATIENT and CATH tables were joined at the laboratory and patient serial numbers; 3,016 records resulted. (Apparently numerous catheterizations were recorded in the CATH table of the database without accompanying entries in the PATIENT table.) The PROCED table contained a list of procedures performed on cardiac patients. The PROCED table was used to provide further assurance that coronary angiography was performed on patients in the data set and that the database system was being utilized properly, i.e., that entries were made in the appropriate tables. The PATIENT–CATH table was joined at the laboratory and catheterization serial numbers to the PROCED table to accomplish this assurance. The resulting table, PATIENT-CATH-PROCED, contained 2,767 records.

Data from the CAD table, which are relevant to this project, are listings of the degree of stenoses found during angiography. We compared the stenosis data in the CAD table with the network supervisory variable from the PATIENT-CATH-PROCED table, which noted > 50% coronary stenosis or not. If any record in the PATIENT-CATH-PROCED table noted normal coronary arteries, i.e. no stenosis greater than 50%, and yet there was an entry corresponding to that record in the CAD table, then the record was eliminated. These few records were removed only as cautionary measures in those cases in which the maximum stenosis listed in the CAD table was indeed 50% or less (not however when the stenosis in the CAD table was listed as > 50%). To be sure, when the listing in the CAD table was 50% or less, both entries in the two tables might have been correct, but there remained the suspicion that one of the entries was incorrect simply because there was an entry in the CAD table. Numerous entries of "0" stenosis (not entries left blank) and other small stenoses that were listed led to our suspicion.

On the other hand if the PATIENT-CATH-PROCED table indicated that the coronaries were not normal, i.e., at least one stenosis greater than 50% was encountered in the angiography, but the CAD table did not confirm a stenosis greater than 50%, then the record was deleted from the file. At this point we counted the number of records submitted from each laboratory, and we removed all records submitted from laboratories that had fewer than 30 entries, i.e., less than one coronary angiography per month for two and one-half years. Twenty five hundred twenty records from 27 laboratories remained in the data set, of which 1,009 records represented female patients. The latter records became the data set for the project (Table 1).

2.2 Artificial Neural Network

Artificial neural networks are a type of artificial intelligence [9]-[11]. The elementary components of artificial neural networks are somewhat analogous to biological neurons. Each element receives numerous inputs; the sum of the inputs on each element and their modification by the nonlinear transfer function determines the output of each element. The transfer functions are logistic in the elements of the network in this chapter. When numerous elements are arranged in layers with weighting, i.e., amplifying connections extending between the elements in adjacent layers, the networks so formed are structurally if not functionally analogous to biological neural networks. The numerous weighting connections affect the processing of data and thus affect the output of the network. Along with the transfer functions, the weights determine the magnitudes of the elemental outputs as the data from input variables are fed forward through a network. The weighting connections are modified during "learning" or "training" according to an algorithm. Indeed, the modifications of the weighting connections are the essence of the training of neural networks.

A widely used training algorithm in artificial neural networks, and the one used in this project, is the backpropagation algorithm [12]. During training, a function of the difference between the output of the network and the desired output or correct answer (from the supervisory variable regarding significant stenosis for each record) is propagated back through the layers of the network. A function of that difference is used to change the weighting connections in the multi-dimensional space created by the network by changing the weights in the directions of minimizing the errors. Momentum was included as a part of the procedure for changing weights of the network during training. Therefore, the function that was used to change weights also included a term derived from the most recent changing of weights. The type of training used in this project, is known as supervised training, meaning that correct answers (1/0 for > 50% stenosis or not) were provided to the network during training in the form of supervisory variables from the data set (Table 1).

A common number of layers in artificial neural networks are three: input, middle (called hidden), and output layers, and three layers were used in this project. The input layer serves only to distribute input data to the middle layer through weighting connections. Therefore, the number of elements in the input layer is simply the number of input (independent) variables in the neural network model, 19 in this project. The number of elements in the output layer is often only one, as was the case in this project. As a result of the output element having a logistic transfer function, the network output was always a decimal number between 0 and 1. Subsequent to the training, the exact magnitude of the output represented the network prediction of whether or not there was significant coronary artery stenosis (> 50% or not) in the patient represented by each record. The magnitude of the output between 0 and 1 that distinguished significant stenosis was determined by using what we have called the cutoff determination file (Figure 1).

The optimum number of elements for the middle layer must often be determined empirically. However, we tested a number of methods in a previous project, Mobley et al. [4], and found that the method of determining the number of middle layer elements recommended by the manufacturer of the neural network software seemed to work best. The number of middle layer elements recommended by the Neuroshell 2 software manufacturer (Ward Systems, Inc., Frederick, MD) is the square root of the number of records in the training file plus half the sum of the input plus output elements, i.e., 30 = (square root of 409 + 0.5 (19 + 1)).

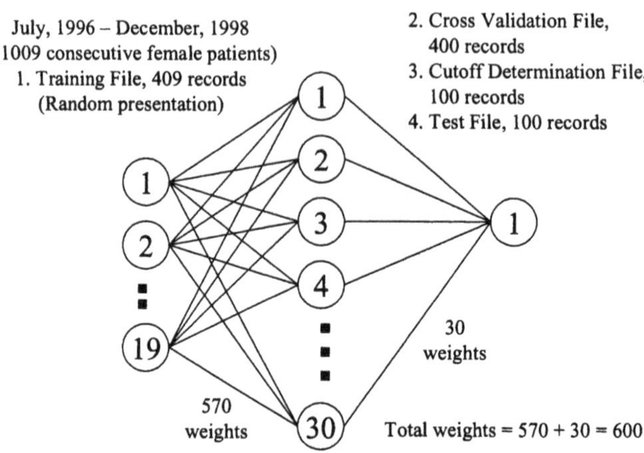

Figure 1. A representation of the artificial neural network used in this project. The 600 weights were changed as each record in the training file was processed. The learning constant, the momentum constant, and the initial maximum for all of the weights were all set at 0.1. The computer was a Gateway 2000, 133MHz (North Sioux City, SD).

2.3 Patient Files

The 1,009 female patient records of the data set were ordered consecutively according to the dates of the angiographies and placed into four files. The first 409 patient records comprised the training file for the neural network. Records from the next 400 patients formed the cross-validation file, i.e. the file on which the performance of the trained network was optimized. Records from the next 100 patients formed the cutoff determination file, the file used to determine the output cutoff of the network between 0 and 1, i.e., the number which was in turn applied to the 100 records of the fourth file, the test file. The training file and the cross-validation file were presented to the network alternately until the training had been optimized on the cross-validation file [4]. Optimization as provided by the software

(Neuroshell 2) was based on the mean squared difference between the supervisory variables and the output variables of the network for all records in the cross-validation file. A million iterations of the network on the training file, with intermittent checks on the cross-validation file, was run subsequent to the achievement of an apparently optimally trained network before finally declaring the network optimally trained. Then the cutoff determination file was presented to the network, and finally the test file was presented for simulating predictions of significant coronary stenosis or not on patients. Modification of the connecting weights of the network was allowed only during the training of the network. During training, the records in the training file were presented to the neural network in random order. A chi square test on the four files showed that the null hypothesis of equal proportions of significant stenosis in the four files could not be rejected ($p = 0.41$).

Table 2
Files and Outcomes*

File name (consecutive patients)	Significant coronary artery stenosis (>50%)	Controls (no significant stenosis)	Total patients
Training	273 (67%)	136	409
Cross-validation	248 (62%)	152	400
Cutoff-determination	60 (60%)	40	100
Test	62 (62%)	38	100
Totals	643 (64%)	366	1009

* P = 0.41 (chi square)

The numerical cutoff from the cutoff determination file was that network output which maximized the specificity while maintaining perfect sensitivity in that file. In effect the cutoff was chosen to identify the maximum number of women who in retrospect may not have needed the angiography because none of their stenoses were deemed significant, while at the same time affirming the need for angiography for all of those who did have at least one significant stenosis. The same numerical cutoff was then applied to the processed records of the test file under the assumption that specificity would be maximized within the constraints of perfect sensitivity for records in that file also.

2.4 Logistic Regression

Logistic regression analysis is most frequently used to examine the risk relationship between disease and exposure, with the ability to test for statistical interaction and control for multi-variable confounding [13]. It is less frequently used to predict the probability of disease using the observed values from multiple risk factors. In this latter role, computer software driven modeling strategies are appropriate in the search for the best predictions [14].

When used alone, the reliability of a logistic regression prediction model is best demonstrated by splitting the data into a training file and a validation file [15]. This technique permits the comparison of the sensitivity and specificity of the training model with the sensitivity and specificity of the predictions resulting from using the same regression equation on the validation file. The validation file will often evidence prediction "shrinkage," i.e., the training model will not perform as well with the new data of the validation file.

However, because the logistic modeling predictions in this chapter are for comparison to the predictive results of a neural network, the first 909 records in the data set were used to determine the best logistic model. This method provided a "best case" logistic model to compare to the neural network results from the last 100 records. The logistic regression formula from the best model (and the cut-off that provided the highest specificity with perfect sensitivity) was used to generate classifications for the last 100 records for comparison with the results of the neural network. Stepwise logistic regression, using a backward variable selection algorithm, was used to generate the model for these comparisons (SPSS 10.0, SPSS Inc., Chicago, IL). At each successive stage of the stepwise procedure, individual predictions of disease probability were generated for the subjects. These data were then used in receiver operating characteristic (ROC) analyses to identify a parsimonious model with the highest specificity and perfect sensitivity in predicting coronary artery stenosis of > 50%.

2.5 ROC Analysis

The distinction between significant coronary artery stenosis or not (> 50% or not, 1 or 0) is a discrete representation by the supervisory variables from the records in the SCA&I data set. However the outputs from artificial neural networks and the logistic regression, which serve to predict significant stenoses or not in the records of the test file, are continuous functions between 0 and 1. ROC analysis is an appropriate means to display sensitivity and specificity relationships when a predictive output for two possibilities is continuous. In its tabular form the ROC analysis displays true and false positive and negative totals and sensitivity and specificity for each listed cutoff value between 0 and 1.

The ROC curve is a plot of the true positive rate (sensitivity) against the false positive rate (1 – specificity) for each possible cutoff. In this chapter we report results after choosing a cutoff, one that maximizes the identification of false positive pre-angiographic records (maximizing the specificity by making the false positives into true negatives), within the constraint of maintaining a sensitivity of 1.0 (no false negatives of the neural network system or logistic regression models). Nonetheless, there remains merit in computing the area of a ROC curve [4]. The area under the ROC curve is a measure of the overall discriminating ability of a system as a whole (the neural network system or the logistic regression models in this chapter). The closer the area under the ROC curve is to a value of 1.0, the better the entire system is at detecting in advance whether there is significant coronary stenosis or not. If a single cutoff could distinguish all of the records with significant stenosis from all of those without significant stenosis, the area under the ROC curve would be 1.0. Whereas if the network or logistic regression assigned significant stenosis or not randomly, the area under the ROC curve would be 0.5. The software used for the ROC analysis was from NCSS Statistical Software (Kaysville, UT).

3 Results

3.1 Neural Network Training and Cross Validation

The learning constant, momentum constant, and initial upper limit for all of the randomly assigned weights in the artificial neural network were all set to 0.1. According to the network performance on the cross validation file of 400 records, the network became optimally trained after 30,266 iterations or 74 epochs of the 409 records in the training file had been processed. However, the training was allowed to proceed for an additional 1,000,005 iterations (2,445 epochs) on records from the training file in order to be sure that the training of the network could not be improved with respect to its performance on the cross validation file.

3.2 Network Application to the Cutoff Determination File

The cutoff determination file was important for determining the network output between 0 and 1 that would be used to distinguish predictions of significant coronary artery stenosis (CAS) in the records of the test file, yet maintaining perfect sensitivity. The assumption was that the network cutoff distinguishing significant stenosis in the test file would also be the same or nearly the same as for the records of the cutoff determination file. The output of the network that became the cutoff was simply the largest network output, which preserved perfect sensitivity by allowing no false negatives. Such an output identified the maximum number of false positives from pre-angiographic testing and correctly classified them as true negatives of the network system, therefore it maximized the specificity, without in effect allowing the network to recommend against angiography for any false negative of the network system. Figure 2 is a scatter plot that shows the network outputs or network predictions on records of the cutoff determination file on the ordinate and the dichotomous indications of significant coronary artery stenosis determined from the angiography on the abcissa (1 indicates stenosis > 50%, 0 indicates no stenosis > 50%).

Obviously the cutoff of network predictions that maximizes specificity while maintaining perfect sensitivity is slightly greater than 0.30. All of the outputs in the SIGNIFICANT CAS = 0 column that were rated less than 0.30 by the trained network were false positives created by the non-invasive pre-angiographic testing, the testing which led to the recommendation for angiography. Because no patient actually having significant stenosis was rated by the network less than 0.30, then there were no false negatives created by the network system when the cutoff was chosen to be 0.30.

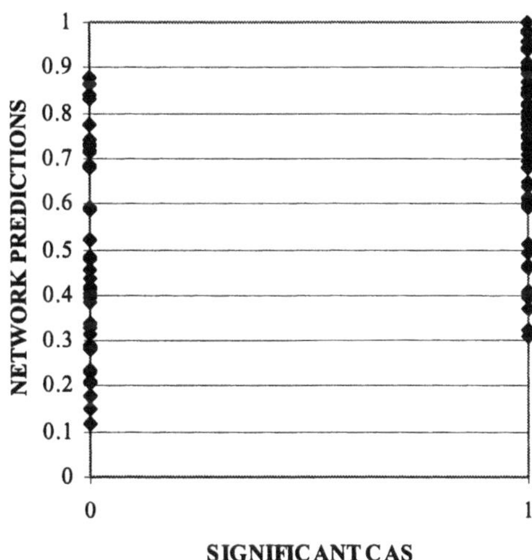

Figure 2. This figure shows a scatter plot of the network predictions on the records in the cutoff determination file with respect to indications of whether or not the patients did or did not have any coronary artery stenosis exceeding 50%.

Table 3 is a tabular ROC analysis of the application of the neural network to the records in the cutoff determination file. The table lists 21 cutoff possibilities at intervals of 0.05 between 0 and 1, and it shows the results of true and false negative and positive results and the sensitivity, specificity and the positive and negative predictive values at each of the cutoff choices. The purpose of the cutoff determination file is to provide a cutoff, which may then be applied to the test file. Therefore it is only of academic interest that a cutoff of 0.30 applied to the cutoff determination file itself identified 12 of 40 (30%) false positive patients from the pre-angiographic testing and placed them in the true negative category, while maintaining a sensitivity of 1.00.

Table 3. ROC Results on the Cutoff Determination File.

Cutoff Predic-tions	True Pos.	False Pos.	False Neg.	True Neg.	Sensi-tivity	Speci-ficity	Pos. Pred. Value	Neg. Pred. Value	Cumulative ROC Area
0.00	60	40	0	0	1.00	0.00	0.60	1.00	0.00
0.05	60	40	0	0	1.00	0.00	0.60	1.00	0.00
0.10	60	40	0	0	1.00	0.00	0.60	1.00	0.00
0.15	60	39	0	1	1.00	0.03	0.61	1.00	0.03
0.20	60	37	0	3	1.00	0.08	0.62	1.00	0.08
0.25	60	32	0	8	1.00	0.20	0.65	1.00	0.20
0.30	60	28	0	12	1.00	0.30	0.68	1.00	0.30
0.35	58	25	2	15	0.97	0.38	0.70	0.88	0.37
0.40	56	23	4	17	0.93	0.43	0.71	0.81	0.42
0.45	54	18	6	22	0.90	0.55	0.75	0.79	0.54
0.50	51	15	9	25	0.85	0.63	0.77	0.74	0.60
0.55	49	14	11	26	0.82	0.65	0.78	0.70	0.62
0.60	47	12	13	28	0.78	0.70	0.80	0.68	0.66
0.65	38	12	22	28	0.63	0.70	0.76	0.56	0.66
0.70	36	10	24	30	0.60	0.75	0.78	0.56	0.69
0.75	30	5	30	35	0.50	0.88	0.86	0.54	0.76
0.80	23	4	37	36	0.38	0.90	0.85	0.49	0.77
0.85	18	2	42	38	0.30	0.95	0.90	0.48	0.79
0.90	13	0	47	40	0.22	1.00	1.00	0.46	0.80
0.95	7	0	53	40	0.12	1.00	1.00	0.43	0.80
1.00	0	0	60	40	0.00	1.00	0.00	0.40	0.80

Area under ROC curve = 0.80, Standard error = 0.04

Figure 3 illustrates the cumulative area under the ROC curve that is presented in Table 3. The cumulative area of 0.80 should be compared to an area of 1.00, which would be the area if there were at least one cutoff value that perfectly distinguished all of the patients with significant stenosis from all of those without significant stenosis.

3.3 Network Application to the Test File

Figure 4 illustrates that the cutoff number of 0.30, which was determined on the cutoff determination file, would serve exceptionally well as a cutoff when applied to the 100 records of the test file. It is clear from the figure that a cutoff of 0.30 maintains perfect sensitivity in the test file, because it prevents the network system from creating any false negatives, which would be patients with significant stenosis

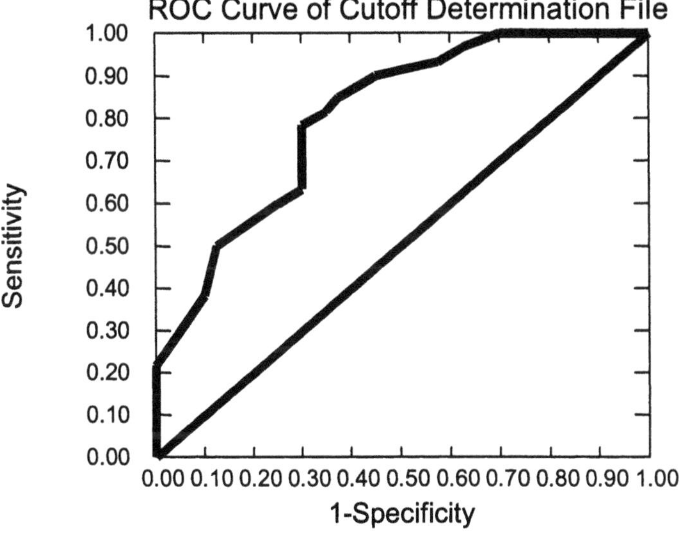

Figure 3. This figure shows an ROC curve representing the results of the application of the trained neural network to all 100 records in the cutoff determination file. The area under the ROC curve is 0.80 with a standard error of 0.04.

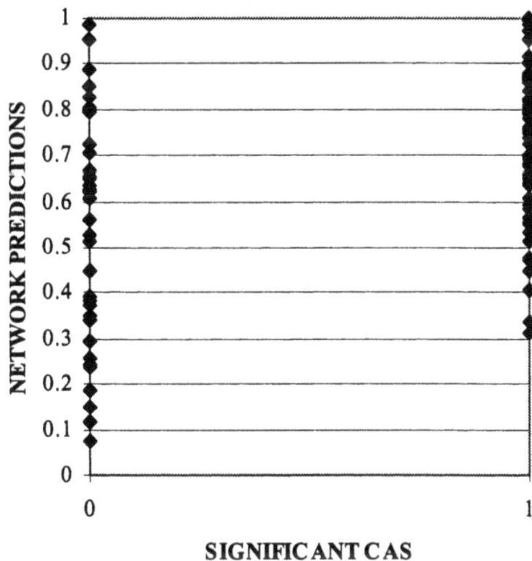

Figure 4. This figure shows a scatter plot of the network predictions on the records in the test file with respect to indications of whether or not the patients did or did not have any coronary artery stenosis exceeding 50%.

but identified by the network as being without significant stenosis. Within that constraint the cutoff of 0.30 maximizes the specificity, by maximizing the identification of false positives created by the pre-angiographic testing.

Table 4 is similar to Table 3 except that it is an ROC analysis depicting the neural network performance on records of the test file. Table 4 reveals exactly how many false positives of the pre-angiographic testing are safely identified by the neural network system. At a cutoff of 0.30, 9 of 38 (24%) women patients without significant stenosis are identified, while the sensitivity of 1.00 is maintained. The 9 women without significant stenosis are true negatives and are identified as such in the table, thus maximizing the specificity within the constraint of maintaining perfect sensitivity.

Table 4. ROC Results on the Test File.

Cutoff Predic-tions	True Pos.	False Pos.	False Neg.	True Neg.	Sensi-tivity	Speci-ficity	Pos. Pred. Value	Neg. Pred. Value	Cumulative ROC Area
0.00	62	38	0	0	1.00	0.00	0.62	1.00	0.00
0.05	62	38	0	0	1.00	0.00	0.62	1.00	0.00
0.10	62	37	0	1	1.00	0.03	0.63	1.00	0.03
0.15	62	35	0	3	1.00	0.08	0.64	1.00	0.08
0.20	62	34	0	4	1.00	0.11	0.65	1.00	0.11
0.25	62	31	0	7	1.00	0.18	0.67	1.00	0.18
0.30	62	29	0	9	1.00	0.24	0.68	1.00	0.24
0.35	60	26	2	12	0.97	0.32	0.70	0.86	0.31
0.40	60	23	2	15	0.97	0.39	0.72	0.88	0.39
0.45	58	23	4	15	0.94	0.39	0.72	0.79	0.39
0.50	56	22	6	16	0.90	0.42	0.72	0.73	0.42
0.55	53	20	9	18	0.85	0.47	0.73	0.67	0.46
0.60	49	19	13	19	0.79	0.50	0.72	0.59	0.48
0.65	44	14	18	24	0.71	0.63	0.76	0.57	0.58
0.70	37	11	25	27	0.60	0.71	0.77	0.52	0.63
0.75	32	9	30	29	0.52	0.76	0.78	0.49	0.66
0.80	24	7	38	31	0.39	0.82	0.77	0.45	0.69
0.85	18	3	44	35	0.29	0.92	0.86	0.44	0.72
0.90	12	2	50	36	0.19	0.95	0.86	0.42	0.73
0.95	9	2	53	36	0.15	0.95	0.82	0.40	0.73
1.00	0	0	62	38	0.00	1.00	0.00	0.38	0.73

Area under ROC curve = 0.73, Standard error = 0.05

Figure 5 shows the graphical ROC analysis that is presented in Table 4. The area of 0.73 under the ROC curve indicates that when judged in its totality as a discriminating system for significant stenosis, the neural network we used may not be outstanding. However, the proposal in this chapter is to utilize the neural network in a limited and conservative way, i.e., to pick one cutoff that maximizes specificity within the constraint of maintaining perfect sensitivity. In that limited respect the neural network performs quite well by potentially and safely sparing from angiography 24% of the women who were recommended for the procedure.

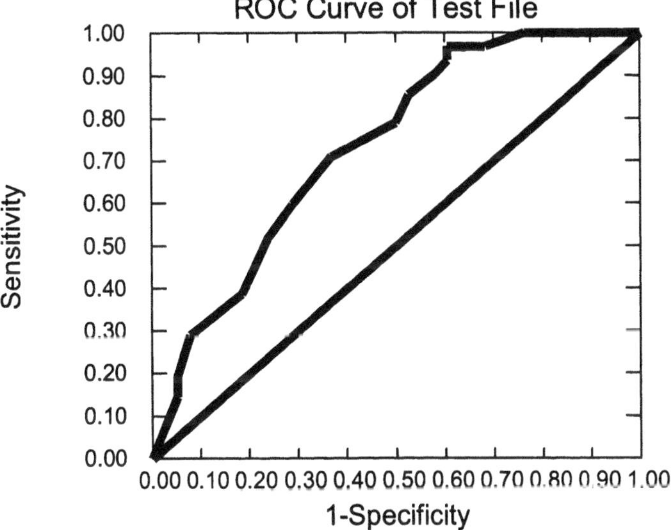

Figure 5. This figure shows an ROC curve representing the results of the application of the trained neural network to all 100 records in the test file. The area under the ROC curve is 0.73 with a standard error of 0.05.

3.4 Relative Weights of the Neural Network

Table 5 shows a ranking of the total weighting associated with each of nineteen input variables to the artificial neural network. There were 60 weights between each input element and the single output element in the network, although 30 of those weights were common to all input variables because they ran from the thirty elements in the hidden layer to the single element in the output layer (see Figure 1). As noted earlier, such rankings of the sum of weights may or may not be important in discerning the relative importance of the variables in determining the decisions of the network. However the fact that the variable of age had the greatest total weight may be indicative of the fact that the sum of weights is indeed relevant to decisions of the network.

Table 5. Independent variable rankings by neural network weights.

Independent variables	Sum of weights
Age	0.1377
Height	0.1163
Weight	0.0774
Stress Test: Positive	0.0675
Race: Other	0.0626
Race: Caucasian	0.0614
Diabetes	0.0570
Unstable / Stable Angina	0.0559
Race: Black	0.0512
High Creatinine	0.0489
Hypertension	0.0483
Rest Pain	0.0460
Peripheral Vascular Disease	0.0364
Body-Mass Index	0.0300
Stress Test : Negative	0.0260
Dialysis	0.0225
CCSC4 / CCSC3	0.0217
Stress Test: Inconclusive	0.0185
Ethnicity: Hispanic	0.0148

3.5 Logistic Regression

Logistic regression analysis was performed to compare its restrained predictive ability (maximized specificity given perfect sensitivity) with the restrained predictive ability of the neural network. The full or complete logistic model had only 17 independent variables. However those 17 variables were defined such that they contained all of the information provided by the 19 variables used in the neural network analysis. The parsimonious model was chosen as the model that used the fewest independent variables in producing the greatest specificity with perfect sensitivity.

Table 6. Logistic regression model with all variables.

Variables in the equation	B	S.E.	Sig.	Odds Ratio	95% C.I.for Odds Ratio	
					Lower	Upper
1. Age (years)	.056	.008	.000	1.058	1.042	1.074
2. Height (inches)	-.198	.155	.201	.821	.606	1.111
3. Weight (pounds)	.010	.028	.728	1.010	.956	1.067
4. BMI (per kg/m^2)	-.093	.160	.563	.911	.666	1.248
5. Unstable/Stable Angina	.155	.175	.378	1.167	.828	1.645
6. Rest Pain	-.263	.290	.364	.769	.436	1.356
7. Stress Test: Positive	.791	.260	.002	2.205	1.325	3.667
8. Stress Test: Inconclusive	-.143	.398	.719	.866	.397	1.890
9. CCSC4/CCSC3	.089	.212	.673	1.093	.722	1.657
10. Diabetes	.932	.216	.000	2.540	1.664	3.876
11. Hypertension	.462	.162	.004	1.587	1.154	2.182
12. Peripheral Vascular Disease	.751	.445	.092	2.118	.885	5.068
13. High Creatinine (>2 mg/dl)	-.001	1.108	.999	.999	.114	8.761
14. Dialysis	.915	1.735	.598	2.497	.083	74.887
Race (Caucasian referent)						
15. Black	-.539	.303	.076	.584	.322	1.058
16. Other	.702	.696	.313	2.019	.516	7.891
17. Hispanic	.630	.347	.070	1.877	.950	3.706
Constant	9.427	9.852	.339			

The variables in thc full logistic model and the variables in the model selected for parsimony, perfect sensitivity and best specificity are shown in Tables 6 and 7. The parsimonious model correctly identified only one more patient who was a false positive of the pre-angiographic testing than the full model did. A comparison of the proportion of patients potentially spared angiography by neural network prediction for the last 100 records and logistic regression prediction for the first 909 records is shown in Table 8. It is readily apparent that the performance of the neural network far exceeds that of the more traditional method of logistic regression. The network is 12.4 times as likely to safely spare a patient from angiography. Even though the predictive quality of the chosen logistic model is likely to be much higher than would be the case if it were used on another data set, it is still no match for the rigorously validated neural network. In fact, a

Table 7. Parsimonious logistic regression model with highest specificity given perfect sensitivity.

Variables in the equation	B	S.E.	Sig.	Odds Ratio	95% C.I.for Odds Ratio	
					Lower	Upper
1. Age (years)	.056	.008	.000	1.058	1.042	1.074
2. Height (inches)	-.145	.032	.000	.865	.812	.921
3. BMI (per kg/m^2)	-.037	.014	.009	.964	.938	.991
4. Unstable/Stable Angina	.176	.172	.307	1.192	.851	1.671
5. Rest Pain	-.190	.243	.434	.827	.514	1.331
6. Stress Test: Positive	.836	.210	.000	2.307	1.530	3.479
7. Diabetes	.940	.215	.000	2.560	1.680	3.900
8. Hypertension	.469	.162	.004	1.598	1.163	2.196
9. Peripheral Vascular Disease	.763	.443	.085	2.144	.900	5.111
Race (Caucasian referent)						
10. Black	-.515	.302	.088	.597	.330	1.080
11. Other	.699	.694	.314	2.013	.516	7.848
12. Hispanic	.630	.347	.070	1.878	.950	3.710
Constant	6.050	2.238	.007			

Table 8. Comparison of logistic prediction for first 909 cases with neural network prediction for last 100 cases (910 to 1009).

	Correctly Identified $\leq$ 50% Stenosis	Incorrectly Identified > 50% Stenosis	Total
Neural Net	9 23.7%	29 76.3%	38
Logistic Regression	8 2.4%	320 97.6%	328

Fisher's exact test, two-tailed $p < .001$
Odds Ratio for neural net's ability to potentially spare patients angiography compared to logistic model = 12.4 ; 95% C.I., 4.5 to 34.6
Cut-off for positive test = 0.117226

matched comparison using both methods on the last 100 records reiterated the performance of the neural network when compared to the logistic model (Table 9). The logistic model spared only 1 of 38 from angiography while the neural network spared that one and 8 others. Due to the limited sample size (100) in the matched comparison and the poor performance of the logistic model, a matched odds ratio could not be determined.

Table 9. Matched comparison of logistic prediction and neural net prediction for last 100 records.

		Logistic Regression Prediction		
		Correctly Identified As ≤ 50% Stenosis	Incorrectly Identified As > 50% Stenosis	Total
Neural Network	Correctly Identified As ≤ 50% Stenosis	1	8	9
	Incorrectly Identified As > 50% Stenosis	0	29	29
	Total	1	37	38

McNemar Test, Exact Binomial Probability = 0.008
Matched odds ratio for neural network's ability to potentially spare patients angiography compared to logistic model = 8/0 = indeterminate
Cut-off for positive test = 0.117226

4 Discussion

4.1 Patients and Data

The patients represented by the data set were all female and predominantly white (90.9%), with a mean age of 64.4 ± 11.3 years. They had a high prevalence of diabetes (23.1%), and hypertension (60.7%). A large proportion also presented with unstable angina (69.7%) as opposed to stable angina, and had positive exercise stress tests (85.1%). As a group they also evidenced borderline obesity (Mean BMI = 29.3 ± 6). These demographics suggest that this group is representative of patient profiles that leave the diagnostician little choice but to perform coronary angiography. These data then provided a test of neural network prediction that is appropriate in application and difficulty.

Mobley, et al, [4] used the variables contained in a database to predict the lack of **any** measurable coronary stenosis. Many of the variables available in that database were laboratory values of blood samples. The independent variables available in the SCA&I database used in this project are oriented to chest pain, and the only supervisory variable offered in that database is significant stenosis (>50% or not). Neither of the two databases was designed originally for our purposes in predicting either any stenosis or significant stenosis. We simply used variables available from existing databases and in the process proposed to show that the artificial neural network can be used as a tool for predicting absence of stenosis [4] or significant stenosis (this chapter). We note however that our long-term goal is to produce a large database in conjunction with a clinical trial in which we can include variables describing the chest pain, variables derived from blood samples and variables from a complete patient history and physical. During such a trial we would propose to record the degree of stenosis from the angiography so that we could potentially assess neural network ability to predict different degrees of stenosis.

4.2 Patient Files

We divided the data set into four files (training, cross-validation, cutoff determination, and test) and ordered the records of those files according to the order in which the patients presented for coronary angiography. The reason for such an ordering was that it allowed simulation of a system in which the neural network could make predictions for actual patients, i.e. the patient records in our test file could simulate patients for whom predictions of stenosis were made. Because the records of angiographies in the data set were submitted through 27 laboratories from across the country, there is probably randomness present even in this method of ordering records.

The total size of the data set (1,009 records) required some compromises in the number of records placed in each of the four files (training, cross-validation, cutoff determination, and test). We chose 100 records, i.e., 10% of the total number of records to be in each of the test and cutoff determination files. Our thinking was to have the training and cross-validation files be approximately equal [4] and be as large as possible without making the cutoff determination and test files so small that results of the network system on a given test file might be suspect.

4.3 Cutoff Determination File

The cutoff determination file, proposed in this chapter, may be essential for building a neural network system that can actually be implemented for the purpose of making predictions of the absence of significant coronary artery stenosis. However, we are aware of no other study, which has attempted to incorporate it into a system. Mobley, et al. [4] acknowledged the necessity of using a cutoff determination file, but there were insufficient patient records in their data set of 763 to allow for the inclusion of such a file. One could pick a cutoff number in advance, apply it to the test file, and thereby eliminate the necessity of including a cutoff determination file in the system. The most likely choice for a cutoff chosen a priori would be 0.5. The results presented in Figures 2 and 4 and Tables 3 and 4 illustrate the problems accompanying such a method of selection, i.e., there would be a significant danger of the introduction of false negatives by the network system.

4.4 Predictive Systems

An advantage of the employment of a neural network system or any formal mathematical or statistical system in predicting the absence of significant coronary artery stenosis is that all of the chosen input variables contribute to a decision by the network regarding the recommendation of angiography for each patient. Furthermore, all of the variables contribute to the decision objectively, i.e., to their trained degree.

It would be appropriate at some time to assess the monetary savings that could result from the implementation of a neural network system which could safely shield some patients from coronary angiography and compare the savings with the cost of operating such a system.

4.5 Network Weights

When assessing the relative importance of neural network variables, one may simply add numeric weights in the paths between each input variable and the output. This method has some validity because the number of separate paths between each input element and the single output is identical. However, each network element modifies by a nonlinear logistic function the sum of each weight times the respective output from elements in the preceding layer, therefore the resulting nonlinear modifications of the weights are not conveyed by the simple addition of weights. Therefore the ranking of weights of the 19 independent variables listed in Table 5 is not necessarily presented because of the insight contained therein. Although the relative importance of the different independent variables in the logistic regression experiments were clear, the logistic regression models were unable to match the predictive ability of the neural network. Neural networks remain a "black box" type device, and the justification for using them in a predictive system must result from their superior predictive ability, as demonstrated in numerous rigorous tests.

5 Conclusions

We have shown that a properly designed and trained artificial neural network system may be effective in identifying a fraction of the women referred for coronary angiography whose coronary artery stenosis is not sufficiently extensive to explain their angina. In addition, such identifications can be made within the constraint that the neural network will also identify everyone whose coronary stenosis is sufficiently extensive to explain the angina.

Acknowledgments

We thank the Society for Cardiac Angiography and Interventions for allowing us to work with their database.

References

[1] Graboys, T.B., Biegelsen, B., Lampert, S., Blatt, C.M., and Lown, B. (1992), "Results of a second-opinion trial among patients recommended for coronary angiography," *JAMA*, vol. 268, pp. 2537-2540.

[2] Ouzan, J., Chapoutot, L., Carre, E.J., Liehn, J.C., and Elaerts, J. (1993), "Multivariate analysis of the diagnostic values of clinical examination, exercise testing and exercise radionuclide angiography in coronary artery disease," *Cardiology*, vol. 83, pp. 197-204.

[3] Guerci, A.D., Spadaro, L.A., Goodman, K.J., Lledo-Perez, A., Newstein, D., Lerner, G., and Arad, Y. (1998), "Comparison of electron beam computed tomography scanning and conventional risk factor assessment for the prediction of angiographic coronary artery disease," *JACC*, vol. 32, pp. 673-679.

[4] Mobley, B.A., Schechter, E., Moore, W.E., McKee, P.A., and Eichner, J.E. (2000), "Predictions of coronary artery stenosis by artificial neural network," *Art. Intell. Med.*, vol. 18, pp. 187-203.

[5] Ayanian, J.Z. and Epstein, A.M., (1991), "Differences in the use of procedures between women and men hospitalized for coronary heart disease," *N. Engl. J. Med.*, vol. 325, pp. 221-225.

[6] Jong, P., Mohammed, S., and Sternberg, L., (1996), "Sex differences in the features of coronary artery disease of patients undergoing coronary angiography," *Can. J. Cardiol.*, vol. 12, pp. 671-677.

[7] Baxt, W.G., (1991), "Use of an artificial neural network for the diagnosis of myocardial infarction," *Ann. Int. Med.*, vol. 115, pp. 843-848.

[8] Baxt, W.G. and Skora, J. (1996), "Prospective validation of artificial neural network trained to identify acute myocardial infarction," *Lancet*, vol. 347, pp. 12-15.

[9] Wasserman, P.D. (1989), *Neural Computing: Theory and Practice*, Van Nostrand, Reinhold, New York.

[10] Caudill, M. and Butler, C. (1990), *Naturally Intelligent Systems*, MIT Press, Cambridge, MA.

[11] Hertz, J., Krogh, A., and Palmer, R.G. (1991), *Introduction to the Theory of Neural Computation*, Addison-Wesley, Redwood City, CA.

[12] Rumelhart, D.E., Hinton, G.E., and Williams, R.J., (1986), "Learning internal representations by error propagation," *Parallel Distributed Processing*, pp. 318-362.

[13] Hosmer, D.W. and Lemeshow, S. (1989), *Applied Logistic Regression*, John Wiley & Sons, New York

[14] Kleinbaum, D.G. (1994), *Logistic Regression: a Self-Learning Text*, Springer-Verlag, New York

[15] Kleinbaum, D.G., Kupper, L.L., and Muller, K.E. (1988), *Applied Regression Analysis and Other Multivariable Methods*, Duxbury Press, Belmont, CA.

Chapter 11

A Modular Neural Network System for the Analysis of Nuclei in Histopathological Sections

C.S. Pattichis, F. Schnorrenberg, C.N. Schizas, M.S. Pattichis, and K. Kyriacou

The evaluation of immunocytochemically stained histopathological sections presents a complex problem due to many variations that are inherent in the methodology. This chapter describes a modular neural network system which is being used for the detection and classification of breast cancer nuclei named Biopsy Analysis Support System (BASS). The system is based on a modular architecture where the detection and classification stages are independent. Two different methods for the detection of nuclei are being used: the one approach is based on a feed forward neural network (FNN) which uses a block-based singular value decomposition (SVD) of the image, to signal the likelihood of occurrence of nuclei.

The other approach consists of a combination of a receptive field filter and a squashing function (RFS), adapting to local image statistics to decide on the presence of nuclei at any particular image location. The classification module of the system is based on a radial basis function neural network. A total of 57 images captured from 41 biopsy slides containing over 8300 nuclei were individually and independently marked by two experts. A five scale grading system, known as diagnostic index, was used to classify the nuclei staining intensities. The experts' mutual detection sensitivity (SS) and positive predictive value (PPV) were found to be 79% and 77% respectively. The overall joint performance of the FNN and RFS modules were 55% for SS and 82% for PPV. The classification module correctly classified 76% of all nuclei in an independent validation set containing 25 images. In conclusion, this study shows that the BASS system simulates the detection and grading strategies of human experts and it will enable the formulation of more efficient standardization criteria, which will in turn improve the assessment accuracy of histopathological sections.

1 Introduction

1.1 The Need of Quantitative Analysis in Diagnostic Histopathology

Diagnostic surgical pathology is a branch of pathology in which an expert attempts to diagnose disease by microscopic examination of tissue biopsies [4], [21]. The histopathologic diagnosis is very crucial for patient care, since it determines the most appropriate method of treatment, and also predicts the course of disease.

In the decision making process a histopathologist, evaluates the microscopical features of tissues and cells, in order to diagnose whether a particular lesion is benign or malignant. This task, which has been traditionally subjective, is being supplemented with an increasing range of methods such as morphometry and image analysis [3], [7], [14], [34], [37] that aim to make this information more objective [48]. Currently clinicians demand more quantitative histopathological assessments, that include in addition to cell morphology, information on the biological features of tumour cells [6]. In this respect image analysis methods have the potential of providing more quantitative data that can supplement the histopathological diagnosis. At the same time such systems detect subtle changes in tissues and cells, therefore their use increase the sensitivity of diagnostic procedures [3]. Currently image morphometry is being extensively applied to the automated screening of cytological smears [7], [28] as well as to supplement the diagnosis and prognosis of different types of cancer [6], [14], [21], [34].

The aim of this chapter is to present a Biopsy Analysis Support System, BASS, which has been primarily designed for carrying out semi-quantitative analysis of prognostic factors in breast cancer biopsies. The system is based on a modular neural network architecture incorporating detection, classification and biopsy scoring modules.

1.2 A Brief Overview of the Use of Artificial Neural Network (ANN) Systems in Diagnostic Histopathology

Accurate histopathological diagnosis of more disorders is often difficult, due to the pathophysiological complexity of the underlying processes that cause disease. In addition reproducibility in diagnosis and prognosis is further influenced by human factors. The advantages of ANN that make them attractive to investigate in this field can be summarized as follows: (i) exhibit adaptation or learning, (ii) pursue multiple hypothesis in parallel, (iii) may be fault tolerant, (iv) may process degraded or incomplete data, (v) make no assumptions about the underlying probability density functions, and (vi) seek answers by carrying out transformations [4], [22].

In recent years an increasing number of investigators are applying ANNs in diagnostic pathology, in order to improve the diagnosis and prognosis of several diseases [9], [17], [19], [30], [38], [41]. Indeed within the discipline of anatomic pathology, areas witnessing a growing use of ANNs, include tumour classification [15], cervical smear analysis [28], [29], and disease prognosis [30]. In particular different computer based systems using ANNs, have been applied to breast cancer diagnosis and prognosis [35], [51], or to estimate the probability of relapse [16].

1.3 Quantitative Analysis in Immunocytochemistry

Breast cancer is the most frequent malignancy in the female population of industrialised countries, with one in eight women being affected [20], [27]. Despite major efforts of early detection, the incidence of breast cancer is rising [36], and it remains the leading cause of death from cancer among women [39].

However, there is a wide variability in the survival of women affected with breast cancer which is influenced by several clinicopathological variables, called prognostic factors [10]. More recently with the advent of immunohistochemistry much attention has been focused on biological factors that can predict the behaviour of different types of tumours including breast cancers.

In order to improve the predictive accuracy of the immunohistochemical data, regarding estrogen and progesterone receptors, several investigators have devised diagnostic schemes such as the H-score [33] or the diagnostic index [46]. These are based on the combined evaluation of two variables, namely the staining intensity of individual tumor nuclei and the percentage of cells that are stained at each intensity class. The aim of these manual diagnostic schemes, is to enable a semi-quantitative assessment of the microscopical images, the interpretation of which is only subjective when carried out routinely. Moreover, there is currently a major effort in standardization and quality assurance in histopathology [47]. In this context commercial image analysis systems, such as SAMBA and CAS, have been developed and are being applied for the quantitation of immunocytochemical images [1], [2], [8], [11], [12]. These systems, although dedicated, do not actually simulate the manual procedure employed by human experts which involves the classification of individual tumor nuclei. CAS and SAMBA use a global approach to classify stained nuclei by introducing threshold levels that distinguish between specific staining and background (non-specific staining). In this respect the development and application of computer-aided systems such as BASS [42]-[45], which reproduce and enhance the experts' ability to detect objects of interest, stained nuclei in this case, on an individual rather than a global basis, may enable a more quantitative assessment of immunohistochemical results and therefore improve their predictive accuracy.

The material used in this study is described in the next section. Then the modular neural network system and the experimental setup for system validation are explained, followed by the results section. Finally, the findings are discussed and future work is presented.

2 Material

Cryostat sections from frozen biopsies of 41 breast cancer patients, were cut at 6 μm and placed on poly-L-lysine coated slides. The sections were fixed and immunolabelled using specific antibodies to estrogen and progesterone receptors (ER-ICA/ PgR-ICA kits, Abbott, Germany). Positive nuclear staining, brown color, was visualized using the strept ABC kit linked to peroxidase (DAKO, Denmark). Subsequently, sections were counterstained with hematoxylin to highlight unlabeled nuclei which stained blue.

The immunocytochemically stained slides were microscopically and independently evaluated by two experts using a five point grading diagnostic scheme as described in Appendix A. Differences in grading certain cases were resolved by reevaluation and consensus agreement between the two experts. Following manual scoring, a medical expert selected up to two regions of interest from each specimen for subsequent digitization. In total 57 images from slides stained for either estrogen or progesterone receptors were digitized for subsequent analysis by BASS. The images were captured at x400 magnification (Zeiss Axiophot microscope, SONY DXC-980P camera) in 24 bit color and 640 · 480 pixel spatial resolution.

3 Modular Neural Network System

To assign a diagnostic index to a biopsy specimen, BASS implements a modular approach which resembles the algorithm used by human experts to perform the task. BASS proceeds to grade a biopsy image by first finding the location of nuclei in the image (see Figure 1, Module I, Nuclei Detection [43], [44]) and then classifying them into one of five classes according to staining intensity (see Figure 1, Module II, Nuclei Classification and Biopsy Scoring [42], [44]). Once the nuclei are graded a five class nuclei proportion vector is computed which is used to determine the diagnostic index according to the manual grading scheme as described in Appendix A. BASS also contains an image database retrieval interface (see Figure 1, Module III, Retrieval Interface [45]), which can be used for content-based biopsy image retrieval from a database of cases.

3.1 Detection of Nuclei: the Receptive Field-Squashing Function (RFS) Module

The RFS algorithm uses a combination of a difference of gaussian on-center-off-surround receptive field and a squashing function to detect nuclei in biopsy images [43], [44]. Receptive fields, well known from biological vision [24] and image processing [32] appear to be a particularly good choice as part of a detector mechanism for nuclei. A detector always needs to decide when an event is 'to be detected' or 'not to be detected'. A squashing function, initialized using image statistics as part of the iterative process of the RFS algorithm, acts as a soft threshold. The function divides the image pixels gradually into background pixels and pixels belonging to nuclei by gradually transforming both sets closer to the extreme values.

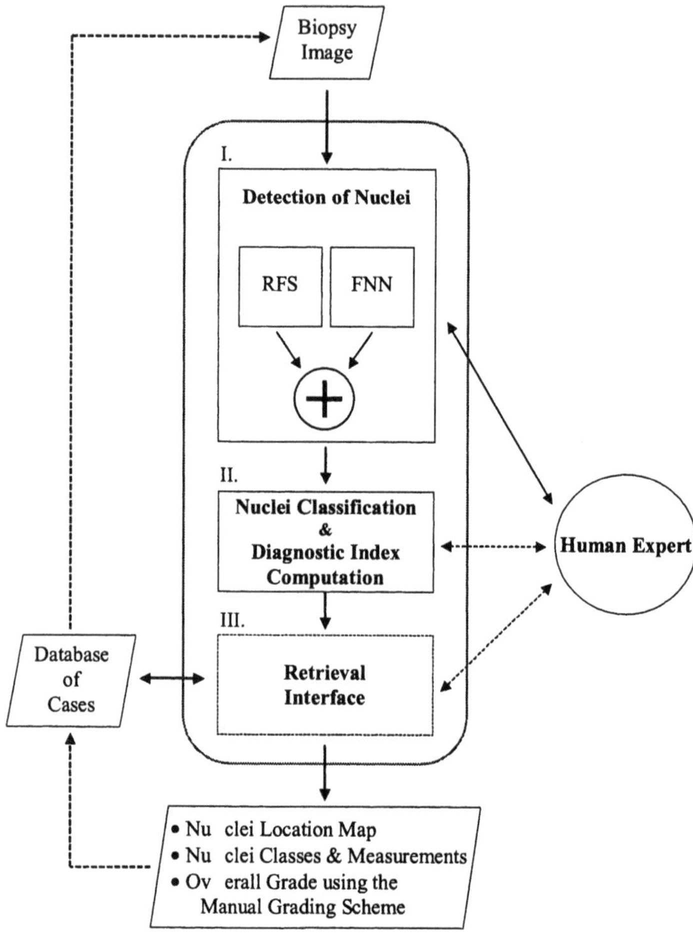

Figure 1. Diagram of the biopsy analysis support system (BASS). (From [44], © 2000 IEEE, with permission).

3.1.1 Step 1: Convert Color Image to Optical Density Image

The original RGB color image is transformed into a scalar array, **I**, using the Y channel (optical density) of the RGB-YIQ transform [26].

$$\mathrm{I}(x,y) = 255 - \begin{bmatrix} 0.299 & 0.587 & 0.114 \end{bmatrix} \cdot \begin{bmatrix} \mathrm{R}(x,y) \\ \mathrm{G}(x,y) \\ \mathrm{B}(x,y) \end{bmatrix} \tag{1}$$

where $0 \leq \mathbf{I}, \mathbf{R}, \mathbf{G}, \mathbf{B} \leq 255$. The Y channel captures the sensitivities of the human visual system with respect to perceived optical density of each color. The scalar array **I** consists of the inverted Y channel to accommodate the detection characteristics of the on-center-off-surround type receptive field.

3.1.2 Step 2: Compute the Receptive Field Filter

The receptive field array is given by the following equation:

$$\mathrm{Rf}(x,y) = B \cdot \left\{ \alpha \cdot \exp\left(-\frac{x^2+y^2}{2\rho_1^2} \right) - J \cdot \exp\left(-\frac{x^2+y^2}{2\rho_2^2} \right) \right\} \tag{2}$$

where B, a, ρ_1, ρ_2, and J are constants. The parameters ρ_1 and ρ_2 determine the sensitivity of the filter regarding the range of object sizes and were set to 2.5 and 7.5. B, a, and J are automatically determined as given in [43].

3.1.3 Step 3: Apply Iteratively the Receptive Field and the Squashing Function

$$\begin{array}{l} \mathrm{WHILE}\ (k < 3)\ \mathrm{DO} \\ \quad \mathrm{ADJUST}\ \mathit{scale},\ \mathit{offset},\ \mathit{incl} \\ \quad \mathbf{I}_{k+1} = \mathrm{Sq}_k(\mathbf{Rf}_k \otimes \mathbf{I}_k) \\ \mathrm{ENDWHILE} \end{array} \tag{3}$$

where $\mathbf{I}_k$ and Sq_k depict the image array and the squashing function at the k-th iteration. At the beginning of each iteration the squashing function parameters *scale*, *offset*, and *incl* are adjusted according to the current image array and the previous parameter values. Then, the image is convolved with the receptive field and then squashed. Within three iterations, the image intensity distribution takes on a bimodal shape, clearly indicating the presence of two pixel classes: (i) background, and (ii) candidate nuclei. Repeated application of the receptive field and the squashing function to the image ensures that object detection is mostly dependent on object geometry and not on object intensity. The squashing function is defined as

$$\mathrm{Sq}(\mathbf{I}_k) = \frac{scale}{1 + \exp\left[-incl \cdot \left(\mathbf{I}_k - offset\right)\right]} \tag{4}$$

where *scale* determines the range of the function, *incl* determines the inclination, and *offset* determines the offset along the abscissa. These parameters are again determined automatically as described in [43].

3.1.4 Step 4: Threshold Bimodal Histogram

The histogram vector of $\mathbf{I}_3$, i.e., the image array after the third iteration, is smoothed with a moving average filter. The threshold value T is set equal to the histogram bin with the minimum amount of pixels between the two modes of the histogram.

$$T = \min_{mode} (\mathrm{hist}(\mathbf{I}) \otimes [a_0, \ldots, a_{0.3 \cdot scale-1}]) \tag{5}$$

where hist(**I**) returns the histogram of **I**, $a_0, \ldots, a_{0.3 \cdot scale-1} = |0.3 \cdot scale|^{-1}$ define the moving average filter coefficients, $\min_{mode}$ returns the minimum between two maxima, which is used to obtain the corresponding threshold intensity, T. The threshold value T is used to segment the sample image into background and candidate nuclei. The candidate nuclei in the image, i.e., all connected sets of pixels, are recorded in an object list.

3.1.5 Step 5: Revise the List of Detected Nuclei

The nuclei center locations are computed by determining the center of gravity of each candidate nucleus and returned.

3.2 Detection of Nuclei: the Feedforward Neural Network (FNN) Module

The algorithm presented in this section detects the locations of nuclei in biopsy images based on a supervised neural network. Block-based processing of the images is adopted, followed by a singular value decomposition, SVD, of each block. The most important singular values are fed as inputs to a neural network classifier which in turn determines the likelihood that the original image block contains a nucleus. The neural network is trained in a supervised mode, thus allowing the knowledge provided by the experts to be included in its design.

The singular values of a matrix have a very important property that of energy compaction [22]. This property makes them suitable as input to the neural network classifier investigated in this study [23]. Each N×N image block is completely described by the corresponding N singular values, while, for example, in the case of the Discrete Cosine Transform (DCT) transform all N^2 coefficients are needed to completely describe the same block. Let $\mathbf{A}_{m\times n}$ be a real rectangular matrix and k denote its rank. Without loss of generality we can assume that $m \geq n$. Then there exist two orthogonal matrices $\mathbf{U}_{m\times m}$, $\mathbf{V}_{n\times n}$ and a diagonal matrix $\Lambda_{m\times n}$ for which the following formula holds:

$$\mathbf{A} = \mathbf{U}\,\Lambda \mathbf{V}^{\mathrm{T}} \tag{6}$$

where $\Lambda = diag(\lambda_1, \lambda_2, \ldots, \lambda_K, 0, 0, \ldots, 0)$, $\lambda_1 > \lambda_2 > \ldots > \lambda_K$, and $(\;)^{\mathrm{T}}$ denotes the transpose of a matrix. Each λ_i^2, $i = 1, 2, \ldots, k$, is an eigenvalue of $\mathbf{AA}^{\mathrm{T}}$, or equivalently $\mathbf{A}^{\mathrm{T}}\mathbf{A}$, and λ_i are the singular values of matrix $\mathbf{A}$. The eigenvectors $\mathbf{u}_i$ of $\mathbf{AA}^{\mathrm{T}}$, related with the eigenvalues λ_i^2, are the columns of matrix $\mathbf{U}$ and the eigenvectors $\mathbf{v}_i$ of $\mathbf{A}^{\mathrm{T}}\mathbf{A}$ are the columns of matrix $\mathbf{V}$.

Given the diagonal matrix and using a column vector $\mathbf{e} = (1, 1, \ldots, 1)^{\mathrm{T}}$ of size n, we can generate the singular value vector (**svv**) of matrix **A** by post-multiplying matrix Λ with vector **e**:

$$\mathbf{svv} = \Lambda \cdot \mathbf{e}. \tag{7}$$

Under the constraint $\lambda_1 > \lambda_2 > ... > \lambda_K$, matrix Λ and vector **svv** are unique for a given matrix **A** [23].

Apart from the above, it should be mentioned that singular values are insensitive to small changes in matrix **A**. Assuming that matrix **A** denotes an image block, **svv** and consequently the neural network detector are insensitive to small changes of pixel values caused by noise or different illumination conditions. Moreover, singular values remain the same even if the image block is rotated, translated, or transposed. The above properties are highly desirable in the detection task, where the position, and not the orientation, of the nuclei are required. The proposed scheme is composed of the following steps:

3.2.1 Step 1: Color Image to Optical Density Image Conversion

This step is identical to that of the RFS module, except that the optical density image is not inverted (see Equation 1).

3.2.2 Step 2: Histogram Stretching and Thresholding

The histogram stretching and thresholding step aims at smoothing the noisy background and increasing, i.e., normalizing, the contrast between the nuclei and the background of the optical density image. In addition, contrast varies between the images, potentially affecting the efficiency of the neural network detector. Thus, contrast variations should be minimized. If **X** is an optical density image whose histogram values are limited in the interval $[a, b]$, where $a \geq 0$, $b \leq c$, then histogram stretching generates an image **Y** whose histogram lies in the extended interval $[0, c]$. Image **Y** is derived by the following transformation:

$$\mathbf{Y}(i,j) = c \cdot \frac{\mathbf{X}(i,j) - a}{b - a}. \quad (8)$$

After histogram stretching, the noisy background is smoothed using an appropriate threshold, T_s, which is a function of image Y. The thresholded image **Z** is given by:

$$\mathbf{Z}(i,j) = \begin{cases} \mathbf{Y}(i,j), & \mathbf{Y}(i,j) < T_s \\ c, & \mathbf{Y}(i,j) \geq T_s \end{cases}. \quad (9)$$

3.2.3 Step 3: SV Expansion and Feedforward Neural Network Identification of Image Blocks

This step is based on a block-oriented architecture. In particular, the preprocessed images are raster scanned both horizontally and vertically using a scanning step of k pixels and a block size of $N \times N$ pixels. Thus, the image is separated into overlapping blocks $\mathbf{B}(i, j)$ of $N \times N$ pixels. These blocks are subsequently transformed using SVD, producing Singular Value (SV) feature vectors. The SV feature vectors are then subjected to dimensionality reduction, to further decrease the complexity of the algorithm and to drop those singular values, which do not significantly contribute to the separability of the dataset. The truncated feature vectors are then fed into the neural network. Consequently, identification is performed in the singular value domain. The analog neural network output values for all blocks form an output intensity image $\mathbf{O}$, the size of which is $(k \times k)$ times less than that of the original image, where k is the scanning step.

3.2.4 Step 4: Calculation of the Exact Nuclei Locations

Ideally, the output of the neural detector would include isolated points indicating the nuclei positions. However, this is not the case, due to the different nuclear sizes and to overlapping nuclei. The output image, in general, contains instead of isolated points clusters of points (i.e., blocks) within, and possibly around, the area of the nuclei. In order to achieve the final nuclei identification of thc blocks, a global threshold T is implemented. Then, the exact positions of nuclei are computed as the local maxima of the output image pixel values as documented in [44].

3.3 Combination of Detection Modules

Since the RFS and FNN modules work on different principles our aim was to test their individual, as well as evaluate their combined performance. In this study, the detection results of the RFS and FNN module were combined based on modified logical OR and AND operators. ORDt (RFS .OR. FNN) and ANDDt (RFS .AND. FNN) modules were evaluated along side the individual RFS and FNN modules. Detected nuclei locations were considered to coincide if the centers were located within a distance of 9 pixels from each other as detected by either of the two modules.

The following rules were adopted for either ORDt or ANDDt: (i) ANDDt's detection results include only those nuclei which coincide regarding the location and have matching class labels, while (ii) ORDt's detection results include all nuclei from both modules, substituting the locations detected by the RFS module and the corresponding labels, in case the class labels do not agree.

3.4 Nuclei Classification and Diagnostic Index Calculation

Following the detection of nuclei, a radial basis function (RBF) neural network [42], [44] is applied to classify each nucleus into one of five staining intensity classes. After the classification of each nucleus, the diagnostic index is computed as illustrated in Figure 4. The RBF neural network classifier is trained based on the nuclei feature vectors from four selected images (with more than 1000 nuclei feature vectors) marked by Expert 2. (Expert 2 turned out to be more consistent in marking cell nuclei, see analysis in Section 4.)

The RBF network structure consisted of a single RBF unit layer which was fully connected to a layer with linear output units (see Figure 2). The four dimensional input vectors (see below) were simultaneously fed to each RBF unit. The output layer contained five linear units which encoded the five possible nuclei staining intensity classes with binary values. The training procedure for RBF networks [13] was carried out using an incremental solver [18], which dynamically adds one RBF neuron per training epoch to the RBF layer and adjusts the weights between the RBF layer and the linear output layer, until either a maximal number of neurons has been added or the sum-squared error falls beneath an error goal. The transfer function (TF) of each RBF neuron has the following form:

$$TF(\mathbf{p}) = RBF\left(\mathrm{dist}(\mathbf{w},\mathbf{p}) * 0.8326/SP\right) \tag{10}$$

where **p** is the input feature vector, **w** is the weight vector, dist is the Euclidean distance measure, SP is the spread constant, and RBF is a Gaussian function. The transfer function TF (Equation 10) takes on its maximal value of unity when its argument becomes zero. TF will return

0.5 when its argument has the value 0.8326. Thus, TF will be unity, if the distance between the vectors **p** and **w** is zero. If, for example, the spread constant SP equals to 0.1, TF would return 0.5 for every vector at a distance of 8.326 from **w**. The main steps of the classification algorithm are summarized as follows in Sections 3.4.1 to 3.4.3.

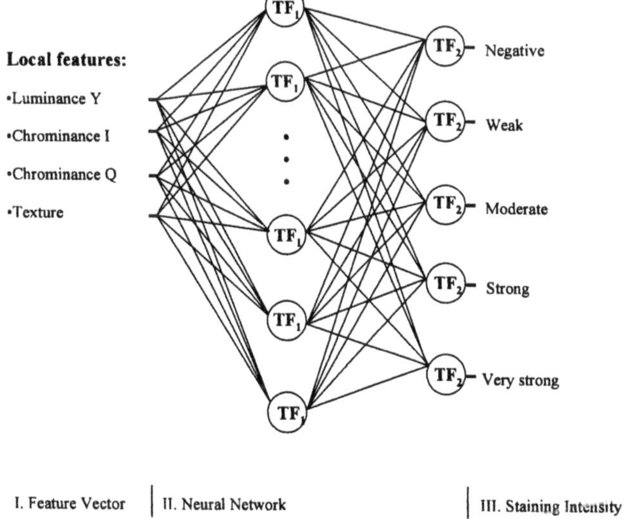

Figure 2. Structure of neural network classifiers: the FF multi-layer networks have sigmoidal transfer functions (TF1 and TF2), while RBF networks consist of Gaussian transfer functions (TF1) in the first layer and linear transfer functions (TF2) in the second layer.

3.4.1 Step 1: Extract Features for Each Nucleus

The following features are extracted: the average of the Y, I, and Q channel as well as a texture measure *Txt* are computed for each nucleus. The texture measure *Txt* is defined as follows:

$$Txt = 1 - \frac{1}{1 + \sigma^2_{\left(Y_{probe}\right)}} \tag{11}$$

where σ^2 is the variance, and Y_{probe} denotes the optical density feature of a nucleus.

3.4.2 Step 2: Classify Each Nucleus

A nucleus is classified into one of five staining intensity classes (negative, weak, moderate, strong, very strong) using the radial basis function neural network classifier.

3.4.3 Step 3: Compute Diagnostic Index

The diagnostic index is computed according to the manual semi-quantitative scheme which is described in Appendix A. The assessment results of the preceding steps are finally stored in the database of cases [45]. This interface may also be used to retrieve biopsy images based on the contents as described by the assessment results. For this purpose special matching operators were implemented which enable the retrieval of images based on similarity or on presence of one or more characteristics.

3.5 System Validation

Ultimately, any system or expert can be said to perform successfully if the nuclei proportions in an image are estimated accurately and reliably, since the diagnostic index is based on this estimate. However, at present it is difficult to evaluate either the performance of the experts or the systems due to the unavailability of universal gold standards at the nuclei detection and nuclei classification, i.e., diagnostic index, levels. Firstly, the system was evaluated at the nuclei detection level by comparing its performance to that of two human experts using as the basis 200 to 300 nuclei per case, which were marked by the experts. Secondly, the system was evaluated at the nuclei classification level by comparing the nuclei classified by BASS to the nuclei classified by the experts. Finally, the diagnostic index, calculated based on BASS' nuclei classification results, was compared to the diagnostic index of each case obtained routinely by human experts.

To create a basis for comparison at the nuclei level, a small circular probe of 9 pixels diameter was placed centrally on top of each nucleus detected by the system modules. Independently, two experts placed the same sized probes manually where they perceived the nuclei centers were located using the mouse. In addition, the probes were color coded depending on the staining intensity class computed by BASS' classification algorithm described in Section 3.4 or assigned by the expert. As a result of this procedure, for each module and each expert, i.e., the combinations, one mask image was created. The diagnostic index for all mask images was automatically computed according to Table 3, using the above described color coded mask images.

A detection event was defined as the set of pixels belonging to one of the probes in the image. If two probes from different mask images overlapped, then the two corresponding detection events were said to coincide and the corresponding nucleus was interpreted to have been detected in both mask images. In the case of one probe touching several other probes, only one coinciding detection event was counted. Two hybrids, called OREx (Ex1 .OR. Ex2) and ANDEx (Ex1 .AND. Ex2), were derived with a modified logical OR and AND operation from the individual experts' masks, as implemented in the ORDt and ANDDt modules (see Section 3.3).

System validation was performed using four methods. In particular, the detection performance and the joint detection and classification performance of the BASS system were assessed with the following methods:

1. Receiver-operator characteristic measures (ROC) were used to analyze individual nuclei detection performance of BASS' detection modules compared to that of two experts. ROC measures are useful to compare the detection performance with respect to individual nuclei, because no assumption about the underlying probability distribution of the detection events is made. Based on the definition of the detection events, two measures, sensitivity (SS) and positive predictive value (PPV), were chosen to characterize the detection performance. Sensitivity is the likelihood that a nucleus will be detected in case it is also marked as a nucleus in the gold standard. It is defined as follows:

$$SS = TP / (TP + FN) \tag{12}$$

   where TP (true positive) are those nuclei marked in both the gold standard and the image, and FN (false negative) are those nuclei which are marked in the gold standard, but not in the image. Positive predictive value (PPV) is the likelihood that the detection of a nucleus is actually associated with a nucleus marked in the gold standard.

$$PPV = TP / (TP + FP) \tag{13}$$

where FP (false positive) are those nuclei which are marked in the image, but not in the gold standard.

2. Confusion matrices were used to evaluate the experts' and BASS' performance in the assignment of diagnostic indices as compared to the diagnostic index assigned routinely (see Appendix A). The average percentage of correctly graded images per diagnostic index class (0, 1+, 2+, 3+, and 4+) CGI (i.e., the average of the diagonal entries in the confusion matrix) summarizes an important performance characteristic and is reported. This part of the study was based on 29 images which were selected by the experts on the basis of clarity of color perception of the digitized images and minimal overlap between nuclear boundaries.

3. Spearman's rank-order correlation coefficient [40] was determined to assess the joint performance of BASS' detection and classification modules compared to that of Expert 1, Ex1, and Expert 2, Ex2, based on the staining intensity class proportions of all biopsy images.

4 Results

A total of 57 images from 41 patients were captured and analyzed using BASS. The actual distribution of images for each diagnostic index category was 17 %, 17 %, 17 %, 35 %, and 14 % starting with the '0' and ending with the '4+' diagnostic index categories respectively.

4.1 Detection Example, Figure 3

Figure 3 shows a subregion of a breast cancer biopsy image marked by all experts and modules and their combinations. Figures 3a and 3b indicate the nuclei marked by Ex1 and Ex2, whereas Figures 3c and 3d present the detection results from OREx and ANDEx. Figures 3e-h show the detection performance for the RFS module, the FNN module, the ORDt module, and the ANDDt module.

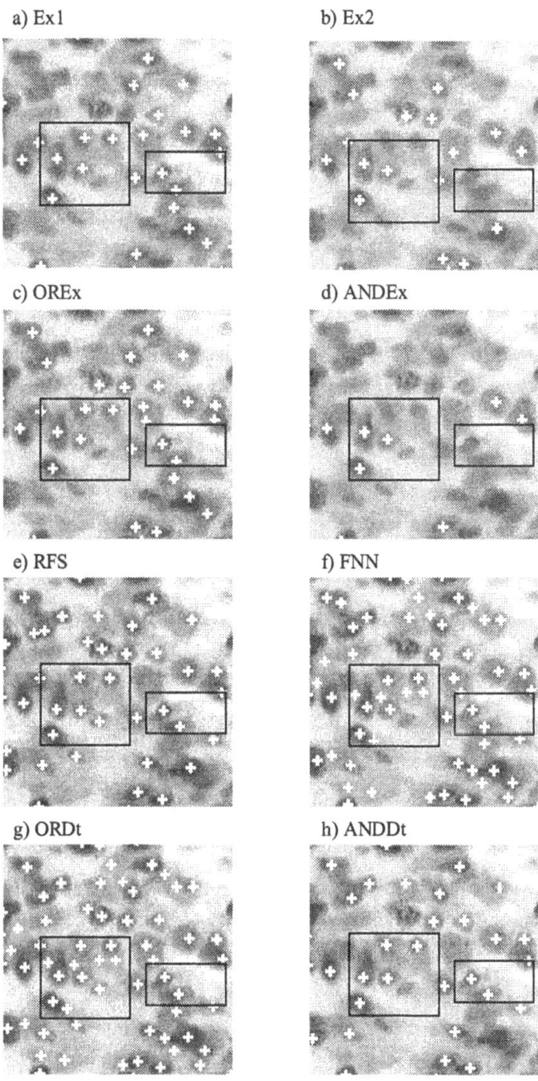

Figure 3. Example for nuclei detection: (a) Ex1, (b) Ex2, (c) OREx, (d) ANDEx, (e) RFS module, (f) FNN module, (g) ORDt, (h) ANDDt. The large square highlights the differences between experts and modules. The small rectangle points to an example of how the OR and the AND operator affect the generation of combined results. (From [44] © 2000 IEEE, with permission).

The large square inserted in each figure, highlights a region of the image which displays the performance of the experts and the modules.

The modules generally detect more nuclei in the biopsy images than both experts. The small rectangle illustrates the performance of the OR and AND operators. In particular, the individual experts and the modules have marked the nucleus towards the left side of the rectangle, but each expert assigned a different nuclear staining intensity class (difference is only visualized in color). In addition, Ex1 detected two more nuclei which Ex2 did not mark. The result is that the OR operator includes one of the locations and the corresponding nuclear class for the nucleus detected by both experts, and the two nuclei not detected by Ex1. In contrast, the AND operator ignores all the nuclei in the smaller rectangle, since there is disagreement regarding nuclei locations as well as their staining intensity classes.

4.2 ROC Analysis, Tables 1 and 2

Table 1 tabulates the detection performance regarding the 57 images in the database. In this table the second component is taken to be the gold standard. It is noted that if component 1 were to be chosen as the gold standard, the numbers would only have to be interchanged due to the definition of SS and PPV. The table consists of five sections ordered by the applied laboratory gold standard. Table 1, section I shows that Ex1 has a SS of 79.2 % and a PPV of 76.5 % with Ex2 as the gold standard. When Ex1 and Ex2 are compared to OREx, their sensitivity value increases to 83.8 % and 81.1 % respectively. For ANDEx the PPV value of Ex1 and Ex2 are just 52.8 and 54.7 % with a high standard deviation of over 20 %. The SS and the PPV values for the individual experts when compared to ANDEx and OREx is by definition 100 %.

Table 1, sections II to V show the detection results of the individual and combined modules. The detection results for the individual modules are presented first, followed by ORDt and ANDDt. The SS and PPV values of the RFS and FNN modules (table sections II and III) when compared to either Ex1 or Ex2 were in general quite similar, varying from 76.7 to 81.4 % and 54.9 to 58.5 % respectively. It is noted that the PPV values of the individual RFS and FNN modules (table sections II and III) remain about 20 % below the 78 % (table section I, Ex1 - Ex2, (79.2+76.5)/2) average level of the human experts. The SS values of the individual modules versus OREx (table section IV) drop only slightly.

The PPV value, however, increases by 8.6 % and 10.8 % for the RFS module and 9.6 % and 8.9 % for the FNN module.

Table 1. Comparison of detection performance for 57 images using sensitivity (SS) and positive predicitive value (PPV). The second component is always considered to be the gold standard. (From [44] © 2000 IEEE, with permission).

	Detector vs. Gold Standard	SS	S.D.	PPV	S.D.
I. Expert	Ex1 - Ex2	79.2	± 13.7	76.5	± 15.3
	Ex 1 - OREx	83.8	± 13.2	100.0	± 0.0
	Ex 2 - OREx	81.1	± 14.5	100.0	± 0.0
	Ex 1 - ANDEx	100.0	± 0.0	52.8	± 20.6
	Ex 2 - ANDEx	100.0	± 0.0	54.7	± 21.7
II. Expert 1	RFS - Ex1	81.4	± 16.3	58.5	± 14.5
	FNN - Ex1	76.7	± 12.6	54.9	± 15.2
	ORDt - Ex1	92.3	± 5.1	48.3	± 11.5
	ANDDt - Ex1	57.3	± 15.2	71.2	± 16.5
III. Expert 2	RFS - Ex2	81.1	± 15.2	56.3	± 17.2
	FNN - Ex2	79.6	± 8.9	55.8	± 17.5
	ORDt - Ex2	92.6	± 5.2	47.8	± 14.8
	ANDDt - Ex2	59.9	± 14.2	72.2	± 16.9
IV. OREx	RFS - OREx	79.6	± 16.1	67.1	± 14.1
	FNN - OREx	75.4	± 11.9	64.7	± 16.8
	ORDt - OREx	91.6	± 5.1	57.6	± 12.9
	ANDDt - OREx	55.4	± 14.4	82.4	± 14.6
V. ANDEx	RFS - ANDEx	84.6	± 14.9	32.0	± 14.9
	FNN- ANDEx	81.3	± 10.8	30.1	± 12.5
	ORDt - ANDEx	94.5	± 5.1	25.7	± 11.4
	ANDDt - ANDEx	62.7	± 15.4	40.7	± 17.0

Combining the RFS and FNN modules via the logical OR operation achieves SS values of over 91.6 %, but the PPV values decrease by a maximum of 8 % below the lowest value of any individual module for all combinations with the experts (ORDt entries in table sections II to V). ANDEx scores over 30 % lower SS values than OREx. However, the PPV values are at least 12.7 % higher than the best performances by any single module or ORDt, reaching 82.4 %, which approaches favorably the performance of the experts.

The modules and their combination perform poorly as regards the PPV value when compared to ANDEx (table section V). ANDDt performs best regarding the PPV, while ORDt perform best regarding the SS. Since the number of nuclei marked and classified in agreement is lower than the numbers from individual experts, the PPV value for both the modules and the individual experts is very low. On average the experts agree only in about 53 % of the cases (table section I, Ex1 - ANDEx, Ex2 -ANDEx, (52.8 + 54.7)/2) on the location, while ANDDt agrees only with a maximum of 40.7 %. It is noted that the definition of the AND operator requires detected nuclei not only to spatially coincide, but also to be labeled identically (Section 3.3).

Table 2 summarizes the results of the second run validation, which was performed by one of the experts on a subset of 29 images (20 patients). The data in Table 2 shows that the PPV increases for the RFS, the FNN, the ORDt, and the ANDDt modules by 8.2, 6.1, 8.4, and 4.7 % respectively. ANDDt achieves the best PPV value of 83.6 %, while ORDt's SS score was highest at 92.2 %.

Table 2. Results of the second run validation procedure: Modules are compared to the orex on a subset of 29 images (20 patients). (From [44] © 2000 IEEE, with permission).

Detector vs. Gold Standard	w/out second run validation				with second run validation			
	SS	**S.D.**	**PPV**	**S.D.**	**SS**	**S.D.**	**PPV**	**S.D.**
RFS - OREx	76.4	± 16.9	66.8	± 14.0	78.1	± 16.3	75.0	± 16.9
FNN - OREx	79.3	± 8.4	56.1	± 13.5	80.7	± 8.3	62.2	± 15.4
ORDt - OREx	91.2	± 4.8	52.7	± 11.7	92.2	± 4.4	61.1	± 15.1
ANDDt - OREx	59.8	± 15.0	78.9	± 14.0	61.1	± 15.0	83.6	± 15.0

4.3 Classification and Diagnostic Index Computation Module, Figure 1 (Module II)

The best radial basis function classifier obtained, consisted of 46 basis function units with a fixed spread factor of 20. In total 85.22 %,

79.57 %, and 76.36 % of the nuclei feature vectors from the training set, the test set, and the validation set were classified correctly. These numbers correspond to an average per nucleus staining intensity class classification accuracy of 82.06 %, 80.46 %, and 65.16 % for the training, test, and validation sets. The total number of nuclei in the training, test and validation sets were 805, 328, 5077, respectively. The distribution of nuclei classes in the combined dataset was 31.37 %, 33.99 %, 23.24 %, 9.4 %, 1.99 % beginning with the 'negative' nuclei class and ending with the 'very strong' nuclei class respectively.

Furthermore, the actual distribution of diagnostic indices was 17 %, 17 %, 17 %, 35 %, 14 % starting with the '0' and ending with the '4+' diagnostic index entries respectively. The confusion matrix analysis emphasizes the per class accuracy rather than the accuracy with respect to the overall dataset. It was shown that Ex1 and Ex2 differ by 9 % in their (average percentage of correctly graded images per diagnostic index class) CGI values, achieving 68 % and 77 % respectively. ANDEx achieved 69 % and OREx scored 84 % CGI. Incorrectly graded images were mainly overscored by both experts, while ANDEx overscored 0, 1+, and 2+ images and underscored 3+ and 4+ images. OREx only overscored in the case of 0, 1+, and 3+ images. The RFS and FNN module scored 69 % and 61 % CGI respectively. There is agreement in their performance for 3+ and 4+ images, while the RFS module performed better than the FNN module for 0, 1+, and 2+ images. The performance of ORDt lies 4 % below the FNN module and 12 % below the RFS module. However, ANDDt performed the same as the RFS module, despite differences in the absolute numbers of detected nuclei per image [44].

Spearman's rank order correlation analysis was carried out for all 57 images. This analysis depends on the detection and classification performance of each expert or module. The experts achieved correlation values of the order of 0.7 to 0.9 depending on the nuclei staining intensity, whereas BASS nuclei detection modules correlated well (> 0.75) among all nuclei intensities [44].

5 Discussion

Histopathological sections of breast cancer nuclei immunocytochemically stained for steroid receptors are routinely reported by experts based on the microscopical evaluation of numbers of nuclei stained at particular intensities of brown color. This study shows that detection and classification of individual nuclei in histopathological sections can be reliably performed by the BASS modular neural network system in an accurate and consistent manner. BASS also facilitates interaction with experts and to this effect the second run validation results indicate that this interaction is constructive, since it was demonstrated that the modules correctly detect considerable numbers of nuclei which were not initially detected by the experts. Moreover, since the system simulates detection and grading strategies of human experts it will enable the formulation of more efficient standardization criteria in the assessment of immunocytochemically stained histopathological sections.

The ANDed RFS-FNN module, ANDDt, leads to the best overall results in terms of detection accuracy for the diagnostic indices. It achieved the highest PPV as compared to OREx after the second run validation (83.6 %) and the highest average accuracy for correctly assigning diagnostic indices to the images (69 %). However, the SS is lower than for any other combination of modules (61.1 %). It should be noted that although the RFS module matches the overall performance of ANDDt for the diagnostic indices, its values for SS and PPV were 78.1 % and 75.0 % respectively.

The present data show that a high PPV value is critical for obtaining a good performance with respect to the diagnostic index, as can be seen when comparing the experts and BASS' combined detection and classification modules. On the other hand our data show that the SS value does appear to be a less important factor and not directly related to BASS' performance in computing diagnostic indices. The experts showed a tendency to overscore as is demonstrated by the diagnostic index confusion matrices, while the combinations of detection and classification modules both overscore and underscore. This tendency of the experts to overscore may be explained by the observation that

Spearman rank correlation values were higher for moderate to very strong nuclei. However, the Spearman rank correlation values for the BASS system (RFS, FNN, ORDt, ANDDt, combined with the classification module) lie above 0.87 (except for the 0.76 correlation value for the weak nuclei regarding RFS-FNN), which entails that the modules and/ or their combinations perform consistently and uniformly. In addition to higher accuracy and greater objectivity image analysis systems should also possess greater speed than that required by human experts to perform similar tasks. BASS is able to perform the analysis of one image on average in less than 20 seconds (500 MHz Intel Pentium II PC, 64 Mbyte RAM). This time span compares favorably with the time needed by human experts to perform similar tasks.

In an attempt to improve objectivity and offer rapid analysis speeds some commercial systems, like CAS and SAMBA, rely on global discrimination of structures of interest, between nuclei in this case, and background. These systems measure percent stained surface area using global thresholding techniques. However, there is disagreement among experts about the optimal selection of global thresholds, the choice being fixed, manual, and automatically set thresholds [5]. BASS avoids the need for global thresholding and area measurements, since it detects, counts, and classifies individual nuclei according to the manual semi-quantitative diagnostic index.

BASS was designed to simulate closely the detection and grading strategies as practiced by histopathologists so that experts may be used to supervise and evaluate the system at the nuclei detection, the nuclei classification, and the diagnostic index levels. It was shown that the combination of the detection modules RFS and FNN performed better than the individual modules. The 'nuclei classification and diagnostic index computation' module performance on the whole dataset (see also [44]) compares favorably to the performance of a neural network classifier utilizing 17 mostly textural features [31]. Neural networks are but one technique to classify image feature vectors. Bibbo *et al.* [5], for example, included a variety of diagnostic clues and detailed prior knowledge in a bayesian belief network to grade prostate lesions, while Mangasarian *et al.* [31] showed that linear programming methods may successfully be applied for breast cancer diagnosis and prognosis based on computer-aided image analysis and other clinical data.

It is difficult to assess the true system performance and also compare it to other systems in the absence of reliable and universal gold standards [49]. All experiments performed here had to be based on laboratory gold standards, i.e., either on the nuclei marking results from the experts or the diagnostic index which was manually derived to ensure consistent classification of the images. Since the confusion matrices also serve as a measure of objectivity and consistency of individual experts, Expert 2 was chosen as the source of supervisory information at the beginning of this study. However, the present data demonstrate that both modules RFS and FNN perform consistently and accurately despite the fact that they were developed using different methodologies. In addition, the BASS system can achieve at least similar results as the human experts. Furthermore, BASS facilitates interaction with experts and this combination, as shown in this study, increases the potential for improving accuracy and objectivity.

6 Future Work

Most of the principal aims of computer aided image analysis include higher accuracy, increased objectivity and greater speed. This study demonstrates that the BASS modular neural network system possesses a good potential for detecting and classifying nuclei accurately and consistently. Subsequently, in addition to expanding the database to increase the accuracy of assigning a diagnostic index, BASS performance will be evaluated in a clinical setting whereby its predicitive and prognostic accuracy will be compared to the clinical status of breast cancer patients. Moreover, grading results based on human experts and BASS regarding the diagnostic index may be combined into a hybrid system in an effort to further improve performance.

References

[1] Alcatel TITN Answare (1993), *IMMUNO 4.00: User's Guide*, 1st ed., Grenoble, France.

[2] Bacus, S. and Flowers, J.L. (1988), "The evaluation of estrogen receptor in primary breast carcinoma by computer-assisted image analysis," *Am. J. of Clinical Pathology*, vol. 90, pp. 233-239.

[3] Bartels, P.H. (1992), "Computer generated diagnosis and image analysis, an overview," *Cancer*, vol. 69, pp. 1636-1638.

[4] Becker, R.L. and Usaf, M.C. (1995), "Applications of neural networks in histopathology," *Pathologica*, vol. 87, no. 3, pp. 246-254.

[5] Bibbo, M., Bartels, P.H., Pfeifer, T., Thompson, D., Minimo, C., and Galera Davidson, H. (1993), "Belief network for grading prostate lesions," *Anal. Quant. Cytol. Histol.*, vol. 15, pp. 124-135.

[6] Biesterfeld, S., Kluppel, D., Koch, R., Schneider, S., Steinhagen, G., Mihalcea, A.M., and Schroder, W. (1998), "Rapid and prognostically valid quantification of immunohistochemical reactions by immunohistometry of the most positive tumour focus," *Journal of Pathology*, vol. 185, no. 1, pp. 25-31.

[7] Birdsong, G.G. (1996), "Automated screening of cervical cytology specimens," *Human Pathology*, vol. 27, pp. 468-481.

[8] Brugal, G. (1985), "Color processing in automated image analysis for cytology," in Mary, J.Y. and Rigaut, J.P. (Eds.), *Quant. Image Analysis in Cancer Cytology and Histology*, Amsterdam: Elsevier, pp. 19-33.

[9] Burke, H.B. (1994), "Artificial neural networks for cancer research. Outcome prediction," *Sem. Surgical Oncology*, vol. 10, pp. 73-79.

[10] Carter, C.L., Allen, C., and Henson, D.E. (1989), "Relation of tumour size, lymph mode status and survival in 24,740 breast cancer cases," *Cancer*, vol. 63, pp. 181-187.

[11] Cell Analysis Systems Inc. (1990), *Cell Analysis Systems: Quantitative Estrogen Progesterone Users Manual*, Application Version 2.0, Catalog Number 201325-00, USA.

[12] Charpin, C., Martin, P.M., DeVictor, B., Lavaut, M.N., Habib, M.C., Andrac, L., and Toga, M. (1988), "Multiparametric study (SAMBA 200) of estrogen receptor immunocytochemical assay in 400 human breast carcinomas," *Cancer Research*, vol. 48, pp. 1578-1586.

[13] Chen, S., Cowan, C.F.N., and Grant, P.M. (1991), "Orthogonal least squares learning algorithm for radial basis function networks," *IEEE Trans. Neural Networks*, vol. 2, no. 2, pp. 302-309.

[14] Cohen, C. (1996), "Image cytometric analysis in pathology," *Human Pathology*, vol. 27, no. 5, pp. 482-493.

[15] Dawson, A.E., Austin Jr., R.E., and Weinberg, D.S. (1991), "Nuclear grading of breast carcinoma by image analysis," *American Journal of Clinical Pathology*, vol. 95 (Suppl. 1), pp. S29-S37.

[16] De Laurentiis, M., De Placido, S., Bianco, AR., Clark, G.M., and Ravdin, P.M. (1999), "A prognostic model that makes quantitative estimates of probability of relapse for breast cancer patients," *Clinical Cancer Research*, vol. 5, no. 12, pp. 4133-4139.

[17] Deligdisch, L., Einstein, A.J., Guera, D., and Gil, J. (1995), "Ovarian dysplasia in epithelial inclusion cysts. A morphometric approach using neural networks," *Cancer*, vol. 76, no. 6, pp. 1027-1034.

[18] Demuth, H. and Beale, M. (1994), *Neural Network Toolbox*, The MathWorks, Inc., Natick, Mass., USA.

[19] Furness, P.N., Levesley, J., Luo, Z., Taub, N., Kazi, J.I., Bates, W.D., and Nicholson, M.L. (1999), "A neural network approach to the biospy diagnosis of early acute renal transplant rejection," *Histopathology*, vol. 35, pp. 461-467.

[20] Garfinkel, L., Boring, C.C., and Heath, C.W. Jr. (1994), "Changing trends. An overview of breast cancer incidence and mortality," *Cancer*, vol. 74, pp. 222-227.

[21] Goldschmidt, D., Decaestecker, C., Berthe, J.V., Gordower, L., Remmelink, M., Danguy, A., Pasteels, J.L., Salmon, I., and Kiss, R. (1996), "The contribution of image cytometry and artificial intelligence-related methods of numerical data analysis for adipose tumor histopathologic classification," *Laboratory Investigation*, vol. 75, no. 3, pp. 295-306.

[22] Haykin, S. (1994), *Neural Networks: a Comprehensive Foundation*, New York, USA: Macmillan, 1994.

[23] Hong, Z.-Q. (1991), "Algebraic feature extraction of image for recognition," *Pattern Recognition*, vol. 24, no. 3, pp. 211-219.

[24] Hubel, D.H. and Wiesel, T.N. (1962), "Receptive fields, binocular interaction and functional architecture in the cat's visual cortex," *J. Physiol., Lond.*, vol. 160, pp. 106-154.

[25] Jagoe, R., Steele, J.H., Vucicevic, V., Alexander, N., van Noorden, S., Wooton, R., and Polak, J.M. (1991), "Observer variation in quantification of immunocytochemistry by image analysis," *Histochemical Journal*, vol. 23, pp. 541-547.

[26] Jain, A.K. (1989), *Fundamentals of Digital Image Processing,* Englewood Cliffs, New Jersey, USA: Prentice Hall, 1989.

[27] Kelsey, J.L. and Horn-Ross, P.L. (1993), "Breast cancer: magnitude of the problem and descriptive epidemiology," *Epidemiological Reviews*, vol. 15, no. 1, pp. 7-16.

[28] Kok, M.R. and Boon, M.E. (1996), "Consequences of neural network technology for cervical screening," *Cancer*, vol. 78, pp. 112-117.

[29] Koss, L.G. (2000), "The Application of PAPNET to Diagnostic Cytology," in Lisboa, P.J.G., Ifeachor, C., and Szczepaniak, P.S. (Eds.), *Artificial Neural Networks in Biomedicine*, Springer-Verlag, London, pp. 51-67.

[30] Lundin, M., Lundin, J., Burke, H.B., Toikkanen, S., Pylkkanen, L., and Joensuu, H. (1999), "Artificial neural networks applied to survival prediction in breast cancer," *Oncology*, vol. 57, pp. 281-286.

[31] Mangasarian, O.L., Street, W.N., and Wolberg, W.H. (1995), "Breast cancer diagnosis and prognosis via linear programming," *Operations Research*, vol. 43, no. 4, pp. 570-577.

[32] Marr, D. and Hildreth, E. (1980), "Theory of edge detection," *Proc. R. Soc. Lond.*, vol. B 207, pp. 187-217.

[33] McCarty Jr., K.S., Miller, L.S., Cox, E.B., Konrath, J., and McCarty Sr., K.S. (1985), "Estrogen receptor analyses. Correlation of biochemical and immunohistochemical methods using monoclonal antireceptor antibodies," *Arch. Pathol. Lab. Med.*, vol. 109, pp. 716-721.

[34] Millot, C. and Dufer, J. (2000), "Clinical applications of image cytometry to human tumour analysis," *Histology Histopathology*, vol. 15, no. 4, pp. 1185-200.

[35] Naguib, R.N., Sakim, H.A., Lakshmi, M.S., Wadehra, V., Lennard, T.W., Bhatavdekar, J., and Sherbet, G.V. (1999), "DNA ploidy and cell cycle distribution of breast cancer aspirate cells measured by image cytometry and analyzed by artificial neural networks for their prognostic significance," *IEEE Trans Information Technology Biomedicine*, vol. 3, no. 1, pp. 61-69.

[36] Newcomb, P.A. and Lantz, P.M. (1993), "Recent trends in breast cancer incidence, mortality, and mammography," *Breast Cancer Research and Treatment*, vol. 28, pp. 97-106.

[37] O'Brien, M.J. and Sotnikov, A.V. (1996), "Digital imaging in anatomic pathology," *American Journal of Clinical Pathology,* vol. 106, no. 4, suppl. 1, pp. S25-S32.

[38] Pantazopoulos, D., Karakitsos, P., Iokim-Liossi, A., Pouliakis, A., Botsoli-Stergiou, E., and Dimopoulos, C. (1998), "Back propagation neural network in the discrimination of benign from malignant lower urinary tract lesions," *Journal of Urology*, vol. 159, no. 5, pp. 1619-1623.

[39] Pisani, P., Parkin, D.M., Bray, F., and Ferlay, J. (1999), "Estimates of the world mortality from 25 cancers in 1990," *International Journal of Cancer*, vol. 83, pp. 18-29.

[40] Press, W.H., Flattery, B.P., Teukovsky, S.A., and Vetterling, W.T. (1988), *Numerical Recipes in C*, Cambridge, UK: University Press.

[41] Ravdin, P.M. and Clark, G.M. (1992), "A practical application of neural network analysis for predicting outcome of individual breast cancer patients," *Breast Cancer Research and Treatment*, vol. 22, pp. 285-293.

[42] Schnorrenberg, F., Pattichis, C.S., Kyriacou, K., Vassiliou, M., and Schizas, C.N. (1996), "Computer-aided classification of breast cancer nuclei," *Technology and Health Care*, vol. 4, no. 2, pp. 147-161.

[43] Schnorrenberg, F., Pattichis, C.S., Kyriacou, K., and Schizas, C.N. (1997), "Computer-aided detection of breast cancer nuclei," *IEEE Trans. Information Technology in Biomedicine,* vol. 1, no. 2, pp. 128-140.

[44] Schnorrenberg, F., Tsapatsoulis, N., Pattichis, C.S., Schizas, C.N., Kollias, S., Vassiliou, M., Adamou, A., and Kyriacou, K. (2000), "Improved detection of breast cancer nuclei using modular neural networks," *IEEE Engineering in Medicine and Biology Magazine*,

Special Issue on Classifying Patterns with Neural Networks, vol. 19, no. 1, pp. 48-63.

[45] Schnorrenberg, F., Pattichis, C.S., Kyriacou, K., and Schizas, C.N. (2000), "Content-based retrieval of breast cancer biopsy slides," *Technology and Health Care,* vol. 8, to appear in Dec.

[46] Störkel, S., Reichert, T., Reiffen, K.A., and Wagner, W. (1993), "EGFR and PCNA expression in oral squamous cell carcinomas: a valuable tool in estimating the patients prognosis," *European Journal of Cancer*, vol. 29B, pp. 273-277.

[47] Taylor, C.R. (1993), "An exaltation of experts: concerted efforts in the standardization of immunohistochemistry," *Applied Immunohistochemistry*, vol. 1, pp. 232-243.

[48] True, L.D. (1996), "Morphometric applications in anatomic pathology," *Human Pathology*, vol. 27, pp. 450-467.

[49] Weinberg, D.S. (1994), "Quantitative immunocytochemistry in pathology," in: Marchevsky, A.M. and Bartels, P.H. (Eds.), *Image Analysis: a Primer for Pathologists*, New York, USA: Raven Press Ltd., pp. 235-260.

[50] Willemse, F., Nap, M., Henzen-Logmans, S.C., and Eggink, H.F. (1994), "Quantification of area percentage of immunohistochemical staining by true color image analysis with application of fixed thresholds," *Analytical and Quantitative Cytology and Histology*, vol. 16, no. 5, pp. 357-364.

[51] Wolberg, W.H., Street, W.N., and Mangassarian, O.L. (1999), "Importance of nuclear morphology in breast cancer prognosis," *Clinical Cancer Research,* vol. 11, pp. 3542-3548.

Appendix A Semi-Quantitative Diagnostic Index

Routinely, biopsy slides of immunocytochemically stained sections are manually assessed and classified by a human expert with the help of a light microscope [46]. The assessment is based on the intensity of staining and the percentage of cells stained as documented in Table 3. These two factors are used to calculate the diagnostic index or the H-score [46]. as illustrated by Figure 4. This derivation of the H-score may induce interobserver and intraobserver variation errors [25]. Despite these limitations, studies have shown that the results obtained from manual biopsy assessment schemes are clinically important. However, due to the semi-quantitative nature of the manual assessment, there is a need to improve the accuracy, even with scoring schemes that apply five classes for classifying the results.

Table 3. Computation of manual semi-quantitative immunocytochemical diagnostic index [46] (see example in Figure 4).

% of Cells Positive	Score	Staining Intensity	Score	Total Score	Diagnostic Index
0	0	Negative	0	0	0
0 - 25 %	1	Weak	1	1 - 4	1+
26 - 50 %	2	Moderate	2	5 - 8	2+
51 - 75 %	3	Strong	3	9 - 12	3+
≥ 76 %	4	Very Strong	4	≥ 13	4+

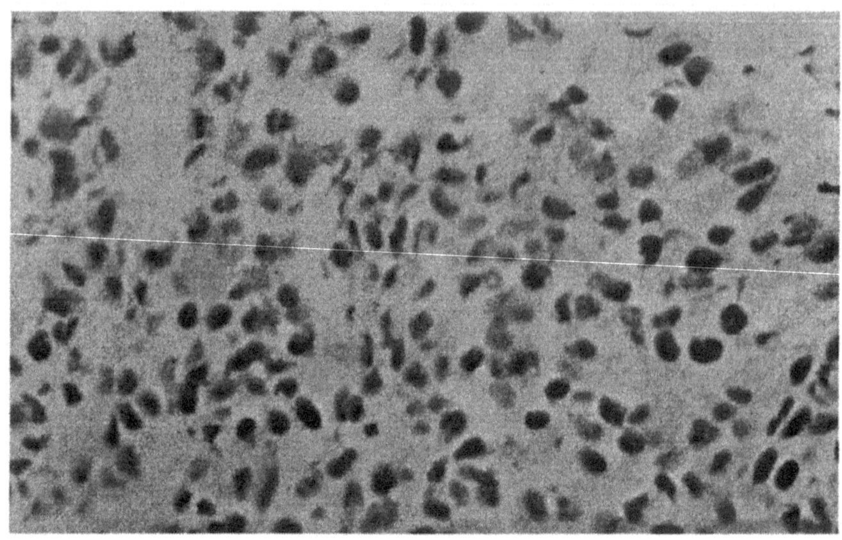

0% of nuclei, score 0, are negative (very light gray, original image: blue), score 0:	0*0 = 0
50% of nuclei, score 2, are weakly stained (light gray, original image: blue/ brown), score 1:	2*1 = 2
25% of nuclei, score 1, moderately stained (medium gray, original image: light brown), score 2:	1*2 = 2
20% of nuclei, score 1, are strongly stained (dark gray, original image: dark brown), score 3:	1*3 = 3
5% of nuclei, score 1, very strongly stained (very dark gray, orig. image: dark brown), score 4:	1*4 = 4
Total Score:	11
Diagnostic index:	3+

Figure 4. Light micrograph showing immunohistochemical staining of breast cancer nuclei for estrogen receptors of diagnostic index 3+ (see Table 3) (localized gray color; original image: brown color). (Magnification x400.)

Chapter 12

Septic Shock Diagnosis by Neural Networks and Rule Based Systems

R. Brause, F. Hamker, and J. Paetz

In intensive care units physicians are aware of a high lethality rate of septic shock patients. In this contribution we present typical problems and results of a retrospective, data driven analysis based on two neural network methods applied on the data of two clinical studies.

Our approach includes necessary steps of data mining, i.e., building up a data base, cleaning and preprocessing the data and finally choosing an adequate analysis for the medical patient data. We chose two architectures based on supervised neural networks. The patient data is classified into two classes (survived and deceased) by a diagnosis based either on the black-box approach of a growing RBF network and otherwise on a second network which can be used to explain its diagnosis by human-understandable diagnostic rules. The advantages and drawbacks of these classification methods for an early warning system are discussed.

1 Introduction

In intensive care units (ICUs) there is one event which only rarely occurs but which indicates a very critical condition of the patient: the septic shock. For patients being in this condition the survival rate dramatically drops down to 40-50% which is not acceptable.

Up to now, there is neither a successful clinical therapy to deal with this problem nor are there reliable early warning criteria to avoid such a situation. The event of sepsis and septic shock is rare and therefore statistically not well represented. Due to this fact, neither physicians can develop well grounded experience in this subject nor a statistical basis for this does exist. Therefore, the diagnosis of septic shock is still made too late, because at present there are no adequate tools to predict the progression of sepsis to septic shock. No diagnosis of septic shock can be made before organ dysfunction is manifest.

The criteria for abnormal inflammatory symptoms (systemic inflammatory response syndrome SIRS) are both non-specific and potentially restrictive [25]. Experience with the ACCP/SCCM Consensus Conference definitions in clinical trials has highlighted the fact that they are unable to accurately identify patients with septic shock who might respond to interventions targeted to bacterial infections and its consequences, identify patients at risk for septic shock and to improve the early diagnosis of septic shock.

Our main goal is the statement of diagnosis and treatment on the rational ground of septic data. By the data analysis we aim to

- help in guideline development by defining sufficient statistical criteria of SIRS, sepsis, and septic shock,
- provide the necessary prerequisites for a more successful conduct of innovative therapeutic approaches,
- give hints which variables are relevant for diagnosis and use them for further research,
- provide new approaches based on the statistical cause and context to sepsis diagnosis implementing cost-effective clinical practice guidelines for improved diagnosis and treatment of septic shock.

It should be underlined that our analysis does not provide medical evidence for the diagnostic rules and therapeutic guidelines obtained in the data mining process but facilitates the discovery of them. It is up to additional, rigorously controlled studies to verify the data mining proposals.

Instead, to assist physicians protecting patient's life, our main concern is not to make a final prognosis about the survival of the patients, but to build up an early warning system to give individual warnings about the patient's critical condition. The principle of such a system is shown in Figure 1.

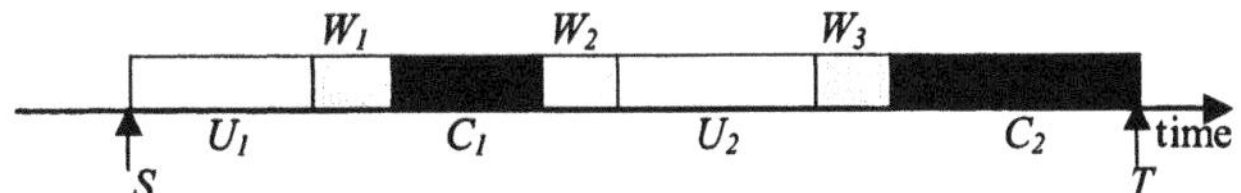

Figure 1. The concept of an early warning system. S = time of admission, T = time of death, shaded time intervals W_1, W_2, W_3: change of state, U_1, U_2 = uncritical period of time, C_1, C_2 = critical period of time

In clinical stay patients may change their state. Let us assume that in the periods of time U_i patients are uncritical, in C_j they are critical. Now, the aim of an early warning system is to give an alarm as early as possible in the transition phases W_k (k = 1,3) and of course in C_j.

Critical illness states are defined as those states which are located in areas of the data showing a majority of measurements from deceased patients, see [16]. By detecting those states we expect to achieve a reliable warning, which should be as early as possible.

2 The Data

Vcry important for mcdical data analysis, cspecially for retrospective evaluations is the preprocessing of the data. In medical data mining, after data collection and problem definition, preprocessing is the third step. Clearly, the quality of the results from data analysis strongly depends on the successful execution of the previous steps. The three steps are an interdisciplinary work from data analysts and physicians and represent often the main work load.

In the following sections, we will show the main problems associated with our data. According to our experience, these problems are typical for medical data and should be taken into account in all approaches for medical data diagnosis. They include the selection of the number and kind of variables, treatment of small sets of mixed-case data with incorrect and missing values, selection of the subset of variables to analyze and the basic statistical proportions of the data.

2.1 The Data Context

Special care has to be taken in selecting and collecting patient data. In our case, the epidemiology of 656 intensive care unit patients (47 with a septic shock, 25 of them deceased) is elaborated in a study made between November 1995 and December 1997 at the clinic of the J.W.Goethe-University, Frankfurt am Main [36]. The data of this study and another study made in the same clinic between November 1993 and November 1995 is the basis of our work.

We set up a list of 140 variables, including readings (temperature, blood pressure, ...), drugs (dobutrex, dobutamin, ...) and therapy (diabetes, liver cirrhosis, ...). We only analyzed the metric ones which represented half of them. Our data base consists of 874 patients; 70 patients of all had a septic shock; 27 of the septic shock patients and 69 of all the patients deceased.

2.2 Data Problems and Preprocessing

There are typical problems associated with medical data preprocessing. The problems and our approaches to maintain data quality are listed below.

- The **data set** is **too small** to produce reliable results. We tried to circumvent this problem by combining two different studies into one data pool.
- The medical **data** from the two different studies had to be **fused**. With the help of physicians we set up a common list of variables. Different units had to be adapted. Some variables are only measured in one of the two studies. It happened that time stamps were not clearly identifiable. Some data entries like *see above* or *zero* were not interpretable. So some database entries had to be ignored. The result is one common study with an unified relational database design including input and output programs and basic visualization programs.
- Naturally, our medical data material is very **inhomogeneous** (*case mix*), a fact that has to be emphasized. Each of the patients has a different period of time staying in the intensive care unit. For each patient a different number of variables (readings, drugs, therapies)

was documented. So we had to select patients, variables and periods of time for the data base fusion. Because different data are measured at different times of day with different frequency (see Table 1), which gave hard to interpret multivariate time series, we used resampling methods to set the measurements in regular 24 hours time intervals.

Table 1. Averages of sampling intervals of four measured variables from all patients without any preprocessing. It is evident that a priori there is no state of the patient where all variables are measured synchronously.

variable	average interval in [days : hours : min]
systolic blood pressure	1: 12: 11
temperature	1: 12: 31
thrombocytes	1: 18: 13
lactate	5: 0: 53

- **Typing errors** were detected by checking principal limit values of the variables. Blood pressure can not be 1200 (a missing decimal point). Typing errors in the date (03.12.96 instead of 30.12.96) were checked with the admission and the discharge day.
- A lot of variables showed a high number of **missing values** (internally coded with -9999) caused by faults or simply by seldom measurements, see Table 2.

Table 2. Available measurements of septic shock patients after 24-hours sampling for six variables.

variable	measurements
systolic blood pressure	83.27 %
temperature	82.69 %
thrombocytes	73.60 %
inspiratorical O_2-concentration	65.81 %
lactate	18.38 %
lipase	1.45 %

The occurrence of faulty or missing values is a problem for many classical data analysis tools including some kinds of networks. The alternative of regularly sampled variables with a constant sample rate is not feasible in a medical environment. Since most of the samples are not necessary for the patient diagnosis or too expensive either in terms of unnecessary labor cost or in terms of high laboratory or device investment charges most of the important variables are measured only on demand in critical situations. Here, the sample rate depends also on the opinion of the supervising physician about the patient's health conditions. Therefore, we have to live with the fact of missing values.

The treatment of missing values in the analysis with neural networks is described in more detail in Section 3.

In conclusion, it is almost impossible to get 100% clean data from a medical data base of different patient records. Nevertheless, we have cleaned the data as good as possible with an enormous amount of time to allow analysis, see Section 2.4.

For our task we heavily rely on the size of the data and their diagnostic quality. If the data contains too much inaccurate or missing entries we have no chance of building up a reliable early warning system even if it is principally possible.

2.3 Selecting Feature Variables

The data base contains about 140 variables of metric and categorical nature. For the small number of patients and samples we have, the number of variables is too high. Here, we encounter the important problem of "curse of dimensionality" [9] which is very hard to treat in the context of medical data acquisition. For a reliable classification the data space has to be sufficiently filled with data samples. If there is only a small number of samples available as in our case of septic shock patients, the training results become influenced by random: the classification boundaries depend on the values and sequence order of the samples.

An important approach to deal with this problem is the selection of a subset of "important" variables.

Which ones are important? There are systematic approaches for feature subset selection based on probabilities, see e.g. [21]. In our case, for analysis the physicians gave us recommendations which variables are the most important ones for a classification, based on their experience. The chosen variable set *V* is composed of n=16 variables: pO_2 (arterial)

[mmHg], pCO_2 (arterial) [mmHg], pH, leukocytes [1000/µl], thromboplastin time (TPZ) [%], thrombocytes [1000/µl], lactate [mg/dl], creatinin [mg/dl], heart frequency [1/min], volume of urine [ml/24h], systolic blood pressure [mmHg], frequency of artificial respiratory [1/min], inspiratorical O_2-concentration [%], medication with antithrombine III AT3 [%], medication with dopamine and dobutrex [µg/(kg·min)].

2.4 Basic Statistical Analysis

Now, we give an impression of the basic statistical properties for our data set. We are aware of the problem that a relative small data set of subjects (in our case only 70 patients) with a septic shock, including missing values in some variables, are not sufficient for excellent results but we can give some hints and first results in the right direction based on the available data.

For the basic statistics, we calculated some statistical standard measures for each of the variables (mean, standard deviation etc.) including all patients or only the septic shock patients combined with all days or comprising only the last day of their stay in the intensive care unit.

Q-Q-plots show that the distributions are usually normal with an huge overlap of values from deceased and survived patients; the pure probability distributions do not show any significant difference. Figure 2 shows two histograms for two variables.

If some variable values are correlated, it will not show up in the distributional plots. So, we checked this case also. A correlation analysis of the data shows high absolute values for the correlations between medicaments and variables, so surely the medicaments complicate the data analysis. Correlations between variables and {survived, deceased} are not high or not significant.

More interesting are the correlations COR(X,Z) calculated one time with the sets X_d, Z_d of samples from deceased and one time with the sets X_s, Z_s of samples from survived patients. The corresponding differences taken from all patients and all days is listed in Table 3. The significance level was calculated with SPSS 9.0. The correlations with significance level 0.01 are printed in bold font.

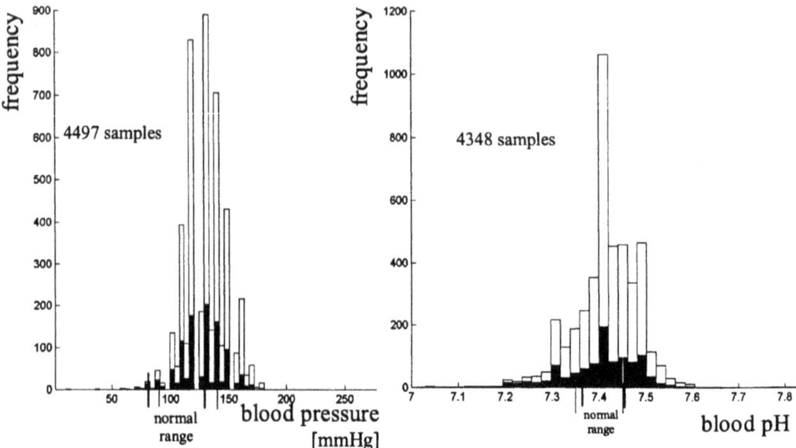

Figure 2. Histograms for a) systolic blood pressure and b) pH value for survived (white boxes) and deceased patients (black boxes). Clearly, the huge overlap of the two sample classes makes a classification very difficult.

Table 3. Correlations between two variables (all patients, all days of hospital stay) with the highest correlation differences ≥ 0.3 between survived and deceased patients and frequency of measurement of each variable ≥ 20%. Significant correlations (level 0.01) are printed in bold letters. GGT is the abbreviation of gammaglutamyltransferase.

variable X	variable Z	COR(X_s,Z_s)	COR(X_d,Z_d)	diff.
inspir. O_2-concentration	pH	-0.03	**-0.39**	0.36
leukocytes	GGT	0.00	**0.32**	0.32
iron (Fe)	GGT	**0.31**	0.01	0.30
(total) bilirubin	urea	**0.26**	-0.07	0.33
urea	creatinin	**0.14**	**0.57**	0.43
fibrinogen	creatinin in urine	0.05	**-0.31**	0.36
arterial pO_2	potassium(K)	**-0.13**	**0.18**	0.31
thromboplastin time TPZ	chloride	**0.24**	-0.07	0.31

Both correlation values for the pairs urea, creatinin and arterial pO_2, potassium are significant (level 0.01), so that the difference could be an indicator for survived or deceased patients. Therefore, these variables should be measured very often to calculate the correlation in a time window during the patients actual stay at hospital. If they turn out to be too high, early warnings could be triggered.

Also, by training a neural network with the correlation values one can find out the exact threshold for a warning based on correlation values or combinations or modifications of such values (for first results see [16]).

Generally, this result seems to be reasonable because physicians reported that the interdependence of variables, measured from critical illness patients, could be disturbed by septic shock [34].

3 The Neural Network Approach to Diagnosis

In the last years many authors contributed to machine learning, data mining, intelligent data analysis and neural networks in medicine (see, e.g., [4], [5], and [23]). For our problem of septic shock diagnosis supervised neural networks have the advantages of nonlinear classification, fault tolerance for missing data, learning from data and generalization ability. The aim of our contribution is not a comparison of statistical methods with neural network results (e.g., see [31]) but to select an appropriate method that can be adapted to our data. Here, our aim is to detect critical illness states with a classification method.

It is widely accepted in the medical community that the septic shock dynamics are strictly nonlinear [32], [34]. After preliminary tests we also concluded that linear classifiers are not suitable for classification in this case. In addition, most nonlinear classification methods also detect linear separability if it exists.

3.1 The Network

The neural network chosen for our classification task is a modified version of the supervised growing neural gas (abbr. SGNG, see [8], [12], and [13])[1]. Compared to the classical multilayer perceptron trained with backpropagation (see [18]) which has reached a wide public, this network achieved similar results on classification tasks, see [19]. The results are presented in Section 3.4.

---

[1] Logistic regression is a statistical alternative to supervised neural networks

The algorithm with our improvements and its parameters is noted in detail in [16]. It is based on the idea of radial basis functions (abbr. RBF, see [18]). The centers of the radial basis functions are connected through an additional graph that is adapted within the learning process, see appendix A. The graph structure allows to adapt not only the parameters (weights, radii) of the best matching neuron but also those of its neighbors (adjacent neurons). Its additional advantage is the ability to insert neurons within the learning process to adapt its structure to the data, see appendix A.

3.1.1 The Network Architecture

The neural network is build by two layers: the hidden layer (representation layer) and the output layer which indicates the classification decision for an input pattern.

The cell structure of the representation layer forms a parametrical graph $P = P(G, S)$ where each node $v_i \in V$ (each neuron) has just one weight vector $w_i \in S$ with $S \subset \mathbb{R}^n$. The neighborhood relations between the nodes are defined by a non-directional graph G (see [7] and [24]) where $G = G(V, E)$ consists of a set of nodes $V = \{v_1, ..., v_m\}$ and a set of edges $E = \{e_1, ..., e_m\}$. An incidence function f maps each edge to an unordered pair $[v_i, v_j]$ of nodes v_i, v_j, the end points or end nodes. The neighbors of a node are defined as those nodes which share an edge with it. For the graph $G = G(V, E)$ the set N_i of neighbors of node i is defined by the equation

$$N_i = \{ v_j / \exists\, e_k : f(e_k) = [v_i, v_j] \}. \tag{1}$$

Each node of the representation layer computes its activity y_i by the RBF activation function

$$y_i = e^{-\frac{\|\mathbf{x}-\mathbf{w}_i\|^2}{\sigma_i^2}} \quad \forall v_i \in G \tag{2}$$

where the width of the Gaussian function, the standard deviation σ_i, is given by the mean edge length s_i of all edges connected to node v_i.

The m output neurons representing m classes are linear, i.e., their activity is computed as

$$z_j = \sum_{v_i \in G} w_{ji}^{out} \cdot y_i \quad \forall v_i \in G \tag{3}$$

using the output layer weight vectors $\mathbf{w}_j^{out} = (w_{j1}^{out},...,w_{jn}^{out})$. The decision for class k is based on the maximal output activity by a winner-takes-all mechanism.

$$C_k = \max_j (z_j + \theta_j) \tag{4}$$

which is influenced by a sensitivity parameter θ_j.

3.1.2 Treatment of Missing Values

Networks like the Supervised Growing Neural Gas (SGNG) present an alternative to dropping samples where only a few number of values are absent. By learning also with a fewer number of values more samples can be used for training and testing.

To achieve knowledge about a patient being in a critical illness condition, we need to classify the vectors $\mathbf{x} = (x_1, ..., x_n)^t$ composed of measurements or drugs x_i, $i = 1, ..., n$ with the outcome y_s (survived) resp. y_d (deceased). For the n-dimensional data vector $\mathbf{x}$, we projected the vector $\mathbf{x}$ such that no missing value is in the projected vector $\mathbf{x}_p := (x_{i_1},..., x_{i_m})^t$, $\{i_1,...,i_m\} \subset \{1,...,n\}$, $m \le n$, $x_{i_1},...,x_{i_m}$ are not missing values. Due to the fact that the SGNG is based only on distance calculations between vectors, it is possible to apply this standard projection argument to the adaptation and activation calculations of the SGNG, so that all calculations are done with the projected vectors $\mathbf{x}_p$. To find the best matching neuron we compute the Euclidean distance d_i by

$$d_i = \frac{1}{|I|}\sqrt{\sum_{l \in I}(x_l - w_{il})^2}\,,\quad I = \{l \mid x_l \text{ exists}\} \tag{5}$$

Here, we take only the existing values, excluding explicitly the missing ones. The computation of the activity y_i in Eq. (2) is done in the same way.

Certainly, there is a probable error involved in the classification when not all values are present, depending on the data set. Preliminary experiments showed that in our case it is not appropriate to project to less than half the variables. Therefore we used only samples containing more than 50% valid variables. This procedure causes a statistical bias, but we believe that it is not high because the most part of the data is missing randomly.

3.2 Training and Diagnosis

It is well known that the training performance of learning systems often does not reflect the performance on unknown data. This is due the fact that the system often adapts well on training to the particularities of the training data. In the worst case a network just stores the training pattern and acts as an associative memory.

3.2.1 The Training and Test Performance

In order to test the real generalization abilities of a network to unknown data, it must be tested by classified unknown data, the *test data*. As we already mentioned in Section 2.3, the numbers of patients and samples are not very high in most medical applications. Therefore, the classical scheme of dividing all available data into training and test data is not possible, because the bigger we choose the training data set the smaller the test data set will be and the test results become vague. Choosing a small training set does no good either, because the trained state becomes also arbitrary, depending on the particularities of the training set composition. Here, special strategies are necessary.

One of the most used methods is the *p-fold cross validation* [14], [37]. Here, the whole data set is divided into p parts of equal size. The training is done in cycles or epochs where in each epoch one part (subset) of the data set is chosen as test set and the remaining $p-1$ parts of the data are used for training. This can be done p times. The test performance is computed as the mean value of all p epoch tests.

The concept can be extended to use all M samples as parts such that the test is done by just one sample. This is known as the *leave-one-out* method [26] and was used in our report [16]. It corresponds to the situation of an *online learning* early warning system trained on a set of patients and asked for the diagnosis for a new arriving patient.

For the results of this paper, we did not use this but simply divided the samples into 75% training and 25% test patterns.

3.2.2 The Problem of Medical Data Partition

There is another problem, especially for training with medical data. We might not distinguish between the data of different patients, treat all samples equal and partition the data set of labeled samples randomly. Thus, data from the same patient appears both in the training and in the test set.

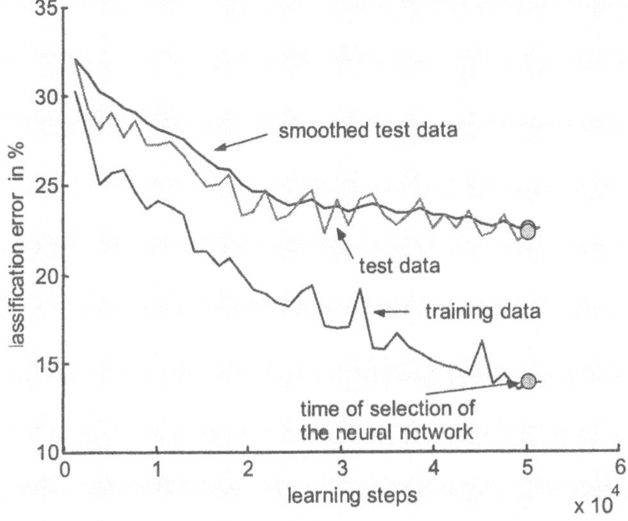

Figure 3. Random division of the data by samples.

This is shown in Figure 3. In contrast to this, the parts can be chosen such that all samples of one patient are either only in the training set or in the test set. The resulting performance is shown in Figure 4.

It turns out that the result with the random partition of samples is much better. But does this result reflect the usage reality of an early warning system? By choosing the random partition, we assume that an early warning system already knows several samples of a patient from the training. This assumption is generally not true in clinical usage.

We have to face the fact that patient data is very individual and it is difficult to generalize from one patient to another. Ignoring this fact would pretend better results than a real system could practically achieve.

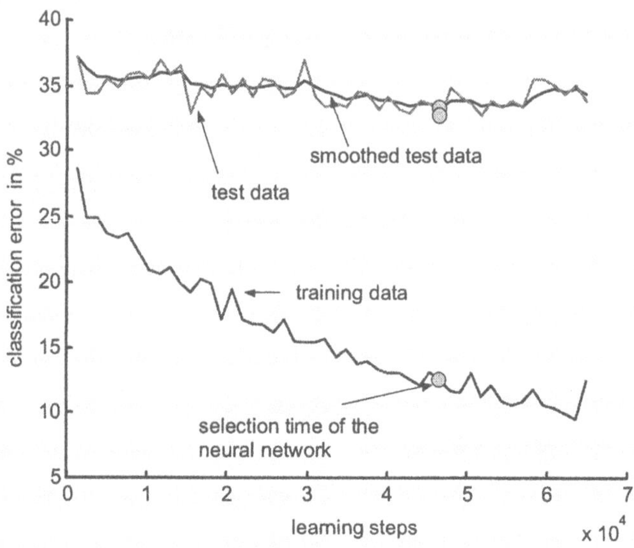

Figure 4. Division of the data by patients.

3.3 Selection and Validation of a Neural Network

One of the important parameters to get a non-overtrained, generalizing network is the time when the training has to be stopped. This time step is obtained by watching the performance of the net on the test set during training. First, the test error decreases in the adaptation process. When the test error increases again, the training should be stopped. Since the samples are randomized, the error should be smoothed in order to be approximately precise. This is shown in Figure 3 at the small circles.

There are three main approaches for selecting a suitable grown network by cross validation:

a) The test set is quite good, but choosing a network by the test set performance makes the choice depend on test set peculiarities. To avoid this, we might choose a third set of independent samples, the validation set. For instance, we might use 50% of the samples for training, 25% for testing and 25% for validation. In the medical environment where we have only a small number of patients and a small number of hand-coded variables, the advantage of independent test and validation becomes obsolete due to the random properties of the very small test and validation sets. The sets differ heavily in their proportions and are no more representative, the stopping criterion and the performance prediction becomes very arbitrary. This can be observed by a high deviation of the performance mean in the *p-fold* cross validation process.

b) The second approach uses the test set both as stopping criterion (choice of the appropriate network) and for validation, i.e., prediction. This improves the performance on the test set, but decreases the prediction performance on unknown data compared to an additional independent validation set. Nevertheless, since we are able to use more of our samples for training, the result becomes closer to the result a real application could achieve.

c) To achieve a maximal training performance in the presence of only a very small number of samples we might use all the samples for training and estimate the best stopping point by the training performance development alone without any explicit test. This includes subjective estimation and does not avoid random deviations of a good state.

The peculiarities of the choice for the sets can be decreased by smoothing the performance results. This can be obtained by taking the moving average instead of the raw value.

In our case we had only 70 patients with the diagnosis "septic shock". The high individual difference between the patients did not encourage us to choose different test and validation sets. Here we chose a test set that contains about 25% of the samples and ensured that all samples in the test set are from patients which are not used in the training set. In another investigation [16], we choose the leave-one-patient-out method to increase the size of the training set and to check each patient under the assumption that all other patients are known.

How reliable is such a diagnostic statement? In classical regression analysis, confidence intervals are given. In cases where there is no probability distribution information available as in our case this is very hard to do, see [17]. There are some attempts to introduce confidence intervals in neural networks [10], [22], [33], but with moderate success. Therefore, we decided to vary the context of testing as much as possible and give as result the deviation, maximum and minimum values additionally to the mean performance.

For the individual case the activity of the classification node of the second layer may be taken as an performance measure for the individual diagnosis [16].

3.4 Results for Septic Shock Diagnosis

Our classification is based on 2068 measurement vectors (16-dimensional samples) from variable set *V* taken from 70 septic shock patients. 348 samples were deleted because of too many missing values within the sample. With 75% of the 1720 remaining samples the SGNG was trained and with 25% samples from completely other patients than in the training set it was tested.

The variables were normalized (mean 0, standard deviation 1) for analysis.

The network chosen was the one with the lowest error on the smoothed test error function. Three repetitions of the complete learning process with different, randomly selected divisions of the data were made. The results are presented in Table 4.

Table 4. Correct classifications, sensitivity, specificity with standard deviation, minimum and maximum in % from three repetitions.

measure	mean value	standard deviation	minimum	maximum
correct classification	67.84	6.96	61.17	75.05
sensitivity	24.94	4.85	19.38	28.30
specificity	91.61	2.53	89.74	94.49

To achieve a generally applicable result ten repetitions would be better, but here it is already clear: with the low number of data samples the results can only have prototypical character, even with more cleverly devised benchmark strategies. Some additional results are reported in [16].

On average we have an alarm rate (= 1 – specificity) of 8.39% for survived patients showing also a critical state and a detection of about 1 out of 4 critical illness states. For such a complex problem it is a not too bad, but clearly no excellent result. An explanation for this low number is grounded in the different, individual measurements of each patient.

To give an impression of the warnings over time we show in Figure 5 the resulting warnings from classification for 7 out of 24 deceased patients with septic shock.

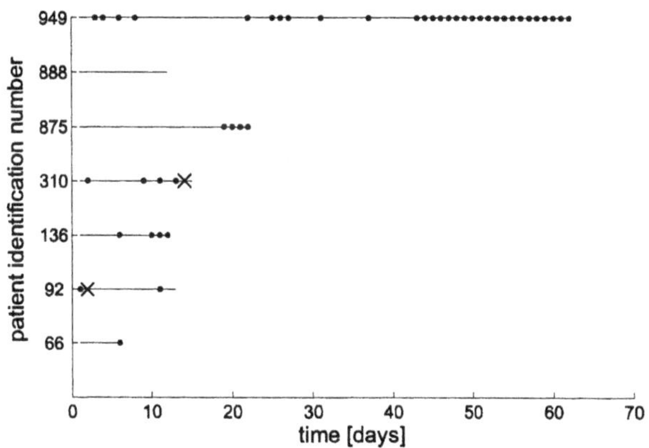

Figure 5. Deceased septic shock patients during their hospital stay with warnings (dot markers). A too high number of missing values causes some missing states (crosses). If there is no marker then no warning will be given.

Not for each deceased patient exists a warning (patient with number 888) and some warnings are given too late (patient with number 66), i.e. the physicians knew already that the patient had become critical. So the ideal time to warn the physician has not yet been found for all patients and remains as future work.

4 The Neuro-Fuzzy Approach to Rule Generation

Results of classification procedures could provide a helpful tool for medical diagnosis. Nevertheless, in practice physicians are highly trained and skilled people who do not accept the diagnosis of an unknown machine (black box) in their routine. For real applications, the diagnosis machine should become transparent, i.e., the diagnosis should explain the reasons for classification. Whereas the explanation component is obvious in classical symbolic expert system tools, neural network tools hardly explain their decisions. This is also true for the SGNG network used in the previous section.

Therefore, as important alternative in this section we consider a classification by learning classification rules which can be inspected by the physician. Actual approaches to rule generation consider supervised learning neuro-fuzzy-methods [14], [20], especially for medical applications [6], [27].

Usually, medical data contain both metric and categorical variables. Here, our data is substantially based on *metric* variables, so in the following we consider the process of rule generation only for metric variables.

We chose an algorithm based on rectangular basis functions for the rule generation approach for metric variables which we apply to the septic shock patient data.

4.1 The Rule Extraction Network

First we describe the fundamental ideas of the algorithm and then we give a detailed description of it. The network structure – as we use it for two classes – is shown in Figure 6.

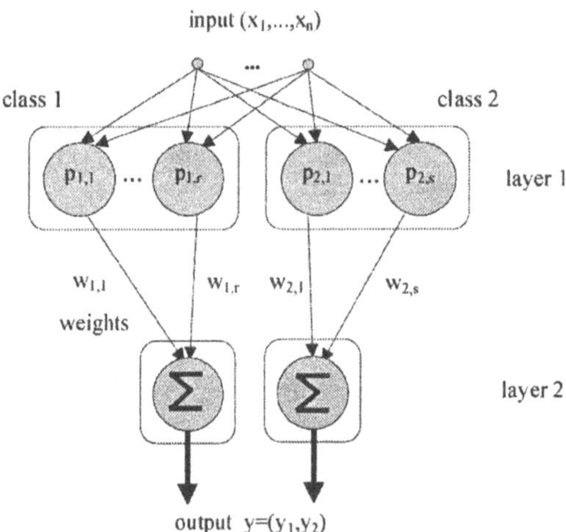

Figure 6. Network structure for two classes. Each class in layer 1 has its individual number of neurons.

The 2-layer network has neurons – separately for every class – in layer 1. The *r* neurons $p_{1,1}$, ..., $p_{1,r}$ belong to class 1 and the *s* neurons

$p_{2,1}, ..., p_{2,s}$ to class 2. The activation functions of the neurons represent rule prototypes using different asymmetrical trapezoidal fuzzy activation functions $R_{1,1}, ..., R_{1,r}$ and $R_{2,1}, ..., R_{2,s}$ with image [0,1].

The algorithm is an improved version of the RecBFN algorithm of Huber and Berthold [20] which in turn is based on radial basis functions [18] with dynamic decay adjustment [2], [3]. During the **learning phase** the input data is passed unmodified to layer 1. Then all neurons are adapted, i.e., the sides of the smaller rectangles (= *core rules*) and the sides of the larger rectangles (= *support rules*) of the fuzzy activation function graph are adapted to the data samples, see Figure 7.

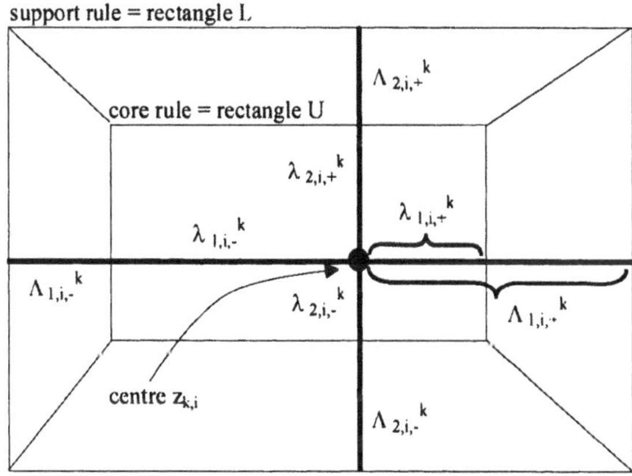

Figure 7. Two-dimensional projection (bird's view) of the trapezoidal function of one neuron with support and core rule and parameters of the algorithm in appendices B and C, representing one fuzzy rule for class k (see Figure 1 in [20]). U is the upper and L the lower rectangle of the trapezoid.

This happens in four phases for every new training data sample vector $\mathbf{x} \in \mathbb{R}^n$ of class k with n as dimension of the data space,

(1) *cover*: if **x** lies in the region of the support rule for all neurons – generated so far – of the same class k as **x**, expand one side of the core rule to cover **x** and increment the weight of the neuron.
(2) *commit*: if no support rule covers **x**, insert a new neuron p with center **x** of the same class k and set its weight to one; the expansions of the sides are initially set to infinite.
(3) *shrink committed neuron*: for a committed neuron shrink the volume of the support and the core rectangle within *one* dimension of the neuron in relation to the neurons belonging to other classes.

(4) *shrink conflict neurons*: for all neurons, belonging to another class not equal to k, shrink the volume of both rectangles within *one* dimension in relation to $\mathbf{x}$.

For details of the main algorithm and the shrinking procedure see appendices B and C.

An advantage of the method is its simplicity that softens the combinatorial explosion in rule generation by its cover-commit-shrink-procedure. By side expansions of the fuzzy activation function to infinite it is possible to find out the variables that are not interesting for a rule, see rules (9) and (10) below. It is also directly possible to integrate a-priori known rules after fuzzification.

Finally, **classification activity** is done by a *winner-takes-all* mechanism, i.e., the calculation of the output $y_k = y_k(\mathbf{x})$ as the sum of the weights multiplied by fuzzy activation for every class $k \in \{1, 2\}$:

$$y_1 := w_{1,1} R_{1,1} + \ldots + w_{1,r} R_{1,r} \tag{6}$$

$$y_2 := w_{2,1} R_{2,1} + \ldots + w_{2,s} R_{2,s} \tag{7}$$

Then, choose class c_{max} as classification result, where c_{max} is the class label of the maximal output:

$$c_{max} := \text{class}\left(\max_k \{y_k(\mathbf{x})\}\right). \tag{8}$$

If the second highest value c_{second} is equal to c_{max} the data is output as *not classified*. It is easy to change the algorithm to function with $c>2$ classes [20]. Usually three to seven epochs are needed for the whole training procedure.

The result of the training procedure are rules of the form (belonging to the core or support rectangle)

if variable 1 **in** $(-\infty, 50)$ **and if** variable 2 **in** $(20,40)$
and if variable 3 **in** $(-\infty,\infty)$ **then** *class l* (9)

in addition with a classification based on (8). Interestingly, in rule (9) variable 3 is not relevant, so variable 3 can be omitted and in such a case we get the simplified rule

if variable 1 **in** $(-\infty, 50)$ **and if** variable 2 **in** (20,40) **then** *class l* (10)

How good are the resulting rules?

The relevance of a rule for a class can be measured by the number of samples of class k that lie in core (resp. support) rule p divided by the number of all samples. This is called the *frequency*. Additionally, the *class confidence* in a class decision is defined as the number of samples of class k that lie in p divided by the number of all the samples that lie in p. Both measures, the frequency and the class confidence of a rule, should always be calculated on test data samples, not on training data samples.

Using these two measures we can expand the rules to a more precise form. The expanded rule (10) becomes rule (11):

if variable 1 **in** $(-\infty, 50)$ **and if** variable 2 **in** (20,40)
then *class* 1 **with** *frequency* 5% **and** *class confidence* 80% (11)

This concludes our tool set for extracting rule based knowledge of a data base.

4.2 Application to Septic Shock Patient Data

Now we present the results of the rule generation process of Section 4.1 with the data set D of Section 2. The data set D is 16-dimensional. The missing values were replaced by random data from normal distributions similar to the original distributions of the variables. So it was assured that the algorithm can not learn a biased result due to biased replacements, e.g., means. We demand a minimum of 10 out of 17 variables measured for each sample, so there remained 1677 samples out of 2068 for analysis.

The data we used in 5 complete training sessions - each with a different randomly chosen training data set - was in mean coming from class 1 with a percentage of 72.10% and from class 2 with a percentage of 27.91%. In the mean 4.00 epochs were needed (with standard deviation 1.73, minimum 3 and maximum 7). Test data was taken from 35 randomly chosen patients for every training session, containing no data sample of the 35 patients in the training data set. In Table 5 the classification results are presented.

Table 5. Mean, standard deviation, minimum and maximum of correct classifications and not classifiable data samples of the test data set. In %.

	mean	standard deviation	minimum	maximum
correct classifications	68.42	8.79	52.92	74.74
not classified	0.10	0.22	0.00	0.48

Average specificity ("deceased classified / all deceased") was 87.96% and average sensitivity ("survived classified / all survived") was 18.15%. The classification result is not satisfying, although similar to the results in Section 3.4 but with the benefit of explaining rules. Samples of deceased patients were not detected very well. Reasons for this can be the very individual behavior of the patients and the data quality (irregularity of measurements, missing values). In this way it seems not possible to classify *all* the patients correctly, but it could be that in some areas of the data space the results are better (*local rules*). So we will present the results of the rule generation. On average 22.80 rules were generated for class survived and 17.80 rules were generated for class deceased.

In Table 6 the core and support frequencies resp. class confidences of the generated rules are shown.

Table 6. Mean of frequency resp. class confidence of support and core rules (calculated on test data). In %. The average values were taken from all repetitions and all rules of every repetition.

performance measure	class survived	class deceased
support frequency	15.93	13.33
core frequency	2.39	0.62
support class confidence	74.37	30.88
core class confidence	59.96	11.70

If no test data sample lies within a rule p, class confidence of p was set conservatively to zero, so that it is possible that the core class confidence could be lower than the support class confidence. All frequency values are in the normal range. Class confidence performance is not high, because there are a lot of *small* rules and a lot of rules containing samples of deceased *and* survived patients.

Despite these results it is possible to give some *single* rules with a better performance, e.g.:

> **if** heart frequency **in** (105.00,∞) **and** systolic blood pressure **in** (130.00,∞) **and** inspiratorical O_2 pressure **in** ($-\infty$, 60.00) **and** frequency of respiratory **in** (19.00,∞) **and** leukocytes **in** ($-\infty$, 16.70) **and** dobutrex **in** ($-\infty$, 1.50) **then** *class* survived **with** *frequency* 9.2% **and** class confidence 91.2% (containing data of 11 different patients)
>
> **if** systolic blood pressure **in** (120.00,∞) **and** leukocytes **in** (24.10,∞) **and** dobutrex **in** (0.00, 6.00) **then** *class* deceased **with** *frequency* 7.6% **and** *class confidence* 69.7% (containing data of 13 different patients)

Considering the latter rule, we can present it to a medical expert in fuzzy notation after *defuzzification* (see [1]):

> **if** systolic blood pressure **is** high **and** (number of) leukocytes **is** high **and** dobutrex **is** given **then** patient **is** in a very critical condition

With the help of such rules, it may be possible for the physician to recommend therapies based on data analysis.

5 Conclusions and Discussion

The event of septic shock is so rare in the clinic routine that no human being has the ability to make a well-grounded statistical analysis just by plain experience. We have presented a data analysis approach for medical data and used it for the important problem of septic shock. The typical problems in analyzing medical data are presented and discussed. Although the special problem of septic shock diagnosis prediction is hard to solve the results of the basic analysis and the more advanced analysis by a growing neural gas are encouraging for the physicians to achieve an early warning system for septic shock patients, but our results are not final. In spite of severe restrictions of the data we achieved good results by using several preprocessing steps.

Our patient data of SIRS, sepsis and septic shock overlap heavily in the low-dimensional subspace we analyzed. Therefore, any prognostic system can not predict always the correct future state but may just give early warnings for the treating physician. These warnings constitute only an additional source of information; the backward conclusion that, if there is no warning there is also no problems, is not true and should be avoided.

Another diagnostic approach by neural networks is adaptive rule generation. By this, we can explain the class boundaries in the data and at the same time find out the necessary variables for the early warning system. By using a special approach of rectangular basis networks we achieved approximately the same classification results as by the growing neural gas. Additionally, the diagnosis was explained by a set of explicitly stated medical rules.

To see how difficult the problem of building an early warning system for septic shock patients is, we asked an experienced senior medical expert to propose an experience-based rule. The following rule was proposed:

> **if** pH **in** $(-\infty,7.2)$ **and** arterial pO_2 **in** $(-\infty,60)$ **and** inspiratorical O_2 concentration **in** $(80,\infty)$ **and** base excess **in** $(-\infty,5)$ **then** *class* deceased

In fact, *no* data point of our data lies in the defined region: There is no data support for this opinion! So a rational data driven machine learning approach to metric rule generation is a great benefit in comparison with subjectively induced rules for the problem of septic shock.

Although the automatic rule generation approach is principally favorable, the number of 40 rules obtained is not much, but too much for daily clinical use. Here, much more research is necessary for selecting the most relevant rules. The performance measures class frequency and class confidence help, but do not solve these problems. In principal, we are faced with a principal problem: how do we get general rules if most of the samples are very individual ones, showing no common aspects? One solution to this fundamental problem is the search for new kinds of similarity. For instance, instead of static correlations or coincidences one might look for a certain dynamic behavior of the variables or their derivatives. In our case, small sampling frequencies and small data bases impeded such an approach.

The alternative to this weak diagnosis lies in the parallel analysis of all variables (in our case: about 140), not only a subspace of 16 in order to get rid of the overlappings and find good class boundaries in hyperspace. But here we encounter the important problem of "curse of dimensionality" [9] which is very hard to treat in the context of medical applications. Two main problems impede a successful approach: the small number of homogeneous patient data and the large number of missing values.

To improve our results we are collecting more data from septic shock patients from 166 clinics in Germany to evaluate our algorithms on this larger amount of patient data.

Generally, for both problems there is only hope if automatic data acquisition and exchange is available which is not the case in most hospitals in Europe. Nevertheless, by the introduction of cost controlling mechanisms (TISS-score etc.) hospital people are forced to enter all available data in the electronic patient record in order to get paid for their efforts. In turn, this may enable better analysis for us in near future by pushing the change from the paper-and-pencil documentation style to electronic data acquisition systems.

There is another problem which should be mentioned here. Even if we have enough good quality data we encounter the problem of combining different kind of variables: metric variables like the one analyzed in this paper and categorical variables like operation and diagnostic code, drug prescription and so on. The transformation of each type into the other causes either an information loss or the introduction of additional, not justified information (noise). The standard approach to avoid this is the construction of an expert for each kind of data and to combine the output of both experts by a meta diagnosis, but there is no unifying approach for the analysis of both kind of data.

In the near future we will try to improve the performance of these results by other methods. Further work will be the extraction of typical time series patterns for medical use. Some results from cluster analysis are presented in [16].

Acknowledgments

This work was partially supported within the DFG-project MEDAN (Medical Data Analysis with Neural Networks). The authors like to thank all the participants of the MEDAN working group especially Prof. Hanisch and all the other persons involved in the MEDAN project for supporting our work. Parts of the results have been published earlier, see [16] and [29]. Section 4 is a contribution of J. Paetz.

References

[1] Berthold, M. (1999), "Fuzzy logic," Chapter 8 in Berthold, M. and Hand, D.J. (Eds.), *Intelligent Data Analysis: an Introduction*, Springer-Verlag, pp. 269-298.

[2] Berthold, M. and Diamond, J. (1995), "Boosting the performance of RBF networks with dynamic decay adjustment," *Advances in Neural Information Processing Systems*, vol. 7, pp. 521-528.

[3] Berthold, M. and Diamond, J. (1998), "Constructive training of probabilistic neural networks," *Neurocomputing*, vol. 19, pp. 167-183.

[4] Brause, R. and Hanisch, E. (2000), *Medical Data Analysis ISMDA 2000*, Springer Lecture Notes in Comp.Sc., LNCS 1933, Springer Verlag, Heidelberg.

[5] Brause, R. (1999), "Revolutionieren neuronale Netze unsere Vorhersagefähigkeiten?" *Zentralblatt für Chirurgie*, vol. 124, pp. 692-698.

[6] Brause, R. and Friedrich, F. (2000), "A neuro-fuzzy approach as medical diagnostic interface," *Proc. ESANN 2000*, De Facto Publ., Brussels, pp. 201-206.

[7] Bruske, J. (1998), "Dynamische Zellstrukturen. Theorie und Anwendung eines KNN-Modells," Dissertation, Technische Fakultät der Christian-Albrechts-Universität, Kiel, Germany.

[8] Bruske, J. and Sommer, G. (1995), "Dynamic cell structure learns perfectly topology preserving map," *Neural Computation*, vol. 7, pp. 845-865.

[9] Bellman, R. (1961), *Adaptive Control Processes: a Guided Tour*, Princeton, NJ: Princeton University Press.

[10] Dybowski, R. (1997), "Assigning confidence intervals to neural network predictions," Technical Report, Division of Infection, UMDS (St Thomas' Hospital), London, 2 March 1997, available at http://www.umds.ac.uk/microbio/richard/nnci.pdf .

[11] Fein, A.M. et al. (Eds.) (1997), *Sepsis and Multiorgan Failure*, Williams & Wilkins, Baltimore.

[12] Fritzke, B. (1994), "Fast learning with incremental RBF networks," *Neural Processing Letters*, vol. 1, no. 1, pp. 2-5.

[13] Fritzke, B. (1995), "A growing neural gas network learns topologies," in: Tesauro, G., Touretzky, D.S., and Leen, T.K. (Eds.), *Proc. Advances in Neural Information Processing Systems* (NIPS 7), MIT Press, Cambridge, MA, pp. 625-632.

[14] Fritzke, B. (1997), "Incremental neuro-fuzzy systems," in: Bosacchi, B., Bezdek, J.C., and Fogel, D.B. (Eds.), *Proc. SPIE, vol. 3165, Applications of Soft Computing*, pp. 86-97.

[15] Geisser, S. (1975), "The predictive sample reuse method with applications," *Journal of The American Statistical Association*, vol. 70, pp. 320-328.

[16] Hamker, F., Paetz, J., Thöne, S., Brause, R., and Hanisch, E. (2000), "Erkennung kritischer Zustände von Patienten mit der Diagnose 'Septischer Schock' mit einem RBF-Netz," Tech. Report, Interner Bericht 04/00, Fachbereich Informatik, J.W. Goethe-Uni-

versität Frankfurt a. M., http://www.cs.uni-frankfurt.de/fbreports/ fbreport04-00.pdf .

[17] Hartung, J. (1993), *Statistik: Lehr- und Handbuch der Angewandten Statistik*, Oldenbourg-Verlag, München.

[18] Haykin, S. (1999), *Neural Networks, a Comprehensive Foundation*, Prentice Hall, 2nd edition, Upper Saddle River, NJ 07458.

[19] Heinke, D. and Hamker, F. (1998), "Comparing neural networks, a benchmark on growing neural gas, growing cell structures, and fuzzy ARTMAP," *IEEE Transactions on Neural Networks*, vol. 9, no. 6, pp. 1279-1291.

[20] Huber, K.-P. and Berthold, M.R. (1995), "Building precise classifiers with automatic rule extraction," *IEEE International Conference on Neural Networks*, vol. 3, pp. 1263-1268.

[21] Inza I., Merino M., Larrañaga P., Quiroga J., Sierra B., and Girala M. (2000), "Feature subset selection using probabilistic tree structures. a case study in the survival of cirrhotic patients treated with TIPS," in [4], pp. 97-110.

[22] Kindermann, L., Lewandowski, A., Tagscherer, M., and Protzel, P. (1999), "Computing confidence measures and marking unreliable predictions by estimating input data densities with MLPs," *Proceedings of the Sixth International Conference on Neural Information Processing (ICONIP'99)*, Perth, Australia, pp. 91-94.

[23] Lavrač, N. (1999), "Machine learning for data mining in medicine," in: Horn, W. et al (Eds.), *Proc. AIMDM'99. LNAI 1620*, Springer-Verlag, Berlin Heidelberg, pp. 47-62.

[24] Martinetz, T.M. and Schulten, K.J.(1994), "Topology representing networks," *Neural Networks*, vol. 7, pp. 507-522.

[25] Members of the American College of Chest Physicians / Society of Critical Care Medicine Consensus Conference Committee (1992), "Definitions for sepsis and organ failure and guidelines for the use

of innovative therapies in sepsis," *Crit. Care Med.*, vol. 20, pp. 864-874.

[26] Mosteller, F. and Tukey, J.W. (1968), "Data analysis, including statistics," in: Lindzey, G. and Aronson, E. (Eds.), *Handbook of Social Psychology 2*, Addison-Wesley.

[27] Nauck, D. (1999), "Obtaining interpretable fuzzy classification rules from medical data," *Artificial Intelligence in Medicine*, vol. 16, no. 2, pp. 149-169.

[28] Neugebauer, E. and Lefering, R. (1996), "Scoresysteme und Datenbanken in der Intensivmedizin – Notwendigkeit und Grenzen," *Intensivmedizin*, vol. 33, pp. 445-447.

[29] Paetz, J., Hamker, F., and Thöne, S. (2000), "About the analysis of septic shock patient data," in [4], pp. 130-137. Also available at http://www.cs.uni-frankfurt.de/~paetz/PaetzISMDA2000.pdf.

[30] Pietruschka, U. and Brause, R. (1999), "Using growing RBF nets in rubber industry process control," *Neural Computing & Applications*, Springer Verlag, vol. 8, no. 2, pp. 95-105.

[31] Schumacher, M., Rößner, R., and Vach, W. (1996), "Neural networks and logistic regression, part I," *Computational Statistics & Data Analysis*, vol. 21, pp. 661-682.

[32] Seely, A. and Christou, N. (2000), "Multiple organ dysfunction syndrome, exploring the paradigm of complex nonlinear systems," *Crit. Care Med.*, vol. 28, no. 7, pp. 2193-2200.

[33] Tagscherer, M., Kindermann, L., Lewandowski, A., and Protzel, P. (1999), "Overcome neural limitations for real world applications by providing confidence values for network predictions," *Proceedings of the Sixth International Conference on Neural Information Processing (ICONIP'99)*, Perth, Australia, pp. 520-525.

[34] Toweill, D., Sonnenthal, K., Kimberly, B., Lai, S., and Goldstein, B. (2000), "Linear and nonlinear analysis of hemodynamic signals

during sepsis and septic shock," *Crit. Care Med.*, vol. 28, no. 6, pp. 2051-2057.

[35] Vach, W., Roner, R., and Schumacher, M. (1996), "Neural networks and logistic regression: part II," *Computational Statistics and Data Analysis*, vol. 21, 683-701.

[36] Wade, S., Büssow, M, and Hanisch, E. (1998), "Epidemiology of systemic inflammatory response syndrome, sepsis and septic shock in surgical intensive care patients," *Chirurg*, vol. 69, pp. 648-655.

[37] Wahba, G. and Wold, S. (1975), "A completely automatic French curve: fitting spline functions by cross-validation," *Communications in Statistics*, vol. 4, pp. 1-17.

Appendix A: The Network Adaptation and Growing

Adaptation of the Layers

Let us input a multidimensional pattern $\mathbf{x}$ into the system. First, all neurons compare their match $\| \mathbf{w}_i - \mathbf{x} \|$ with that of the neighbors. That node b with the highest similarity, i.e. the smallest Euclidean distance between its weight vector and the input vector, will win the competition by its high activity y_i (*winner-takes-all*). There is also as second winner a node s with the second best match. Then, the weight vectors $\mathbf{w}_i$ in the neighborhood of the best matching node b are adapted by

$$\begin{aligned} \Delta \mathbf{w}_b &= \eta_b \cdot (\mathbf{x} - \mathbf{w}_b) \\ \Delta \mathbf{w}_c &= \eta_c \cdot (\mathbf{x} - \mathbf{w}_c) \qquad \forall c \in N_b \end{aligned} \qquad \eta_b=0.1,\ \eta_c=0.01 \tag{12}$$

as centers of Radial Basis Functions with the "step size" parameters η_b and η_c. In order to avoid rapid changes the new width $\sigma_i(k)$ of the bell-shaped functions are computed at time step k as shifted mean of the old values $\sigma_i(k-1)$ and the actual distances s_i

$$\sigma_i(k) = \gamma \cdot \sigma_i(k-1) + (1-\gamma) \cdot s_i \qquad \forall\, v_i \in G \quad \gamma=0.8 \tag{13}$$

There is an error associated with each classification. This is defined as the Euclidean distance between the m-dimensional output vector **z** and the desired class vector **u** which has a one at dimension k if class k is desired as output and zero otherwise.

$$d(\mathbf{u},\mathbf{x}) = \left\| \mathbf{u} - \mathbf{z}(\mathbf{x}) \right\| \tag{14}$$

The adaptation of the output weights is based on the delta rule [18] to decrease the error

$$\Delta w_{ji}^{\text{out}} = \eta_o (u_j - z_j)\, y_i \; ; \quad \forall\; j \in \{1,\ldots,m\}, \quad \forall\; v_i \in G \quad \eta_0 = 0.01 \,. \tag{15}$$

Additionally, there is an error counter variable τ_i associated to every node v_i. The best matching neuron b stores the computed error of the output if the error is not marginal and exceeds a certain threshold θ_C.

$$\Delta\tau_b = \begin{cases} d(\mathbf{u},\mathbf{x}) & \text{if } d(\mathbf{u},\mathbf{x}) > \theta_C \\ 0 & \text{else} \end{cases}, \qquad \theta_C = 0.2 \,. \tag{16}$$

All other error counters are exponentially decreased by

$$\Delta\tau_i = -\alpha\tau_i \,, \qquad \forall\; v_i \in G \,, \; \alpha = 0.995 \,. \tag{17}$$

Growing of the Representation Layer

In order to reduce the output error not only by adaptation but also by structural change, we insert a new neuron (new node) in the graph of the first layer. To do this, the node p with the highest error counter value is selected after a certain number (here: 100) of adaptation steps. Between this node and its direct neighbor q with the highest error counter value a new node r is inserted. This new neuron receives a certain fraction β of the error of node p and the errors of p and q are decreased by β.

$$\begin{aligned} \tau_r &:= \beta\,\tau_p \\ \tau_p &:= (1-\beta)\,\tau_p \,, \qquad \beta = 0.5 \,. \\ \tau_q &:= (1-\beta)\,\tau_q \end{aligned} \tag{18}$$

This cell growing allows us to start with a very small network and let it grow appropriately to the needs of the application. In comparison with other growing RBF nets (e.g., [30]) there is also a neighbor topology of edges. Each edge has an attribute called "age." According to this age, the edges may be deleted and update the topology of the graph.

- Increment the age of all edges [b , .] from the winner *b* by one.
- Reset the age of the edge between *b* and *s* to zero. If no edge between these nodes exists, create a new one with age zero.
- Delete all edges with age $\geq \theta_{age}$; $\theta_{age} = 60$.
- Delete all nodes without an edge.

By insertion and center adaptation we control the construction of the network: regions with high error are increased while regions with no activity are decreased.

Appendix B: The Main Rule Building Algorithm

The parameters of the algorithm are (see Figures 6 and 7):

$w_{k,i}$	weight of class k that is connected to neuron i,
$\mathbf{z}_{k,i}$	center of i-th rule prototype (= neuron) of class k,
$\lambda_{n,i,-}^{k}$	negative expansion of upper rectangle U,
$\lambda_{n,i,+}^{k}$	positive expansion of upper rectangle U,
$\Lambda_{n,i,-}^{k}$	negative expansion of lower rectangle L,
$\Lambda_{n,i,+}^{k}$	positive expansion of lower rectangle L

with *n* as data dimension, $i = 1, ..., m_1$ with $m_1 = r$ for class $k = 1$ and $i = 1, ..., m_2$ with $m_2 = s$ for class $k = 2$.

Reset weights:

for $c = 1$ **to** 2
  for $i = 1$ **to** m_c **do** $w_{c,i} := 0$; $\lambda_{n,i,+-} := 0$; $\Lambda_{n,i,+-} := \infty$; **end**
end

Training of one epoch:
for each data sample $\mathbf{x}$ of class k **do**
  if $p_{k,i}$ covers $\mathbf{x}$ // i.e. $\mathbf{x}$ lies in L
   then // $\mathbf{x}$ is covered by $p_{k,i}$ (*cover*)
     $w_{k,i} := w_{k,i} + 1$;
     adjust $\lambda_{n,i,+-}^{k}$, so that U covers $\mathbf{x}$;
     if $\mathbf{x}$ lies in a core rule U of a prototype of class $c \neq k$
      then set all $\lambda_{n,i,+-}^{k} := 0$; **end** // to prohibit over-
          lapping core-rules, additional to [20]

  Insert new neuron (commit):
   else
     $m_k := m_k + 1$;
     $w_{k,i} := 1.0$; // with $i = m_k$
     $\mathbf{z}_{k,i} := \mathbf{x}$; // $\mathbf{x}$ is center of the new rule
     $\lambda_{n,i,+-}^{k} := 0$;
     $\Lambda_{n,i,+-}^{k} := \infty$;

  Shrink committed neuron:
     for $c \neq k$, $1 \leq j \leq m_c$ **do**
      shrink $p_{k,i+1}$ by $\mathbf{z}_{c,j}$, i.e., shrink($p_{k,i+1}$, $\mathbf{z}_{c,j}$); // see app. C
     end
  end

  Shrink conflict neurons:
  for $c \neq k$, $1 \leq j \leq m_c$ **do**
    if $\mathbf{x}$ lies in support region L of p_j^c
      then shrink $p_{c,j}$ by $\mathbf{x}$, i.e., shrink($p_{c,j}$, $\mathbf{x}$); // see app. C
    end
  end
end

Appendix C: The Rule Shrinking Procedure

In the shrinking procedure, we added a threshold $\Lambda_{n,bestinfinite,+-}$ because $\Lambda_{n,min,+-}$ does not always exist. The original algorithm [20] can not be used with our real world data because the algorithm crashes, if not for all $n = 1, ..., m_c$ $\Lambda_{n,min,+-}$ exists, i.e., if for all n the relation $N < \sigma_{n,min}$ holds. If $\lambda > \Lambda$ for one of the λ's within a shrink procedure, set $\lambda := \Lambda$.

shrink(p,x) :

p one rule prototype,
x data sample,
$z_{n,+-}$ center of the rule prototype (each dimension n is considered), left and right expansions are considered separately,
$\sigma_{n,min}$ usually set to 0.1 (prohibits too small areas within one dimension)

- *minimal volume loss principle*:
  calculate M for all *finite* Λ:
  $M := \min\{ \,|\, z_{n,+-} - x_n \,|\, \mid \text{for all } n \neq c \text{ and } \ldots$
  $\ldots |\, \Lambda_{n,+-} - |\, z_{n,+-} - x_n \,| \,/\, \Lambda_{n,+-} \,| \leq |\, \Lambda_{c,+-} - |\, z_{c,+-} - x_c \,| \,/\, \Lambda_{c,+} \};$
  if M exists **then** $\Lambda_{n,min,+-} := M$;
  if $M \geq \sigma_{n,min}$ **then** $\Lambda_{n,bestfinite,+-} := M$; **end**
  end

- calculate for all *infinite* expansions:
  $N := \max\{ \,|\, z_n - x_n \,|\, \mid \text{for all } n\};$
  if N exists **then** $\Lambda_{n,max,+-} := N$;
  if $N \geq \sigma_{n,min}$ **then** $\Lambda_{n,bestinfinite,+-} := N$; **end**
  end

- Calculate a new $\Lambda_{n,+-}$ for *p*, i.e., a shrink in one dimension of the expansion:
  if $\Lambda_{n,bestfinite,+-}$ exists
  then $\Lambda_{n,+-} := \Lambda_{n,bestfinite,+-}$;
  else
  if $\Lambda_{n,bestinfinite,+-}$ exists **and** (($\Lambda_{n,bestinfinite,+-} \geq \Lambda_{n,min,+-}$) ...
  ... **or** ($\Lambda_{n,min,+-}$ does not exist))
  then $\Lambda_{n,+-} := \Lambda_{n,bestinfinite}$;
  else
  if $\Lambda_{n,min,+-}$ exists
  then $\Lambda_{n,+-} := \Lambda_{n,min,+-}$;
  else $\Lambda_{n,+-} := \Lambda_{n,max,+-}$;
  end
  end
  end

Chapter 13

Monitoring Depth of Anesthesia

J.W. Huang, X.-S. Zhang, and R.J. Roy

This chapter examines the use of complexity analysis, approximate entropy, wavelet transforms, artificial neural networks, fuzzy logic, and neuro-fuzzy method (adaptive network-based fuzzy inference systems) to determine the depth of anesthesia (DOA) of a patient by analyzing mid-latency auditory evoked potentials (MLAEP) and electroencephalograms (EEG). Comparisons are made of the success and computational efficiency of each technique using the data of experimental dogs with different anesthetic modalities.

1 Introduction

Currently there is no direct means of assessing the depth of anesthesia (DOA) of a patient during surgery. An anesthesiologist therefore makes heuristic decisions on the DOA and adjusts the anesthetic dosage by integrating meaningful changes in vital signs with their experience. The traditional signs may include changes in blood pressure or heart rate, lacrimation, muscular movement, and spontaneous breathing. However, these anesthetic adjustments cannot always account for the variability in patient responses to anesthesia or changes in anesthetic requirements through the surgical procedure. Therefore, overdosing, underdosing and intraoperative awareness still complicate general anesthesia today [1], and still present an unresolved medical problem. Central to this problem is a poor understanding of the complex levels of consciousness during anesthesia and an inability to assess the DOA. Reliable and noninvasive monitoring of DOA would be highly desirable.

Since a target site of anesthetic action is the brain, it is reasonable to monitor its activity by examining brain waves, the electroencephalograms (EEG) and evoked responses such as the mid-latency auditory evoked potentials (MLAEP). These waveforms quantitatively measure the anesthetic effects. The raw EEG and MLAEP are difficult to interpret clinically, therefore, the DOA-related information contained in the brain waves, should be condensed and simplified to parameters that strongly correlate with the DOA in different aspects by various advanced signal processing techniques. These derived parameters, as input variables, are very important for building accurate DOA estimation models. The emerging computational intelligence techniques just fit the modeling requirements in this field. In Table 1, these techniques are summarized and compared to the practices of an anesthesiologist in managing depth of anesthesia.

The goal of DOA estimation is to effectively control the DOA during surgery. Accurate DOA control requires accurate DOA estimation. Many control strategies for hemodynamic regulation, drug delivery, and DOA control under anesthesia have been developed: adaptive and intelligent control [2], long-range adaptive predictive control [3], optimal control [4], rule-based control algorithm [5]. However, the promising approach is by fuzzy logic based control [6]-[9]. Since fuzzy logic creates a control surface by combining rules and fuzzy sets, it allows designers to build controllers even when their understanding of the mathematical behavior of the system is incomplete. This capacity is especially practical for assessing physiological systems, which are mostly ill-defined with uncertainties in the state descriptions. Therefore, fuzzy logic based control is superior to other control strategies in designing control scheme for anesthesia administration. An automated closed-loop control system [10] has been constructed at Rensselaer based on their previous studies of fuzzy logic in multiple drug hemodynamic control [11]-[14]. The computational intelligence based DOA monitoring methods proposed in this chapter will find further applications in the fuzzy control of the depth of anesthesia.

Computational intelligence (CI) focuses on the use of: (1) fuzzy logic, where imprecise linguistic statements are used for computation; (2) ANN, where simple models of "brains" are constructed to provide useful stimulus-response functions; (3) evolutionary algorithms, where the metaphor of evolution serves as a means for searching for optimal solutions to complex problems; and (4) the combination of above components with each other and/or with other traditional approaches. CI methods have been applied in many medical specializations to assist the human expert, enhance one's senses and skills, acquire new knowledge, and automate procedures. In this chapter, fuzzy logic, ANN, and neuro-fuzzy are used to construct intelligent system for monitoring DOA.

This chapter begins by briefly introducing the CI techniques to be used for DOA monitoring. It also illustrates the application of fuzzy logic in emulating the practices of an anesthesiologist in monitoring and controlling the DOA by using clinical signs (indirect indicators of DOA). As pointed out, the indirect signs of DOA cannot always be reliable for indicating changes in the level of patient consciousness. Therefore, monitoring of the DOA requires additional parameters not easily influenced by the common procedures of the anesthesiologists and surgeons during the operation, while representing a direct indicator of the consciousness. To this end, two more sophisticated DOA models based on the CI techniques are proposed and validated in dog experiments: (1) an ANN based model using MLAEP wavelet transformed coefficients; (2) neuro-fuzzy based model using EEG complexity measure, regularity measure, and spectral entropy. Finally, some general discussions on the CI methods implemented and the performances of the two systems are presented.

2 Computational Intelligence (CI) for DOA

Traditional artificial intelligence has explored the possibilities of creating programs that can compete with humans by means of heuristics, encapsulated knowledge, and symbolic logic. In contrast, computational intelligence has explored the potential for creating intelligent machines by modeling behaviors and mechanisms that underlie biologically intelligent organisms. Nonlinear models, such as those given by ANN and fuzzy logic, have established a good reputation for medical data analysis as being the computational and logical counterparts to statistical methods. The combination of neural learning together with fuzzy logical network interpretations, and neuro-fuzzy methods further provides enhanced problem-solving capabilities. These techniques are introduced briefly and all treated as defining the mapping functions between the input variables (direct or indirect DOA indicators) and the output variable, the DOA.

2.1 Fuzzy Logic Assessment

Fuzzy logic systems are directly based on expert-knowledge. This sort of standard inference method is most suitable in applications where the expert knowledge is directly available for being encoded into the fuzzy inference system by using rules with linguistic labels, such as systems designed for blood pressure control, artificial heart pump rate adjustment, chest computed tomography segmentation, and automatic differentiation of diffuse liver diseases. However, it is usually a time-consuming and laborious process during the design phase when tuning member functions that quantitatively defined those linguistic labels.

Models of awareness are poorly understood (i.e. being a typical ill-defined complex and nonlinear system) and describable only in natural language terms. Variables such as those traditional signs of DOA are not deterministic and there is not a direct (1:1) correlation between any of these variables and the DOA. Estimations of these variables are therefore required due to the complex interactions in an unknown system with unpredictable physiological delays. An anesthesiologist assesses the DOA and controls the anesthetic titration level based on a set of observable measurements of state variables such as hemodynamics, body temperature, spontaneous breathing, and other signs of the DOA. The decision-making process during the assessment ultimately leads to changes in the anesthetic titration level. It is a complex process that very much relies on the experience and knowledge of the anesthesiologist in interpreting those state variables. A fuzzy logic system can thus be substituted for the operation of anesthesia management where the anesthesiologist's knowledge is transcribed and modeled as fuzzy rules for the task of state variable transformation into estimation for controlling actions. The flow of such fuzzy estimation and control process is illustrated in Figure 1.

The *x's* are the signs of DOA measured or secondarily computed, obtained via the sensors placed on the patient. An input variable of the *x's* can be any of the current state (arterial blood pressure), state error (change in arterial blood pressure), state error derivative (rate of change in arterial blood pressure), and state error integral. The output of the fuzzy system changes the current level of anesthetic titration as necessary based on the fuzzy inference process relating the *x's* to that of the anesthetic needs. This process emulates the thought processes of an anesthesiologist in determining the need for changing the titration level based on a collection of observable parameters describable in fuzzy terms as in Figure 2 without being numerically deterministic.

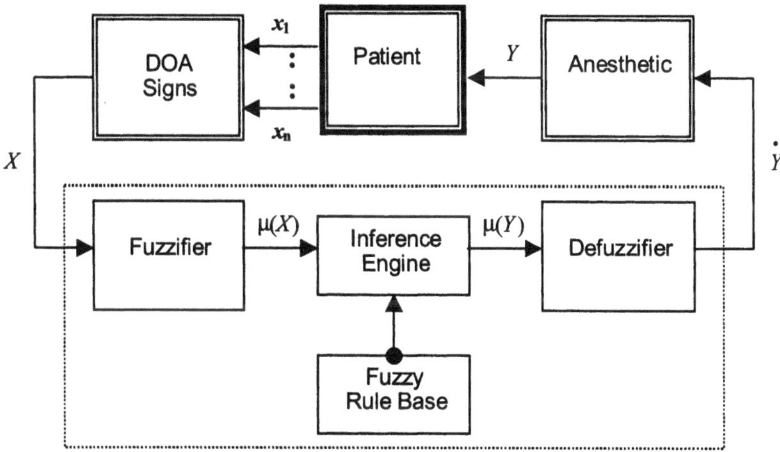

Figure 1. Basic architecture of a fuzzy logic system based on some physician knowledge model. The DOA signs may be any of the "traditional observable inputs" such as hemodynamics, body temperature, patterns of spontaneous breathing, and other indications of awareness. The fuzzy rule base stores the empirical knowledge of the anesthesiologists relating anesthetic titration requirements to changes in DOA signs. The inference process enclosed in the box is the act of DOA assessment.

2.1.1 Fuzzy Inference Process

In such a fuzzy knowledge model, the individual-rule based inference process is conducted by computing the degree of match between the fuzzified input value X and the fuzzy sets describing the meaning of the rule-antecedent as prescribed in the fuzzy ruleset. The fuzzy ruleset contains a series of if-then rules transcribed from an anesthesiologist (expert knowledge). The primary format for each rule is n numbers of "if" conditions as antecedents, which are the fuzzy linguistic DOA signs described earlier, and one or several "then" outcomes as the consequents. The fuzzy consequents are the fuzzy linguistic actions that an anesthesiologist will normally take for changing the anesthetic titration based on the conditions of the antecedents.

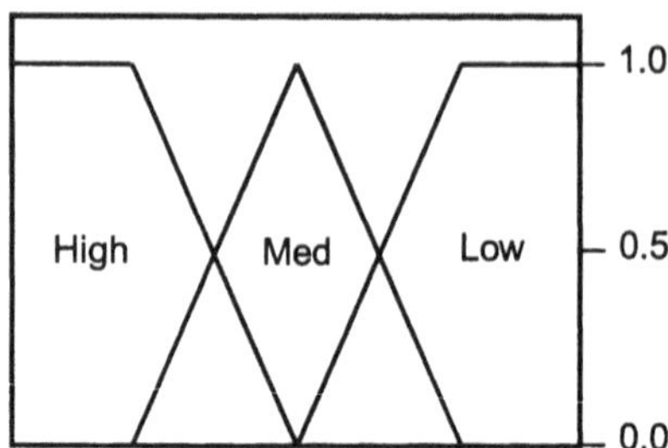

Figure 2. A typical fuzzy membership term set with three membership functions: High, Medium, and Low, can be used semantically in the ruleset to represent knowledge. This term set is therefore applied for fuzzifying the input and output variables in the ruleset. A finer term set with more membership gradation can possibly provide a finer control, however, it also depends on the number of inputs and the size of the ruleset.

Each output is represented by one membership in the ruleset, but in order to reduce the number of rules, an input may have a range of fuzzy memberships. The output μ is produced by clipping the fuzzy member describing the rule-consequent to the degree to which the rule-antecedent has been matched by X. The possibility distribution function is then found by finding the minimal of all μ's:

$$\pi(x_1, \cdots, x_n) = \mu(x_1) \wedge \cdots \wedge \mu(x_n) \tag{1}$$

The minimized value of all μ's therefore determines the degree of applicability for each rule. As π's are aggregated on the fuzzy anesthetic depth term set, the value of the overall output $\dot{Y}$ can then be determined. The rule-consequent is then inferred on the fuzzy anesthetic depth term set. In this example, the defuzzification process utilizes the standard center of gravity method (COG):

$$DEFUZ_{COG}(X) = \frac{\int \mu(x)dx}{\int xdx} \tag{2}$$

The $DEFUZ_{COG}(X)$ determines the output $\dot{Y}$, which is the abscissa of the center of gravity of the area describing the output of the inference engine in the fuzzy anesthetic depth term set.

2.1.2 Why Not Fuzzy?

This fuzzy logic model is based on the states and the changes of various indirect indicators of the DOA, which are variant in a nonlinear system and constantly influenced by unpredictable external events. The application of a muscle relaxant paralyzes patients and ceases any observable muscular movement. Infusions of vasoactive and inotropic drugs diminish the correlations between anesthetic dosage and hemodynamic variability. Furthermore, surgical events and external disturbances reduce the significances of other indirect indicators of the DOA such as breathing patterns and bodily temperatures. Alternatively, studies have shown that the electroencephalogram (EEG) generated from within the central nervous system is an effective sign of the DOA since it provides a graded change associated with an increasing concentration of anesthetics [15]. The EEG can be collected passively or through evoking. Each of the two methods contains different types of information relevant to the graded changes induced by the anesthetic. In the following sections, the computational techniques applied to analyze and process the passive and evoked EEG's under anesthesia are described and compared for their efficiency and effectiveness.

2.2 Artificial Neural Networks

Similar to the human nervous system, an ANN [16] is composed of virtual neurons ("nodes"), axons, and dendrites ("interconnections" and "weights"). Each node contains a simple processing unit (neural element (NE) (see Figure 3a) that uses inputs to calculate an output, just as a biological neuron creates an activation potential based on inputs from its dendrites. This sum is then passed through a threshold unit, whose transfer function usually has the form of a continuous sigmoid shaped gain. Each neural element mathematically defines an N dimensional plane in the N dimensions defined by the inputs. The slope of the plane is defined by the signed weights. A particular set of input values defines a point in this N dimensional space. This point could lie on either side of the plane, or on the plane. The purpose of the threshold unit is to make a decision as to where this point lies. This plane is the decision boundary.

Table 1. Comparison of the anesthesiologist practices and computational intelligence techniques in monitoring the depth of anesthesia.

ANESTHESIOLOGIST		
Input Signals	**Processing**	**Knowledge**
Traditional signs of the DOA such as hemodynamic changes, facial grimacing, muscular movement, lacrimation, spontaneous breathing, and diaphoresis.	Heuristically determined based on the relationship, and the changes among the input signals.	Professional training in anesthesiology and clinical experiences.

COMPUTATIONAL INTELLIGENCE SYSTEMS		
Input Signals	**Processing**	**Knowledge**
Quantified levels and measurable changes in hemodynamic and other monitored parameters of DOA.	Fuzzy Inference	Fuzzy Rule Set
MLAEP	Wavelet Transform	Artificial Neural Network (ANN)
EEG	Complexity Measure	Adaptive Network-based Fuzzy Inference System (ANFIS)
	Approximate Entropy	
	Spectral Entropy	

For complex decision boundaries, a multi-layer neural network is used. Theoretically, a three layer neural network can generate any nonlinear or closed boundary decision function, but often a four-layer network, with two hidden layers, can facilitate the training procedure. Figure 3b shows a four-layer ANN with 2 hidden layers, which consists of an input layer of m neural elements, a hidden layer 1 of n neural elements, a hidden layer 2 of p neural elements, and an output layer of q neural elements. The interconnections between nodes have numerical weights that determine how one node influences the other node. The process of architecture determination (i.e., the selection of the number of hidden layer and the size of each layer) is a delicate process involving issues such as performance, availability of data, and rate of convergence during training (see [17] and [18]). The most widely used training technique of the ANN during the learning phase is the back-propagation. The output error between the actual and the desired outputs is traced back through the network to adjust the weights of the individual neural elements through an error gradient procedure [16]. After training, the obtained parameters of the ANN model can form complex boundaries in the input space for decision-making.

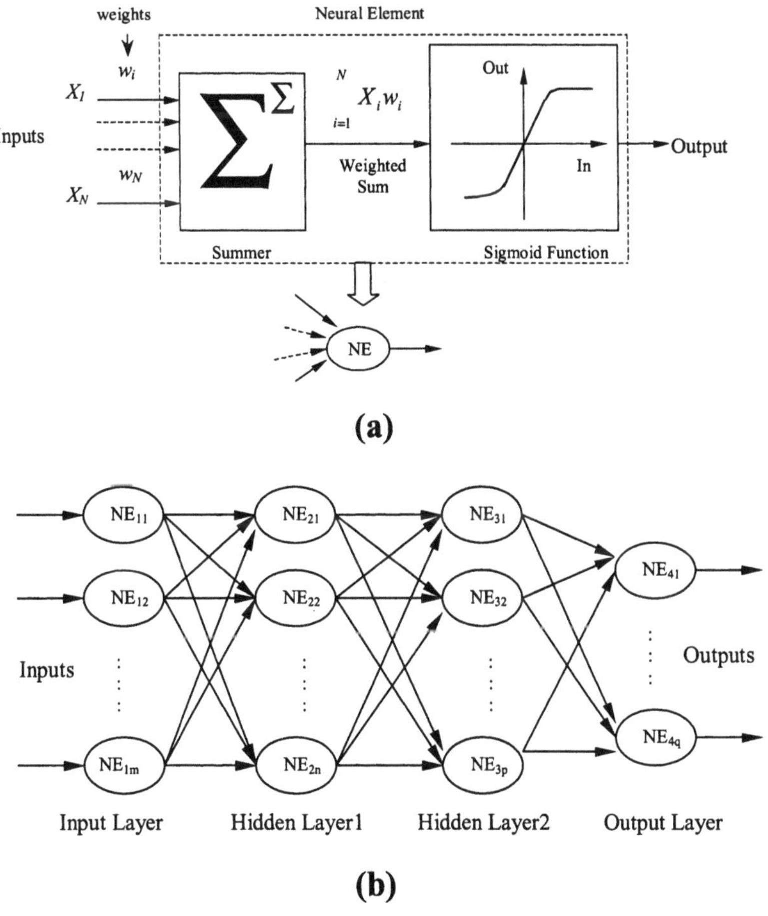

Figure 3. (a) An individual neural element (NE);
(b) A four-layer artificial neural networks (ANN)(m-n-p-q).

In this chapter, the ANN will be used as a classifier for determining the DOA. We intend to leverage the capability of an ANN in learning the features that most discriminate between the states of awake and sleep.

A learned ANN effectively models those multi-dimensional changes in the feature input space, which captures the dynamics of the DOA as the anesthetic level is varied.

2.3 Neuro-Fuzzy Modeling

For the applications where the expert knowledge is not directly available, the neuro-fuzzy methods are used for knowledge acquisition, knowledge refinement, and knowledge interpretation for building effective fuzzy inference systems. Examples are, fuzzy adaptive learning control network (FALCON) [19], adaptive-network-based fuzzy inference system (ANFIS) [20], and adaptive fuzzy neural networks (FuNN) [21]. These systems all utilize multi-layer feed-forward network adaptive architecture, however, relying on different learning algorithms.

Adaptive network based fuzzy inference system (ANFIS), as a neuro-fuzzy method, combines fuzzy logic and neural-nets into a five-layer adaptive network architecture. Details about the structure and learning procedure of ANFIS are in reference [20].

Compared with FALCON and FuNN, ANFIS demonstrates the following advantages: fewer parameters, faster and more accurate learning, better performance in modeling, and better generalization capabilities. These advantageous features made ANFIS a more popular and ideal fuzzy inference system for real-time applications, especially in the medical field. Examples of such applications are: myocardial viability assessment, lung sounds analysis, artificial heart cardiac output estimation, physiological parameters prediction, functional electrical stimulation for locomotion control, and cardiac arrhythmia classification.

To build a derived fuzzy knowledge model based on ANFIS for estimating the DOA, two types of tuning (i.e. model structure tuning and parameter tuning) are required. Structure tuning concerns the structure of the rules: input and output variables selection, variable universe of discourse partition, linguistic labels determination, and type of logical operation to compose each rule. Parameter tuning mainly concerns the fine adjustment of the position of all membership functions together with their shape controlled by *premise parameters* and the Takagi-Sugeno type [6] if-then rules to be extracted controlled by the *consequent parameters*.

Assume, c, a, and e are the input variables of the model, used to estimate the desired output y. Their universes of discourse are defined, respectively, as

$$\begin{aligned} C &= \{c \mid c_l \leq c \leq c_u\} \\ A &= \{a \mid a_l \leq a \leq a_u\} \\ E &= \{e \mid e_l \leq e \leq e_u\} \end{aligned} \tag{3}$$

Where, c_l, c_u, a_l, a_u, e_l, $e_u \in \Re$ are constants representing the upper and lower bounds of input variable operating range. Two fuzzy sets are defined on each of the input spaces, corresponding to linguistic *Small* and *Large* for each variable, and labeled C_k, A_k, and E_k, respectively, with $k = 1, 2$.

The *input space X* is defined as the Cartesian product of the C, A, and E spaces

$$X = \mathrm{C} \times \mathrm{A} \times \mathrm{E} \tag{4}$$

The *output space Y* of the model is defined as

$$Y = \{y \mid y_l \leq y \leq y_u\} \tag{5}$$

Thus, the process to be modeled may be viewed as a mapping from the input space to the output space, which maps C, A, and E to Y. The membership functions, stipulating the linguistic labels of fuzzy sets, take the form of a generalized bell shaped function.

ANFIS employs an efficient hybrid learning procedure that combines gradient descent method and the least squares estimation to tune the parameters both of the membership functions and the Takagi-Sugeno type rules [6]. Each epoch of the learning procedure is composed of a forward pass and a backward pass. In the forward pass, the input data and functional signals go forward to calculate each node output while the *premise parameters* are fixed, and the *consequent parameters* are optimized via least-squares estimation. After the optimum *consequent parameters* are found, the functional signals keep going forward until the output of the network is calculated and the error measure is estimated. Then the backward pass starts. In this stage, the output error propagates from the output end toward the input end while *consequent parameters* are fixed, and the *premise parameters* are optimally updated by the gradient method via a standard back-propagation algorithm. Not only can this hybrid learning procedure decrease the dimension of the search space in the gradient method, but, in general, it will also cut down substantially the convergence time. The least-squares method is, actually, the major driving force that leads to fast training. As a result, ANFIS can usually generate satisfactory results immediately after the first epoch of training, that is, after the first application of the least-squares method. Since the least-squares method is computationally efficient, it can be used for on-line application. Compared with the back-propagation algorithm coupled with forgetting used by FuNN [21] and the hybrid method combining supervised and unsupervised algorithms in FALCON [19], the ANFIS [20] has a learning procedure that is much faster and effective.

Before training, the *consequent parameters* of the ANFIS are all set to zero. As a conventional way of setting parameters in a fuzzy system, the *premise parameters* are set in a way that the membership function can cover the universe of discourse completely, with sufficient overlapping.

After the model being trained, eight Takagi-Sugeno type if-then rules are obtained with fuzzy antecedents, but a crisp consequence, which is a linear combination of the input values plus a constant term, e.g.,

$$(c \in \mathrm{C}_1) \Lambda (a \in \mathrm{A}_1) \Lambda (e \in \mathrm{E}_1) \Rightarrow (y_1 = a_1{*}c + a_2{*}a + a_3{*}e + a_0) \quad (6)$$

The final output y of the model is the weighted average of each rule's output y_i ($i = 1, 2, \ldots, 8$).

3 ANN-Based CI Model for MLAEP

One way to evoke for response is through the auditory pathway since this has been determined to be the most important sensory channel for sensory information processing during general anesthesia [22]. The auditory evoked potentials (AEP) consist of a series of waves that represent processes of transduction, transmission, and processing of auditory information from the cochlea to the brain stem, the primary auditory cortex, and the frontal cortex. The first 10-60 milliseconds of the AEP is known as the mid-latency auditory evoked potentials (MLAEP) and consist of overlapping activation in different structures of the primary auditory cortex. The attenuation of the peak amplitudes and the increase in the latency of the AEP are correlated with increasing blood concentrations of anesthetics, see, e.g., Figure 4.

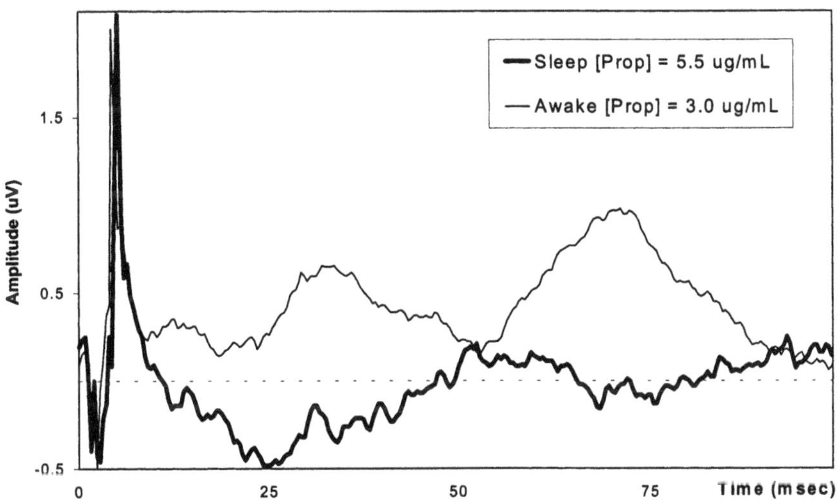

Figure 4. The MLAEP waveforms obtained from a dog that was responsive and not responsive to tail-clamping as a stimulation under the influence of the anesthetic Propofol.

3.1 MLAEP-Derived Parameter Extraction

Features in the MLAEP waveform can be extracted by various time and frequency domain signal analysis techniques. In the time domain, attempts have been made by measuring the amplitude and the lags of the maximum peaks in the MLAEP. However, signal noise and inter-patient variability often render such allometric methods less effective. However, in the frequency domain, Fourier transformation of the MLAEP waveform that decomposes a signal into complex exponential functions of different frequencies has shown to provide gradation of the DOA as the patient shifts from the state of complete awareness to deep sleep. Since the brain has been found to operate around 40 Hz, we can expect predominate power to be found around this band. Selecting those Fourier spectra most sensitive to changes in the DOA is a direct mean of providing features for classification. However, conventional Fourier analysis defined below is most effective for evaluating stationary signals:

$$X(f) = \int_{-\infty}^{+\infty} x(t) \cdot e^{-j2\pi ft} dt \tag{7}$$

since the signal *x(t)*, is multiplied with an exponential term, at some certain frequency *f*, and then integrated over all the times. Therefore, no matter where in time the component with frequency *f* appears, it will affect the result of the integration as well. Since the MLAEP is a non-stationary signal and the variation of MLAEP over different anesthetic dosages brings non-stationary changes in the signal, the Fourier analysis is not the optimal parameter extraction method, as most of the signal strength may be directed toward cancellation. Parameter extraction methods such as the wavelet transformation, specialized for analyzing non-stationary signals, will be more effective.

3.1.1 Wavelet Transformation

The function $\psi(t)$ satisfying the permittable condition $\int_{-\infty}^{+\infty} \psi(t)dt = 0$ is called a *mother wavelet*. The so-called *wavelet* is a family of functions $\{\psi_{a,b}(t)\}$ built by dilating and translating $\psi(t)$:

$$\psi_{a,b}(t) = |a|^{-1/2} \psi(\frac{t-b}{a}), \quad a,b \in R, \quad a \neq 0 \tag{8}$$

where, a and b are scale and position parameter, respectively. The definition of the wavelet transform is

$$W_f(a,b) = < f, \psi_{a,b} > = \frac{1}{\sqrt{|a|}} \int_{-\infty}^{+\infty} f(t) \psi(\frac{t-b}{a}) dt \tag{9}$$

Alternatively, the above formula can be further rewritten in the form of convolution: $W_f(a,b) = f(t) * (1/\sqrt{|a|}) \psi(-t/a)$. As can be seen, $W_f(a,b)$ is the output of a filter with the transfer function $\sqrt{|a|}\hat{\psi}(-a\omega)$, with signal *f(t)* as input. Therefore, wavelet transform is essentially equivalent to using a family of band-pass filters to make multi-passband filtering (one scale a corresponding to one pass-band) for the signal, thus analyzing the information of the signal in different bands.

The central frequency and the bandwidth of the band-pass filter is inversely proportional to the scale a, with the bandwidth automatically being adjusted with the change of the central frequency. The lower the central frequency the narrower the bandwidth, while the higher the central frequency the wider the bandwidth. This embodies the adaptive and zooming ability of the wavelet transform. Thus, the wavelet transformation possesses such advantages over the classical Fourier transformation in signal processing.

Wavelet analysis of the MLAEP is therefore more suitable than Fourier analysis because of the MLAEP's non-stationary nature as its frequency response varies in time. It is a technique more sensitive to the increases in the latency and the decreases in the frequency of the signal as the anesthetic level is raised. In the CI model designed for MLAEP, the discrete time wavelet transformation (DTWT) is used. It is an orthogonal representation of the signal computed by convolution with a series of Finite Impulse Response filters. The filter coefficients derived in [23] lead to a prototype wavelet whose dilation and translation forms an orthogonal set of basis vectors. The DTWT decomposes a given signal into overlapping frequency bands determined by the filters with impulse response $h_{u,b}(t)$. DTWT is mathematically expressed as

$$y_h^1(n)=\sum_{k=0}^{N-1} x(k)h(2n-k) \quad (10)$$

$$y_g^1(n)=\sum_{k=0}^{N-1} x(k)g(2n-k) \quad (11)$$

where N is the sample size equal to some positive integer with a power of 2. The filtered signals are sub-sampled by a factor of two to reduce redundant information, where the time and frequency resolutions are reduced by half. The output of the lowpass filter $g(n)$ is subjected to filtering and sub-sampling in subsequent steps with

$$y_g^i(n)=\sum_{k=0}^{N-1} y_g^{i-1}(k)g(2n-k) \quad (12)$$

$$y_h^i(n)=\sum_{k=0}^{N-1} y_g^{i-1}(k)h(2n-k) \quad (13)$$

where i denotes the iteration number. The highpass filter outputs obtained at each step form the DTWT coefficients.

3.2 System Design Based on ANN for MLAEP

In Figure 5, the flow process of a system [10] incorporating the CI techniques of ANN using the DTWT of the MLAEP waveform is shown. In this system, a 256-point discrete time wavelet transformation of each MLAEP was obtained by using the pyramidal algorithm with a 20th order Daubechies filter [24]. Since a neural network to be trained by a fixed number of training data can be designed by keeping the input feature dimension small and the number of hidden nodes at a minimum, it will tend to have better generalization properties on the testing data. To reduce feature dimensions, stepwise discriminant analysis (SDA) based on Wilk's Lambda Criterion [25] was applied on all data points in deriving the coefficients that have maximum power for discriminating between the responders and the non-responders. The wavelet coefficients 4, 6, 8, and 13 were found to provide maximum discriminating power using such analysis. These wavelets were then used as the inputs to a neural network classifier. The magnitude differences of those SDA-selected wavelet coefficients enable the neural network in classifying the state of the subject under anesthesia.

A four-layer perceptron feedforward network with two hidden-layers is used as the architecture of this neural network [16]. The number of nodes in the first hidden layer was determined experimentally by trial and error while the number of nodes in the second hidden layer was determined by the number of clusters in the input sample space. The training of the neural networks was accomplished by back-propagation by minimizing the mean squared error between the desired output (DOA grading after tail clamping) and the actual output (corresponding neural network output). The training tolerance was arbitrarily set at 0.2 while the testing tolerances was set at 0.49 to reflect the binary nature of the decision process. To train the ANN, 90% of the available 113 data points were randomly chosen for training and the remaining 10% were used for testing as a method of measuring the testing performance. The training process stopped after all training data has been trained within a 0.2 error tolerance, and an accuracy of 100% was achieved for testing that remaining 10% of the data. The weights of each connection in the ANN are now fixed and would be used in future fully automated experiments.

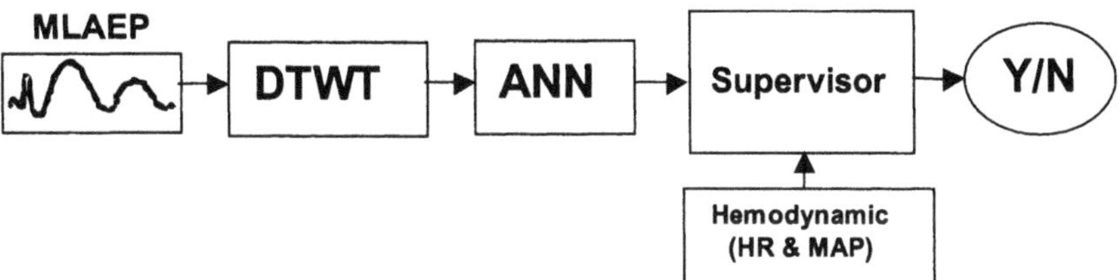

Figure 5. System diagram of the CI model designed for MLAEP. In this process, the acquired MLAEP is passed through discrete wavelet transformation. An artificial neural network based classifier is then used to determine the DOA from the wavelet transformed MLAEP. A supervisor determines whether to increase (Y) or decrease (N) the anesthetic infusion rate based on the DOA and the current hemodynamic state. The system considers the hemodynamic state such as the heart rate (HR) and mean arterial pressure (MAP) to ensure that the non-responsiveness in the subject is not caused by anesthetic overdosing which may lead to life-threatening complications.

The drug agent used in the experiment is the intravenous anesthetic, Propofol. Both the training and the testing data were gathered by incrementing or decrementing the Propofol setpoint ($[Prop]_{sp}$) in steps of 0.5 μg/ml between 2 μg/ml and 13 μg/ml. At each level of $[Prop]_{sp}$, a minimum stabilization period of 10 minutes or more was allowed until the Propofol concentrations in the plasma ($[Prop]_{plas}$) and the effect-site ($[Prop]_{eff}$) have stabilized and equilibrated as estimated by a mathematical model. The MLAEP, HR, MAP, and $[Prop]_{eff}$ were recorded at the end of each stabilization period before a 30-second tail clamp, considered as a supra-maximal stimulus in dogs, was applied to assess the awareness. During each tail clamp, the determination for grading a positive responder (depth zero) or a negative responder (depth one) was estimated.

3.3 ANN System: Experiment Results

Subsequent fully automated experiments were conducted in animal subjects for the validation of the ANN based CI system designed for the MLAEP. The results obtained from an automated experiment on a 16-kg 100-cm male dog are shown in Figure 6.

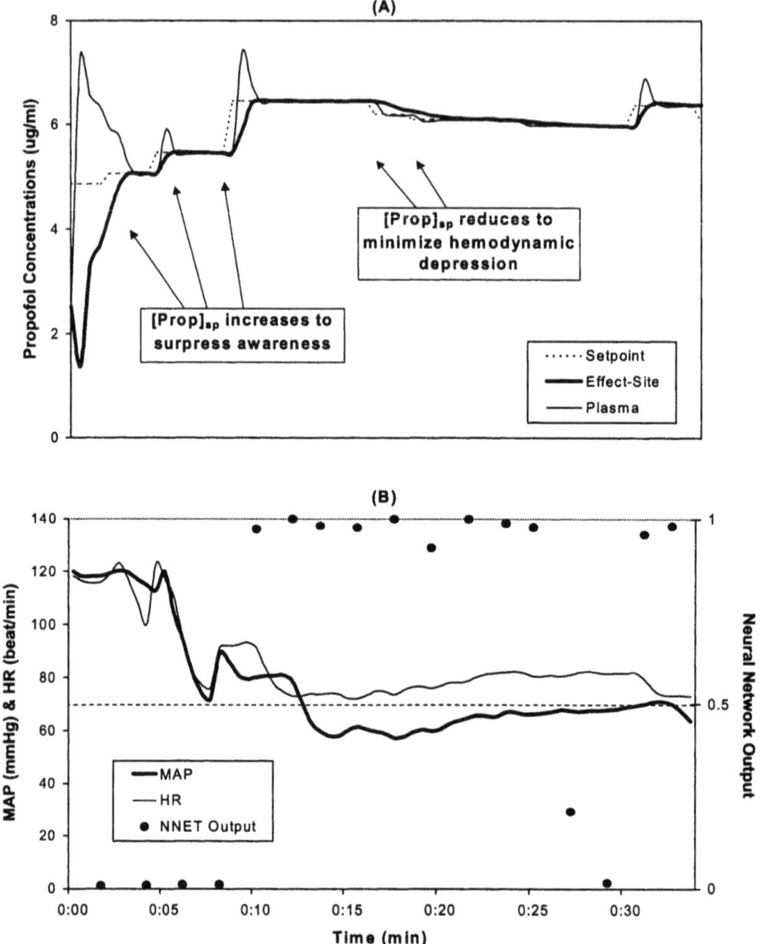

Figure 6. (a) The Propofol setpoint, effect-site and plasma concentrations vary with time during the experiment. The variable step incrementation of Propofol was designed to ensure sufficient anesthetic is delivered while minimizing the responsive period. At various instants, the supervisor performed Propofol decrementation by assessing the hemodynamic states. (b) The neural network outputs, the MAP and the HR are shown. Those states with the neural network outputs of less than 0.5 are presumed to be responsive. Those states are non-responsive when the neural network outputs are greater than 0.5.

In Figure 6a, the system pre-determined the initial $[Prop]_{sp}$ to be 4.87 ug/ml based on the patient profile. The system intervened every two minutes by extracting the MLAEP for processing while the MAP and the HR were being sampled every 30 seconds. Since the system has determined the subject to be responsive for several consecutive terms, the $[Prop]_{sp}$ was step-incremented up to 6.47 ug/ml to guarantee non-responsiveness in the subject. Between the 16-minute and the 25-minute interventions, the slightly depressed hemodynamic states (Figure 6b) have triggered the supervisor to lower the $[Prop]_{sp}$ during several interventions. After 26.5 minutes into the experiment, the system has determined that the anesthesia might be insufficient in the subject; however, the supervisor decided to hold. During the 28.5-minute intervention, a 0.4 ug/ml increase in the $[Prop]_{sp}$ was executed as the system has confirmed the current anesthesia level to be inadequate. The $[Prop]_{sp}$ will oscillate at this level as long as the hemodynamic states are depressed. Nevertheless, the incrementation has the precedence over the decrementation of the $[Prop]_{sp}$ if the classifier has determined the anesthesia to be insufficient, unless the system has determined that the further increases in the $[Prop]_{sp}$ may be detrimental to the hemodynamic states.

4 Neuro-Fuzzy Based CI Model for EEG

Time-domain, frequency-domain, time-frequency domain, and bispectral domain techniques [26]-[28] have been used for processing the EEG to develop a method, which will measure DOA. None of these techniques has found wide clinical acceptance and application. This partly may be attributed to the fact that commonly used signal analysis is based on the assumption that the EEG signal arises from a linear and stationary process. In reality, the nonlinear nature of brain neuronal activity contributes to the formation of the EEG with very complex dynamics [29]. In certain cases, the nonlinearity is an important factor that should be taken into consideration during processing of the EEG [30]. Moreover, the EEG may not be simply generated by a purely deterministic or stochastic process, but rather by some combination of both. The EEG does not change in a linear or monotonic fashion with changes in DOA, and different EEG-derived parameters are not equally useful in estimating DOA. The derived parameters should be used in combination and each method weighted differently as the EEG changes nonlinearly with various levels of stimulation and from light to deep anesthesia. The emerging computational intelligence, the neuro-fuzzy method, can act as a promising modeling candidate.

4.1 EEG-Derived Parameter Extraction

Through nonlinear quantitative analysis, two EEG-derived parameters, complexity measure *C(n)* [31] and approximate entropy *ApEn* [32], are extracted from the raw EEG signals and merged together with the spectral entropy *SE* [35] for estimating the DOA. *C(n)* and *ApEn* quantify the complexity and regularity of the EEG dynamic patterns in a manner consistent with our intuition, as well as being model-independent statistics. Recent studies [34]-[36] indicate their usefulness as relevant features for DOA estimation.

4.1.1 Complexity Analysis

Complexity is a common characteristic of many phenomena, especially for biological systems, with the brain often described as the most complex biological system [29]. Its electrical activity (EEG) exhibits significant complex behavior, which is generated by numerous neuroelectrical events within the brain's structure. The complexity measure $C(n)$, proposed by Lempel and Ziv [33], is extremely well suited for characterizing different spatiotemporal patterns with chaotic temporal components and their development in high-dimensionality nonlinear systems. Compared with other types of complexity measures, the computation of $C(n)$ is simpler, faster, and more suited to real-time EEG analysis [34]. Complexity measures the number of distinct patterns that must be copied to reproduce a given string. The only computer operations considered in constructing a string are copying old patterns and inserting new ones. Briefly described, a string $S = s_1s_2\ldots s_n$ is scanned from left to right, and a complexity counter $c(n)$ is increased by one unit every time a new sub-string of consecutive characters is encountered in the scanning process. After normalization, the complexity measure $C(n)$ reflects the rate of new patterns arising with the increase of string length n. Detailed algorithms for $C(n)$ can be found in [31] and [33].

4.1.2 Regularity Analysis

Approximate Entropy (*ApEn*) is developed to quantify the amount of regularity in the data without any *a priori* knowledge about the system generating them [32]. It is a nonnegative number that will distinguish among data sets, with larger numbers indicating more irregularity, unpredictability, and randomness. *ApEn* is nearly unaffected by low level noise, is robust to occasional very large or small artifacts, gives meaningful information with a reasonable number of data points, and is finite for both stochastic and deterministic processes. These features are useful for quantitatively characterizing changes in the evolving regularity of the EEG. While applying *ApEn* to the EEG, a particular model form is not being sought, such as deterministic chaos, but the intent is to distinguish among the EEG data sets collected under different anesthesia conditions on the basis of regularity. Such regularity can be seen in both deterministic and/or random (stochastic) processes, similar to brain activity. Detailed algorithms for *ApEn* can be found in [32].

4.1.3 Spectral Entropy Analysis

Spectral entropy (*SE*) [35] is selected as the third derived parameter. This measure quantifies the spectral complexity of the EEG signal. The power spectral density (PSD) $\hat{P}(f)$ can be obtained from the EEG signal by a fast Fourier transformation (FFT). The normalization of $\hat{P}(f)$, with respect to the total spectral power, will yield a normalized density function. Application of Shannon's channel entropy gives an estimation of the spectral entropy (*SE*) of the underlying EEG process, where entropy is given as

$$H = \sum_f p_f \ln(1/p_f) \qquad (14)$$

where p_f is the normalized density function value at frequency f. Heuristically, the entropy has been interpreted as a measure of uncertainty about the event at f.

4.2 ANFIS – "Derived" Fuzzy Knowledge Model

By using the ANFIS method, fuzzy if-then rules are obtained to express the complex relationship between the three derived parameters and anesthesia states. These rules are then used to construct a derived fuzzy knowledge model for providing a single variable to represent the DOA [36].

For example, by only using propofol EEG data pairs (*i.e.*, EEG-derived parameters and anesthesia states) for training, the Takagi-Sugeno type rules obtained from the ANFIS are listed in Table 2.

Table 2. Extracted rules by ANFIS for estimating DOA under propofol regimen.

If C(n) is	and ApEn is	and SE is	Then $DOA = a_1 * C(n) + a_2 * ApEn + a_3 * SE + a_0$			
			a_1	a_2	a_3	a_0
Small	*Small*	*Small*	0.26	-0.95	1.14	-2.38
Small	*Small*	*Large*	-1.51	0.47	2.88	-9.46
Small	*Large*	*Small*	-7.32	7.48	-0.75	-0.77
Small	*Large*	*Large*	-0.73	0.06	0.03	0.46
Large	*Small*	*Small*	-158.52	206.01	-68.93	171.67
Large	*Small*	*Large*	105.13	-52.98	3.07	-33.00
Large	*Large*	*Small*	58.49	1.22	4.72	-58.15
Large	*Large*	*Large*	-1.75	-4.08	0.30	5.26

These eight extracted fuzzy rules along with the three input parameters can construct a "derived" fuzzy knowledge model for estimating DOA under propofol regimen. Two membership *Small* and *Large* functions are associated with each input, so the input space is partitioned into eight fuzzy subspaces, each of which is governed by a fuzzy if-then rule. The *premise* part of a rule delineates a fuzzy subspace, while the *consequent* part specifies the output within this fuzzy subspace. The weighted average of the outputs of these eight fuzzy subspaces, i.e. the final output of the model, is a *DOA* index between 0.0 and 1.0, which represents the degree of depth of anesthesia.

The total number of fitting model parameters is 50 (18 *premise* and 32 *consequent* parameters). After being trained, the ANFIS only need perform forward computing for estimating *DOA*. The time needed is about one millisecond. Such a model has the potential to improve real-time DOA estimation accuracy under a propofol regimen while still retaining the structural knowledge.

In the same way, specific "derived" models for other anesthetic regimens, isoflurane and halothane, or a general model across three regimens can also be constructed and justified (see results in Section 4.4).

The meaning of the word "derived" is triple-fold: (1) the input parameters are derived from the EEG by signal processing, not like the hemodynamic parameters, heart rate and blood pressure; (2) the fuzzy knowledge is derived with the help of ANFIS, not directly from experts; (3) the final model and the *DOA* index are derived, not from published data or experience.

4.3 System Design Based on ANFIS for EEG

The designed DOA estimation system (Figure 7) consists of two paths: a dashed line path for off-line training of the ANFIS before the system is put into operation and a solid one for on-line DOA estimation. These two parts contain similar function blocks: EEG collection, Parameters Extraction, and ANFIS. Before the system goes into operation, a Specific Raw EEG Database must be first built for off-line training of the ANFIS. The complexity *C(n)*, regularity *ApEn*, and spectral entropy *SE* are extracted from the raw EEG and form an input feature vector for training the ANFIS. After training, the derived fuzzy if-then rules can be used for on-line DOA estimation. During the on-line application, the recorded EEG is also stored in the Specific Raw EEG Database for updating. Thus, at every period (Δt) the ANFIS is retrained using the newly updated Specific Raw EEG Database and then the new *premise* and *consequent* parameters are sent to the Trained ANFIS for updating its fuzzy if-then rules. In so doing, the system is dynamic not static, and can be continuously refreshed. In addition, a Specific Raw EEG Database for different anesthetic regimens can be constructed, such as for propofol, isoflurane, or halothane. Thus, regimen-specific or general-purpose DOA estimation systems can be easily built.

Adding the times needed to calculate *C*(*n*), *ApEn*, and *SE*, as well as the ANFIS forward computing time, the total time needed to estimate the *DOA* is: 94 + 3911 + 7 + 1 = 4015 ms = 4.015 s. Therefore, the proposed scheme is computationally fast, feasible and suitable for real-time on-line application, where every 10 s one DOA estimation is enough. One example of the results in continuously estimating DOA during a dog experiment is shown in Figure 8 (see Section 4.4).

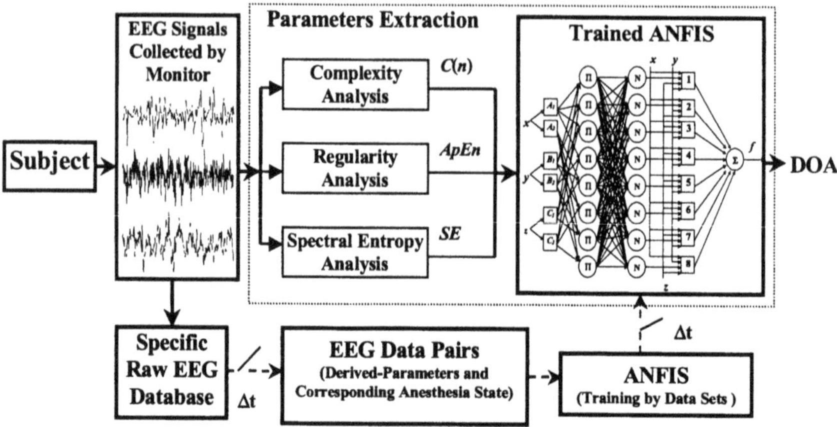

Figure 7. The system diagram for estimating DOA during anesthesia by integrating the complexity, regularity, and spectral entropy information of EEG via ANFIS: dashed flow line for off-line training ANFIS and solid flow line for on-line estimating DOA. Δt denotes that every certain period (Δt) the ANFIS are retrained using the updated Specific Raw EEG Database. The "derived" fuzzy knowledge model is encircled by the dotted rectangle.

4.4 ANFIS System: Experiment Results

Thirty experiments using 15 dogs undergoing anesthesia with three different anesthetic regimens (propofol, isoflurane, and halothane) were performed and a database was obtained. The database consists of EEG recordings and the associated, clinically derived anesthesia states. Totally, 134 EEG recordings were obtained from the propofol experiments, 109 recordings from isoflurane experiments, and 64 recordings from halothane experiments.

To verify and justify the performance of the model in discriminating awake and asleep states and test the applicability for practical use of the model under different anesthetic regimens, the EEG data sets collected under a specific regimen were used to train and test the model. The test results are listed in Table 3 via a "leave-one-out" approach [28].

In order to test the generalization ability of the proposed scheme in detecting anesthesia states, the EEG data sets collected under the three kinds of regimens were mixed and the model was trained and tested by the "leave-one-out" procedure (results see Table 4).

Table 3. Test results by the derived fuzzy knowledge model using the "leave-one-out" procedure for three kinds of regimens, respectively (*i.e.* training and test data sets from the same kind of regimen).

Anesthetic Regimen	**State**	**Sensitivity (%)**	**Specificity (%)**	**Accuracy (%)**
Propofol	Awake	92.3	88.4	90.3
	Asleep	88.4	92.3	90.3
Isoflurane	Awake	89.6	95.1	92.7
	Asleep	95.1	89.6	92.7
Halothane	Awake	82.1	94.4	89.1
	Asleep	94.4	82.1	89.1

Table 4. Test results by the derived fuzzy knowledge model using the "leave-one-out" procedure for studying the generalization ability of the proposed DOA estimation model (*i.e.*, training and test data set from all of the three regimens).

Anesthetic Regimens	**State**	**Sensitivity (%)**	**Specificity (%)**	**Accuracy (%)**
Propofol* + *Isoflurane* + *Halothane	Awake	84.4	87.3	85.9
	Asleep	87.3	84.4	85.9

Using the fuzzy if-then rules listed in Table 2, a DOA estimation system (Figure 7) for the propofol regimen, for example, was obtained. Figure 8 shows the output of this system used during one dog experiment under propofol anesthesia. The windows for calculating the three parameters ($C(n)$, *ApEn*, and *SE*) move forward 500 data points (5 s) for each *DOA* estimation.

During the experiment, the anesthesia state is being changed by adjusting the propofol concentration setpoint (Cs). The estimated *DOA* index (with gradual scaling) continuously tracks the anesthesia state transitions with definite response. Moreover, the value of *DOA* is consistent with the dog's true anesthesia states assessed by the clinician at observation points. The deeper the depth of anesthesia, the higher the value of the output of the model. The system works well for on-line use in real time.

Clinically the *DOA* number tells, in a timely manner, how awake and asleep the patient is. This means that the *DOA* number can help directly assess a patient's level of anesthesia. Currently, anesthesiologists make a subjective guess on DOA according to the observed vital signs and their experience.

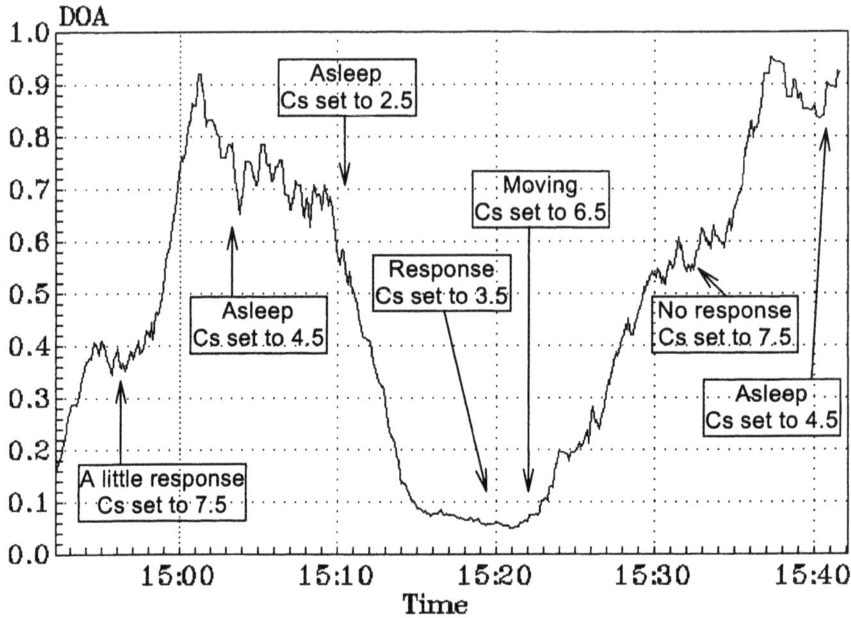

Figure 8. The continuously estimated DOA index by the derived fuzzy knowledge model *versus* time under different anesthesia situations during part of one dog experiment using propofol. Cs denotes the propofol concentration (μg/ml) setpoint at the site of drug effect. The annotation on the figure indicates the state and the concentration set at that observation point.

5 Discussions

5.1 ANN versus ANFIS

Comparing the ANN and the ANFIS in modeling the DOA, the model derived in the ANN is a complete "black box", which does not guarantee convergence. The capability in utilizing linguistic information is specific only to fuzzy inference systems, and is not always available in the ANN. Moreover, the convergence speed can be very slow during training. Therefore, run-time updates of the parameters when new data is available, will be difficult to implement especially when the training process is highly supervised. Another drawback of the ANN is the lack of a direct means in determining the number of nodes in a hidden layer. In contrast, the ANFIS is a useful tool in eliciting knowledge from the training input-output data pairs for building the DOA model. The derived numerical quantitative features from EEG by signal processing, such as *C(n)*, *ApEn*, and *SE*, contain the relevant information about the DOA, but the anesthesiologists have no direct knowledge and expertise using them for assessing DOA. After training ANFIS, the information is derived as knowledge in the form of fuzzy if-then rules. This process will help anesthesiologists in using these EEG-derived parameters in their practice and understand the process of the inference system. From the extracted if-then rules (as listed in Table 2), the anesthesiologists can obtain the knowledge about the relationship between these derived parameters and the DOA and on how to use these derived parameters to get the *DOA* number (which is automatically produced as the output from the model). Fuzzy rules are used here as a framework for knowledge representation. The final output of the model is just one single *DOA* number between 0.0 and 1.0. The number 0.0 represents full awake and 1.0 denotes a flat line of EEG, or complete EEG suppression. The *DOA* number quantitatively tells anesthesiologists the depth of anesthesia. Therefore, readability of the model is at both input and output sides of the model, and somewhat inside.

5.2 EEG versus MLAEP

Although MLAEP has the potential in generating a higher signal-to-noise ratio, the use of the EEG for DOA estimation is preferred clinically. The MLAEP is evoked by rarefaction auditory clicks (e.g. 0.1-ms duration and 70 dB greater than the normal hearing level) presented binaurally with a stimulation frequency (e.g. 9.3 Hz). Thus, given the patient has normal hearing function, one MLAEP recording is generated from 1000 responses with a stimulation rate of 9.3 Hz represented approximately a two-minute period. This means an MLAEP is produced only about every two minutes in providing a single DOA estimation. This may not be enough for the continuous monitoring of the DOA. On the other hand, in the EEG technique, DOA estimation can be performed at 10-second intervals without any special patient requirement. The ease of application, faster response time, and less intrusiveness to the patients, make the EEG technique more appealing as the modality of choice in clinical settings.

5.3 Performance Issues

The successes demonstrated by these proposed DOA models (ANN based MLAEP model and ANFIS based EEG model) depend heavily on the following two considerations: input variable selection and model method selection. It is well known that different signal processing techniques can derive different parameters from MLAEP and EEG to represent different underlying information. The parameters that mostly correlate with the DOA and most suitable for MLAEP and EEG's should be used as the inputs of the model. For this reason, after wavelet transformation on MLAEP, four DTWT coefficients are selected to combine with the universal approximator, which is the ANN. This special combination results in the better performance of the ANN based MLAEP model. For the ANFIS based EEG model, the spectral entropy *SE* measure, complexity measure *C(n),* and regularity measure *ApEn* are selected, since they provide insight into the nonlinear dynamic mechanisms underlying brain activity and allow insight into the evolution of complexity and regularity of the EEG. The advantages of ANFIS over other neuro-fuzzy methods have made ANFIS the tool of choice for mapping the relationship between the three parameters and the *DOA* index. The combinatory approach in using complexity analysis, regularity analysis, spectral entropy analysis (for deriving input variables) and ANFIS (for mapping) enable superior performance in deriving the *DOA* index.

Table 3 shows that the neuro-fuzzy based model has an accuracy in the range of 90% for detecting awake and asleep states under different specific anesthetic regimen. This also demonstrates the capability of the *DOA* index in quantitatively characterizing the level of anesthesia is clinically acceptable. Table 4 further shows that the model has generalization ability with an accuracy of 85.9% in discriminating anesthesia states across the three regimens tested. The DOA number correlates well with the level of anesthesia. Moreover, the DOA number is subject independent (i.e., not sensitive to the large intra- and inter-individual variability), therefore, calibration will not be necessary for any specific individual to be monitored.

The proposed derived fuzzy knowledge model demonstrates the following advantages:

(1) Gradual scaling: model output *DOA* index scaling the depth of anesthesia gradually from 0.0 (full awake) to 1.0 (complete EEG suppression). As such, the *DOA* index quantitatively characterizes the depth of anesthesia, not just qualitatively.
(2) Definite response to the change of anesthesia states.
(3) Accurate discrimination of awake and asleep states.
(4) Independence from the subject to test.
(5) Generalization ability across different anesthetic techniques.
(6) Predictive for the appearance of clinical signs of an inadequate anesthesia.
(7) Readability of the model at the input and output sides, even inside to some extent.
(8) Real-time feasibility for online clinically providing *DOA* index in a timely manner, which is very important for implementation in a real-world monitor.

The output of the proposed model offers all the desirable features for a DOA monitoring index, therefore, this makes the proposed fuzzy knowledge model a promising candidate as an effective tool for continuous assessment of the depth of anesthesia.

The models proposed here do not touch upon another CI technique, evolutionary algorithms, which cover the fields of genetic algorithms, evolution strategies and evolutionary programming. These methods can be used to optimize technical problems and designed methods [37]. The combination of evolutionary algorithms and fuzzy logic or neuro-fuzzy may be the future direction of CI in the field of DOA monitoring.

Acknowledgments

This work was supported by the National Science Foundation under Grant BES-9522639 and by the Whitaker Foundation.

References

[1] Domino, K.B., Posner, K.L., Caplan, R.A., and Cheney, F.W. (1999), "Awareness during anesthesia," *Anesthesiology*, vol. 90, pp. 1053-1061.

[2] Linkens, D.A. (1992), "Adaptive and intelligent control in anesthesia," *IEEE Control Systems*, pp. 6-11.

[3] Kwok, K.E., Shah, S.L., Clanachan, A.S., and Finegan, B.A. (1995), "Evaluation of a long-range adaptive predictive controller for computerized drug delivery systems," *IEEE Trans. Biomed. Eng.*, vol. 42, pp. 79-86.

[4] Wada, D.R. and Ward, D.S. (1995), "Open loop control of multiple drug effects in anesthesia," *IEEE Trans. Biomed. Eng.*, vol. 42, pp. 666-677.

[5] Valcke, C.P. and Chizeck, H.J. (1997), "Closed-loop drug infusion for control of heart-rate trajectory in phamacological stress tests," *IEEE Trans. Biomed. Eng.*, vol. 44, pp. 185-195.

[6] Sugeno, M. and Kang G.T. (1988), "Structure identification of fuzzy model," *Fuzzy Sets and Systems*, vol. 28, pp. 15-33.

[7] Oshita, S., Nakakimura, K., and Sakabe, T.(1994), "Hypertension control during anesthesia: fuzzy logic regulation of nicardipine infusion," *IEEE Eng Med and Bio Mag*, vol. 13, no. 5, pp. 667-670.

[8] Hao, Y. and Sheppard, L.C. (1994), "Regulation mean arterial pressure in postsurgical cardiac patients: a fuzzy logic system to control administration of sodium nitroprusside," *IEEE Eng Med and Bio Mag*, vol. 13, no. 5, pp. 671-677.

[9] Guignard, B., Menigaux, C., Dupont, X., and Chauvin, M. (1998), "Fuzzy logic closed loop system for propofol administration using bispectral index and hemodynamics," *Anesthesiology*, vol. 89(3A), p. 1218.

[10] Huang, J.W., Lu, Y.-Y., Nayak, A., and Roy, R.J. (1999), "Depth of

anesthesia estimation and control," *IEEE Trans. Biomed. Eng.*, vol. 46, no. 1, pp. 71-81.

[11] Huang, J.W. and Roy, R.J. (1998), "Multiple-drug hemodynamic control using fuzzy decision theory," *IEEE Trans. Biomed. Eng.*, vol. 45, no. 2, pp. 213-228.

[12] Held, C.M., and Roy, R.J. (1995), "Multiple drug hemodynamic control by means of a supervisory-fuzzy rule-based adaptive control system: validation on a model," *IEEE Trans. Biomed. Eng.*, vol. 42, no. 4, pp. 371-385.

[13] Huang, J.W., Held, C.M., and Roy, R.J. (1999), "Hemodynamic management with multiple drugs using fuzzy logic," in Teodorescu, H.-N., Kandel, A., and Jain, L.C. (Eds.), *Fuzzy and Neuro-Fuzzy Systems in Medicine*, CRC Press (Boca Raton, London, New York, and Washington DC), chapter 11, pp. 319-340.

[14] Nayak, A. and Roy, R.J. (1998), "Anesthesia control using midlatency auditory evoked potentials," *IEEE Trans. Biomed. Eng.*, vol. 45, no. 4, pp. 409-421.

[15] Gibbs, G.A., Gibbs, E.L., and Lennox, W.G. (1937), "Effect on the electro-encephalogram of certain drugs which influence nervous activity," *Arch. Int. Med.*, vol. 60, pp. 154-166.

[16] Haykin, S. (1994), *Neural Networks: a Comprehensive Foundation*, New York: Macmillan College Publishing Company.

[17] Mehrotra, K.G., Mohan, C.K., and Ranka, S. (1991), "Bounds on the number of samples needed for neural learning," *IEEE Trans. on Neural Networks*, vol. 2, pp. 548-558.

[18] Mirchandani, G. and Cao, W. (1989), "On hidden nodes for neural nets," *IEEE Trans. on Circuits and Systems*, vol. 36, pp. 661-664.

[19] Lin, C.-T. and Lee, C.S.G. (1991), "Neural-network-based fuzzy logic control and decision system," *IEEE Trans. Comp.*, vol. 40, no. 12, pp. 1320-1336.

[20] Jang, J.-S.R. (1993), "ANFIS: adaptive-network-based fuzzy

inference system," *IEEE Trans. On Systems, Man, and Cybernetics*, vol. 23, no. 3, pp. 665-684.

[21] Kasabov, N.K. (1996), "Learning fuzzy rules and approximate reasoning in fuzzy neural networks and hybrid systems," *Fuzzy Sets and Systems*, vol. 82, pp. 135-149.

[22] Goldmann, L. (1988), "Information processing under general anaesthesia: a review," *J. R. Soc. Med.*, vol. 81, pp. 224-227.

[23] Daubechie, I. (1990), "The wavelet transform, time-frequency localization and signal analysis," *IEEE Trans. Info. Theory*, vol. 36, pp. 961-1005.

[24] Daubechie, I. (1988), "Orthonormal bases of compactly supported wavelets," *Comm. on Pure and Appl. Math.*, vol. 41, pp. 909-996.

[25] Klecka, W.R. (1980), *Discriminant Analysis*, London, Sage Publications.

[26] Nayak, A., Roy, R.J., and Sharma, A. (1994), "Time-frequency spectral representation of the EEG as an aid in the detection of depth of anesthesia," *Annals of Biomed. Eng.*, vol. 22, pp. 501-513.

[27] Katoh, T., Suzuki, A., and Ikeda, K. (1998), "Electro-encephalographic derivatives as a tool for predicting the depth of sedation and anesthesia induced by sevoflurane," *Anesthesiology*, vol. 88, pp. 842-650.

[28] Olofsen, E. and Dahan, A. (1999), "The dynamic relationship between end-tidal sevoflurane and isoflurane concentrations and bispectral index and spectral edge frequency of the electroencephalogram," *Anesthesiology*, vol. 90, pp. 1345-1353.

[29] Koch, C. and Laurent, G. (1999), "Complexity and the nervous system," *Science*, vol. 284, pp. 96-98.

[30] Micheloyannis, S., Flitzanis, N., Papanikolaou, E., Bourkas, M., and Terzakis, D. (1998), "Usefulness of non-linear EEG analysis," *Acta Neurol. Scand.*, vol. 97, pp. 13-19.

[31] Kaspar, F. and Schuster, H.G. (1987), "Easily calculable measure for the complexity of spatiotemporal patterns," *Phys Rev A*, vol. 36, pp. 842-848.

[32] Pincus, S.M., Gladstone, I.M., and Ehrenkranz, R.A. (1991), "A regularity statistic for medical data analysis," *J Clin Monit*, vol. 7, pp. 335-345.

[33] Lempel, D. and Ziv, J. (1976), "On the complexity of finite sequences," *IEEE Trans. on Info. Theory*, vol. 22, pp 75-81.

[34] Zhang, X.-S. and Roy, R.J. (1999), "Predicting movement during anesthesia by complexity analysis of the EEG," *Med. & Biol. Eng. & Comp.*, vol. 37, pp. 327-334.

[35] Rezek, I.A. and Roberts, S.J. (1998), "Stochastic complexity measures for physiological signal analysis," *IEEE Trans. Biomed. Eng.*, vol. 45, no. 9, pp. 1186-1191.

[36] Zhang, X.-S. and Roy, R.J. (2000), "Fuzzy knowledge model for estimating the depth of anesthesia," *Proc. of Int. Conf. of Artif. Neural Networks in Eng. (ANNIE) 2000*, St. Louis, MI.

[37] Fathi-Torbaghan, M. and Hildebrand, L. (1999), "Complex system analysis using CI methods," *Proc. of the SPIE*, vol. 3722, pp. 330-341.

Chapter 14

Combining Evolutionary and Fuzzy Techniques in Medical Diagnosis

C.A. Peña-Reyes and M. Sipper

In this chapter we focus on the Wisconsin breast cancer diagnosis (WBCD) problem, combining two methodologies—fuzzy systems and evolutionary algorithms—to automatically produce diagnostic systems. We present two hybrid approaches: (1) a fuzzy-genetic algorithm, and (2) Fuzzy CoCo, a novel cooperative coevolutionary approach to fuzzy modeling. Both methods produce systems exhibiting high classification performance, and which are also human-interpretable. Fuzzy CoCo obtains higher-performance systems than the standard fuzzy-genetic approach while using less computational effort.

1 Introduction

A major class of problems in medical science involves the diagnosis of disease, based upon various tests performed upon the patient. When several tests are involved, the ultimate diagnosis may be difficult to obtain, even for a medical expert. This has given rise, over the past few decades, to computerized diagnostic tools, intended to aid the physician in making sense out of the welter of data.

A prime target for such computerized tools is in the domain of cancer diagnosis. Specifically, where breast cancer is concerned, the treating physician is interested in ascertaining whether the patient under examination exhibits the symptoms of a benign case, or whether her case is a malignant one.

A good computerized diagnostic tool should possess two characteristics, which are often in conflict. First, the tool must attain the highest possible *performance*, i.e., diagnose the presented cases correctly as being either *benign* or *malignant*. Second, it would be highly beneficial for such a diagnostic system to be human-friendly, exhibiting the so-called *interpretability*. This means that the physician is not faced with a black box that simply spouts answers (albeit correct) with no explanation; rather, we would like for the system to provide some insight as to *how* it derives its outputs.

In this chapter we present the combination of two methodologies—fuzzy systems and evolutionary algorithms—to automatically produce systems for breast cancer diagnosis. The major advantage of fuzzy systems is that they favor interpretability, however, finding good fuzzy systems can be quite an arduous task. This is where evolutionary algorithms step in, enabling the automatic production of fuzzy systems, based on a database of training cases. There are several recent examples of the application of fuzzy systems and evolutionary algorithms in the medical domain [28][1], though only a few combine both methodologies in a hybrid way—as we do in this chapter.

This chapter is organized as follows: In the next section we provide an overview of fuzzy modeling, evolutionary computation, and evolutionary fuzzy modeling. In Section 3 we describe the Wisconsin breast cancer diagnosis (WBCD) problem, which is the focus of our interest herein. Section 4 then describes a fuzzy-genetic approach to the WBCD problem. Section 5 presents Fuzzy CoCo, our cooperative coevolutionary approach to fuzzy modeling, and its application to the WBCD problem. Finally, we present concluding remarks in Section 6.

2 Background

2.1 Fuzzy Modeling

Fuzzy logic is a computational paradigm that provides a mathematical tool for representing and manipulating information in a way that resembles human communication and reasoning processes [43]. It is based on the assumption that, in contrast to Boolean logic, a statement can be *par-*

[1]This article provides over one hundred references to works in the medical domain using evolutionary computation.

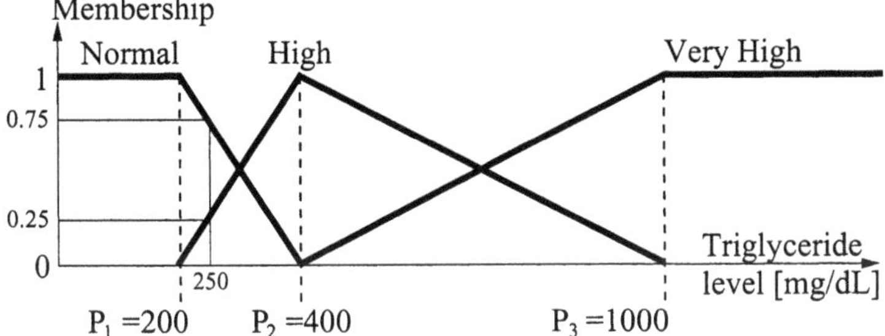

Figure 1. Example of a fuzzy variable: *Triglyceride level* has three possible fuzzy values, labeled **Normal**, **High**, and **Very High**, plotted above as degree of membership versus input value. The values P_i, setting the trapezoid and triangle apices, define the membership functions. In the figure, an example input value 250 mg/dL is assigned the membership values $\mu_{Normal}(250) = 0.75$, $\mu_{High}(250) = 0.25$, and $\mu_{VeryHigh}(250) = 0$. Note that $\mu_{Normal}(250) + \mu_{High}(250) + \mu_{VeryHigh}(250) = 1$.

tially true (or false), and composed of imprecise concepts. For example, the expression "I live near Geneva," where the fuzzy value "near" applied to the fuzzy variable "distance," in addition to being imprecise, is subject to interpretation. A *fuzzy variable* (also called a *linguistic variable*; see Figure 1) is characterized by its name tag, a set of *fuzzy values* (also known as *linguistic values* or *labels*), and the membership functions of these labels; these latter assign a membership value $\mu_{label}(u)$ to a given real value $u \in \Re$, within some predefined range (known as the universe of discourse). While the traditional definitions of Boolean-logic operations do not hold, new ones can be defined. Three basic operations, **and**, **or**, and **not**, are defined in fuzzy logic as follows:

$$\mu_{A\mathbf{and}B}(u) = \mu_A(u) \wedge \mu_B(u) = min\{\mu_A(u), \mu_B(u)\},$$
$$\mu_{A\mathbf{or}B}(u) = \mu_A(u) \vee \mu_B(u) = max\{\mu_A(u), \mu_B(u)\},$$
$$\mu_{\mathbf{not}A}(u) = \neg\mu_A(u) = 1 - \mu_A(u),$$

where A and B are fuzzy variables. Using such fuzzy operators one can combine fuzzy variables to form fuzzy-logic expressions, in a manner akin to Boolean logic. For example, in the domain of control, where fuzzy logic has been applied extensively, one can find expressions such as: **if** room temperature **is** Warm, **then** increase slightly the ventilation-fan speed.

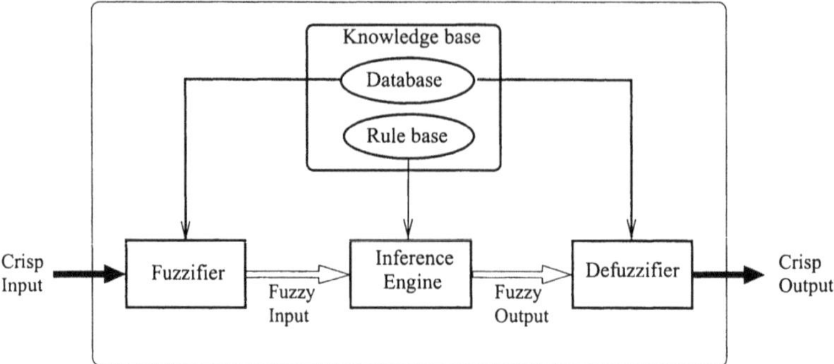

Figure 2. Basic structure of a fuzzy inference system.

A *fuzzy inference system* is a rule-based system that uses fuzzy logic, rather than Boolean logic, to reason about data [43]. Its basic structure consists of four main components, as depicted in Figure 2: (1) a fuzzifier, which translates crisp (real-valued) inputs into fuzzy values; (2) an inference engine that applies a fuzzy reasoning mechanism to obtain a fuzzy output; (3) a defuzzifier, which translates this latter output into a crisp value; and (4) a knowledge base, which contains both an ensemble of fuzzy rules, known as the rule base, and an ensemble of membership functions known as the database.

The decision-making process is performed by the inference engine using the rules contained in the rule base. These fuzzy rules define the connection between input and output fuzzy variables. A fuzzy rule has the form:

$$\textbf{if } antecedent \textbf{ then } consequent,$$

where *antecedent* is a fuzzy-logic expression composed of one or more simple fuzzy expressions connected by fuzzy operators, and *consequent* is an expression that assigns fuzzy values to the output variables. The inference engine evaluates all the rules in the rule base and combines the weighted consequents of all relevant rules into a single fuzzy set using the *aggregation* operation. This operation is the analog in fuzzy logic of the average operator in arithmetic [42] (aggregation is usually performed with the *max* operator).

Fuzzy modeling is the task of identifying the parameters of a fuzzy inference system so that a desired behavior is attained [42]. Note that the

Table 1. Parameter classification of fuzzy inference systems.

Class	Parameters
Logical	Reasoning mechanism Fuzzy operators Membership function types Defuzzification method
Structural	Relevant variables Number of membership functions Number of rules
Connective	Antecedents of rules Consequents of rules Rule weights
Operational	Membership-function values

fuzzy-modeling process has to deal with an important trade-off between the *accuracy* and the *interpretability* of the model. In other words, the model is expected to provide high numeric precision while incurring as little a loss of linguistic descriptive power as possible. With the *direct* approach a fuzzy model is constructed using knowledge from a human expert. This task becomes difficult when the available knowledge is incomplete or when the problem space is very large, thus motivating the use of *automatic* approaches to fuzzy modeling. There are several approaches to fuzzy modeling, based on neural networks [14], [22], [41], evolutionary algorithms [2], [7], [26], and hybrid methods [35], [37]. Selection of relevant variables and adequate rules is critical for obtaining a good system. One of the major problems in fuzzy modeling is the *curse of dimensionality*, meaning that the computation requirements grow exponentially with the number of variables.

The parameters of fuzzy inference systems can be classified into four categories (Table 1) [26]: logical, structural, connective, and operational. Generally speaking, this order also represents their relative influence on performance, from most influential (logical) to least influential (operational).

In fuzzy modeling, logical parameters are usually predefined by the designer based on experience and on problem characteristics. Typical choices for the reasoning mechanism are Mamdani-type, Takagi-Sugeno-

Kang (TKS)-type, and singleton-type [42]. Common fuzzy operators are min, max, product, probabilistic sum, and bounded sum. The most common membership functions are triangular, trapezoidal, and bell-shaped. As for defuzzification, several methods have been proposed, with the Center of Area (COA) and the Mean of Maxima (MOM) being the most popular [19], [42].

Structural, connective, and operational parameters may be either predefined, or obtained by synthesis or search methodologies. Generally, the search space, and thus the computational effort, grows exponentially with the number of parameters. Therefore, one can either invest more resources in the chosen search methodology, or infuse more *a priori*, expert knowledge into the system (thereby effectively reducing the search space). The aforementioned trade-off between accuracy and interpretability is usually expressed as a set of constraints on the parameter values, thus complexifying the search process.

2.2 Evolutionary Computation

The domain of evolutionary computation involves the study of the foundations and the applications of computational techniques based on the principles of natural evolution. Evolution in nature is responsible for the "design" of all living beings on earth, and for the strategies they use to interact with each other. Evolutionary algorithms employ this powerful design philosophy to find solutions to hard problems.

Generally speaking, evolutionary techniques can be viewed either as search methods, or as optimization techniques. As written by Michalewicz [21]:

> Any abstract task to be accomplished can be thought of as solving a problem, which, in turn, can be perceived as a search through a space of potential solutions. Since usually we are after 'the best' solution, we can view this task as an optimization process.

The first works on the use of evolution-inspired approaches to problem solving date back to the late 1950s [4], [5], [8], [10], [11]. Independent and almost simultaneous research conducted by Rechenberg and Schwefel on *evolution strategies* [34], [36], by Holland on *genetic algo-*

rithms [13], and by Fogel on *evolutionary programming* [9] triggered the study and the application of evolutionary techniques.

Three basic mechanisms drive natural evolution: *reproduction*, *mutation*, and *selection*. The first two act on the *chromosomes* containing the genetic information of the *individual* (the *genotype*), rather than on the individual itself (the *phenotype*) while selection acts on the phenotype. Reproduction is the process whereby new individuals are introduced into a *population*. During sexual reproduction, *recombination* (or *crossover*) occurs, transmitting to the offspring chromosomes that are a melange of both parents' genetic information. Mutation introduces small changes into the inherited chromosomes; it often results from copying errors during reproduction. Selection, acting on the phenotype, is a process guided by the Darwinian principle of survival of the fittest. The fittest individuals are those best adapted to their environment, which thus survive and reproduce.

Evolutionary computation makes use of a metaphor of natural evolution, according to which a problem plays the role of an environment wherein lives a population of individuals, each representing a possible solution to the problem. The degree of adaptation of each individual (i.e., candidate solution) to its environment is expressed by an adequacy measure known as the *fitness function*. The phenotype of each individual, i.e., the candidate solution itself, is generally encoded in some manner into its *genome* (genotype). Evolutionary algorithms potentially produce progressively better solutions to the problem. This is possible, thanks to the constant introduction of new "genetic" material into the population, by applying so-called genetic operators which are the computational equivalents of natural evolutionary mechanisms.

There are several types of evolutionary algorithms, among which the best known are *genetic algorithms*, *genetic programming*, *evolution strategies*, and *evolutionary programming*; though different in the specifics they are all based on the same general principles. The archetypal evolutionary algorithm proceeds as follows: An initial population of individuals, $P(0)$, is generated at random or heuristically. Every evolutionary step t, known as a *generation*, the individuals in the current population, $P(t)$, are *decoded* and *evaluated* according to some predefined quality criterion, referred to as the fitness, or fitness function. Then, a subset of individuals, $P'(t)$—known as the *mating pool*—is selected to reproduce, with selection of individuals done according to their fitness. Thus, high-fitness ("good") individuals stand a better chance of "reproducing," while low-fitness ones are more likely to disappear.

Selection alone cannot introduce any new individuals into the population, i.e., it cannot find new points in the search space. These points are generated by altering the selected population $P'(t)$ via the application of crossover and mutation, so as to produce a new population, $P''(t)$. Crossover tends to enable the evolutionary process to move toward "promising" regions of the search space. Mutation is introduced to prevent premature convergence to local optima, by randomly sampling new points in the search space. Finally, the new individuals $P''(t)$ are introduced into the next-generation population, $P(t+1)$; usually $P''(t)$ simply becomes $P(t+1)$. The termination condition may be specified as some fixed, maximal number of generations or as the attainment of an acceptable fitness level. Figure 3 presents the structure of a generic evolutionary algorithm in pseudo-code format.

begin EA
  t:=0
  Initialize population $P(t)$
  while not done **do**
    Evaluate $P(t)$
    $P'(t)$:= Select[$P(t)$]
    $P''(t)$:= ApplyGeneticOperators[$P'(t)$]
    $P(t+1)$:= Introduce[$P''(t)$,$P(t)$]
    t:=t+1
  end while
end EA

Figure 3. Pseudo-code of a standard evolutionary algorithm.

As they combine elements of directed and stochastic search, evolutionary techniques exhibit a number of advantages over other search methods. First, they usually need a smaller amount of knowledge and fewer assumptions about the characteristics of the search space. Second, they can more easily avoid getting stuck in local optima. Finally, they strike a good balance between *exploitation* of the best solutions, and *exploration* of the search space. The strength of evolutionary algorithms relies on their population-based search, and on the use of the genetic mechanisms described above. The existence of a population of candidate solutions entails a parallel search, with the selection mechanism directing the search to the most promising regions, the crossover operator encouraging the exchange of information between these search-space regions, and the mutation operator enabling the exploration of new directions.

The application of an evolutionary algorithm involves a number of important considerations. The first decision to take when applying such an algorithm is how to encode candidate solutions within the genome. The representation must allow for the encoding of all possible solutions while being sufficiently simple to be searched in a reasonable amount of time. Next, an appropriate fitness function must be defined for evaluating the individuals. The (usually scalar) fitness value must reflect the criteria to be optimized and their relative importance. Representation and fitness are thus clearly problem-dependent, in contrast to selection, crossover, and mutation, which seem *prima facie* more problem-independent. Practice has shown, however, that while standard genetic operators can be used, one often needs to tailor these to the problem as well.

We noted above that there are several types of evolutionary algorithms. The distinction is mainly due to historical reasons and the different types of evolutionary algorithms are in fact quite similar. One could argue that there is but a single general evolutionary algorithm, or just the opposite– that "there are as many evolutionary algorithms as the researchers working in evolutionary computation" [31]. The frontiers among the widely accepted classes of evolutionary algorithms have become fuzzy over the years as each technique has attempted to overcome its limitations, by imbibing characteristics of the other techniques. To design an evolutionary algorithm one must define a number of important parameters, which are precisely those that demarcate the different evolutionary-computation classes. Some important parameters are: representation (genome), selection mechanism, crossover, mutation, size of populations P' and P'', variability or fixity of population size, and variability or fixity of genome length.

2.3 Evolutionary Fuzzy Modeling

Evolutionary algorithms are used to search large, and often complex, search spaces. They have proven worthwhile on numerous diverse problems, able to find near-optimal solutions given an adequate performance (fitness) measure. Fuzzy modeling can be considered as an optimization process where part or all of the parameters of a fuzzy system constitute the search space. Works investigating the application of evolutionary techniques in the domain of fuzzy modeling had first appeared about a decade ago [15], [16]. These focused mainly on the tuning of fuzzy inference systems involved in control tasks (e.g., cart-pole balancing, liquid-level system, and spacecraft rendezvous operation). Evolutionary fuzzy modeling has since been applied to an ever-growing number of domains, branching into areas as diverse as chemistry, medicine, telecommunications, biology, and geophysics. For a detailed bibliography on evolutionary fuzzy modeling up to 1996, the reader is referred to [1], [6].

Depending on several criteria—including the available *a priori* knowledge about the system, the size of the parameter set, and the availability and completeness of input/output data—artificial evolution can be applied in different stages of the fuzzy-parameter search. Three of the four categories of fuzzy parameters in Table 1 can be used to define targets for evolutionary fuzzy modeling: structural parameters, connective parameters, and operational parameters [26]. As noted in Section 2.1, logical parameters are usually predefined by the designer based on experience.

Knowledge tuning (operational parameters). The evolutionary algorithm is used to tune the knowledge contained in the fuzzy system by finding membership-function values. An initial fuzzy system is defined by an expert. Then, the membership-function values are encoded in a genome, and an evolutionary algorithm is used to find systems with high performance. Evolution often overcomes the local-minima problem present in gradient descent-based methods. One of the major shortcomings of knowledge tuning is its dependency on the initial setting of the knowledge base.

Behavior learning (connective parameters). In this approach, one assumes that expert knowledge is sufficient in order to define the membership functions; this determines, in fact, the maximum number of rules [42]. The genetic algorithm is used to find either the rule consequents, or an adequate subset of rules to be included in the rule base.

As the membership functions are fixed and predefined, this approach lacks the flexibility to modify substantially the system behavior. Furthermore, as the number of variables and membership functions increases, the curse of dimensionality becomes more pronounced and the interpretability of the system decreases rapidly.

Structure learning (structural parameters). In many cases, the available information about the system is composed almost exclusively of input/output data, and specific knowledge about the system structure is scant. In such a case, evolution has to deal with the simultaneous design of rules, membership functions, and structural parameters. Some methods use a fixed-length genome encoding a fixed number of fuzzy rules along with the membership-function values. In this case the designer defines structural constraints according to the available knowledge of the problem characteristics. Other methods use variable-length genomes to allow evolution to discover the optimal size of the rule base.

Both behavior and structure learning can be viewed as rule-base learning processes with different levels of complexity. They can thus be assimilated within other methods from machine learning, taking advantage of experience gained in this latter domain. In the evolutionary-algorithm community there are two major approaches for evolving such rule systems: the Michigan approach and the Pittsburgh approach [21]. A more recent method has been proposed specifically for fuzzy modeling: the iterative rule learning approach [12]. These three approaches are briefly described below.

The Michigan approach. Each individual represents a *single* rule. The fuzzy inference system is represented by the *entire population*. Since several rules participate in the inference process, the rules are in constant competition for the best action to be proposed, and cooperate to form an efficient fuzzy system. The cooperative-competitive nature of this approach renders difficult the decision of which rules are ultimately responsible for good system behavior. It necessitates an effective credit-assignment policy to ascribe fitness values to individual rules.

The Pittsburgh approach. Here, the evolutionary algorithm maintains a population of candidate fuzzy systems, each individual representing an *entire* fuzzy system. Selection and genetic operators are used to produce new generations of fuzzy systems. Since evaluation is applied to the entire system, the credit-assignment problem is eschewed. This approach allows to include additional optimization criteria in the fitness function, thus affording the implementation of multi-objective optimization. The main shortcoming of this approach is its computational cost, since a population of full-fledged fuzzy systems has to be evaluated each generation.

The iterative rule learning approach. As in the Michigan approach, each individual encodes a single rule. An evolutionary algorithm is used to find a single rule, thus providing a partial solution. The evolutionary algorithm is used iteratively for the discovery of new rules, until an appropriate rule base is built. To prevent the process from finding redundant rules (i.e., rules with similar antecedents), a penalization scheme is applied each time a new rule is added. This approach combines the speed of the Michigan approach with the simplicity of fitness evaluation of the Pittsburgh approach. However, as with other incremental rule-base construction methods, it can lead to a non-optimal partitioning of the antecedent space.

As mentioned before, the accuracy-interpretability trade-off faced by fuzzy modelers implies the assumption of constraints acting on the parameter values, mainly on the membership-function shapes. The following semantic criteria represent conditions driving fuzzy modeling toward human-interpretable systems [26], [30]:

- *Distinguishability.* Each linguistic label should have semantic meaning and the fuzzy set should clearly define a range in the universe of discourse. In the example of Figure 1, to describe variable *Triglyceride level* we used three meaningful labels: **Normal, High**, and **Very High**. Their membership functions are defined using parameters P_1, P_2, and P_3.
- *Justifiable number of elements.* The number of membership functions of a variable should be compatible with the number of conceptual entities a human being can handle. This number should not exceed

the limit of 7 ± 2 distinct terms. The same criterion is applied to the number of variables in the rule antecedent.

- *Coverage.* Any element from the universe of discourse should belong to at least one of the fuzzy sets. That is, its membership value must be different than zero for at least one of the linguistic labels. Referring to Figure 1, we see that any value along the x-axis belongs to at least one fuzzy set; no value lies outside the range of all sets.
- *Normalization.* Since all labels have semantic meaning, then, for each label, at least one element of the universe of discourse should have a membership value equal to one. In Figure 1, we observe that all three sets **Normal**, **High**, and **Very High** have elements with membership value equal to 1.
- *Orthogonality.* For each element of the universe of discourse, the sum of all its membership values should be equal to one (as in the example in Figure 1).

3 Fuzzy Systems for Breast Cancer Diagnosis

In this section we present the medical-diagnosis problem which is the object of our study, and the fuzzy system we propose to solve it with.

3.1 The WBCD Problem

Breast cancer is the most common cancer among women, excluding skin cancer. The presence of a breast mass[2] is an alert sign, but it does not always indicate a malignant cancer. Fine needle aspiration (FNA)[3] of breast masses is a cost-effective, non-traumatic, and mostly non-invasive diagnostic test that obtains information needed to evaluate malignancy.

The Wisconsin breast cancer diagnosis (WBCD) database [20] is the result of the efforts made at the University of Wisconsin Hospital for accu-

[2]Most breast cancers are detected as a lump or mass on the breast, by self-examination, by mammography, or by both [18].

[3]Fine needle aspiration is an outpatient procedure that involves using a small-gauge needle to extract fluid directly from a breast mass [18].

rately diagnosing breast masses based solely on an FNA test [17]. Nine visually assessed characteristics of an FNA sample considered relevant for diagnosis were identified, and assigned an integer value between 1 and 10. The measured variables are as follows:

1. Clump Thickness (v_1);
2. Uniformity of Cell Size (v_2);
3. Uniformity of Cell Shape (v_3);
4. Marginal Adhesion (v_4);
5. Single Epithelial Cell Size (v_5);
6. Bare Nuclei (v_6);
7. Bland Chromatin (v_7);
8. Normal Nucleoli (v_8);
9. Mitosis (v_9).

The diagnostics in the WBCD database were furnished by specialists in the field. The database itself consists of 683 cases, with each entry representing the classification for a certain ensemble of measured values:

case	v_1	v_2	v_3	$\cdots$	v_9	*diagnostic*
1	5	1	1	$\cdots$	1	*benign*
2	5	4	4	$\cdots$	1	*benign*
$\vdots$	$\vdots$	$\vdots$	$\vdots$	$\ddots$	$\vdots$	$\vdots$
683	4	8	8	$\cdots$	1	*malignant*

Note that the diagnostics do not provide any information about the degree of benignity or malignancy.

There are several studies based on this database. Bennet and Mangasarian [3] used linear programming techniques, obtaining a 99.6% classification rate on 487 cases (the reduced database available at the time). However, their solution exhibits little understandability, i.e., diagnostic decisions are essentially black boxes, with no explanation as to how they were attained. With increased interpretability in mind as a prior objective, a number of researchers have applied the method of extracting Boolean

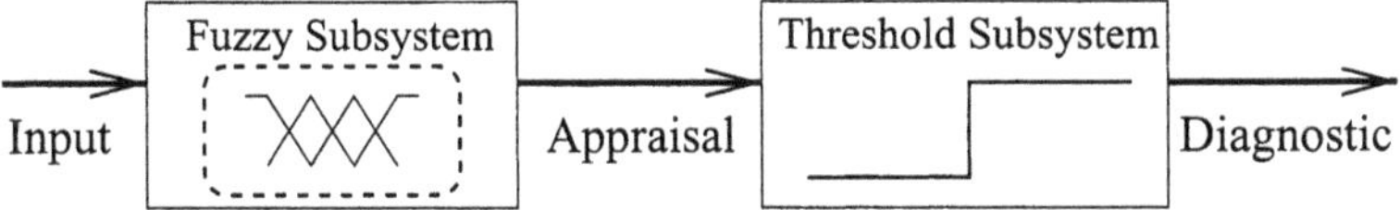

Figure 4. Proposed diagnosis system. Note that the fuzzy subsystem displayed to the left is in fact the entire fuzzy inference system of Figure 2.

rules from neural networks [38], [39]. Their results are encouraging, exhibiting both good performance and a reduced number of rules and relevant input variables. Nevertheless, these systems use Boolean rules and are not capable of furnishing the user with a measure of confidence for the decision made. Our own work on the evolution of fuzzy rules for the WBCD problem has shown that it is possible to obtain diagnostic systems exhibiting high performance, coupled with interpretability and a confidence measure [24]-[27].

3.2 Fuzzy-System Setup

The solution scheme we propose for the WBCD problem is depicted in Figure 4. It consists of a fuzzy system and a threshold unit. The fuzzy system computes a continuous appraisal value of the malignancy of a case, based on the input values. The threshold unit then outputs a *benign* or *malignant* diagnostic according to the fuzzy system's output.

Our previous knowledge about the WBCD problem represents valuable information to be used for our choice of fuzzy parameters (Table 1). When defining our setup we took into consideration the following three results concerning the composition of potential high-performance systems: (1) small number of rules; (2) small number of variables; and (3) monotonicity of the input variables [26]. Some fuzzy models forgo interpretability in the interest of improved performance. Where medical diagnosis is concerned, interpretability is the major advantage of fuzzy systems. This motivated us to take into account the five semantic criteria presented in Section 2.3, defining constraints on the fuzzy parameters: (1) distinguishability, (2) justifiable number of elements, (3) coverage, (4) normalization, and (5) orthogonality.

Referring to Table 1, and taking into account these five criteria, we de-

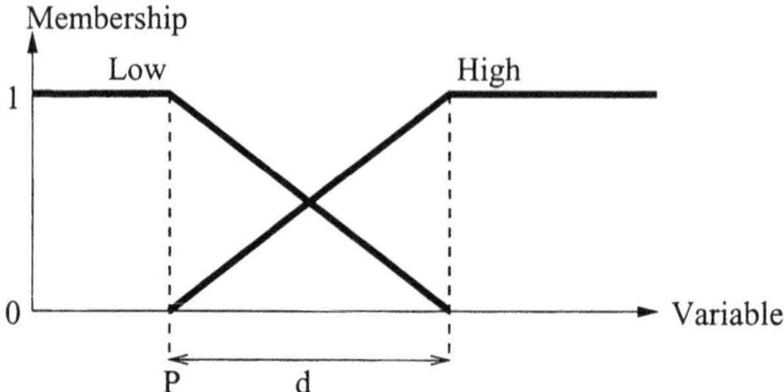

Figure 5. Input fuzzy variables for the WBCD problem. Each fuzzy variable has two possible fuzzy values labeled **Low** and **High**, and orthogonal membership functions, plotted above as degree of membership versus input value. *P* and *d* define the start point and the length of membership function edges, respectively. The orthogonality condition means that the sum of all membership functions at any point is one. In the figure, an example value u is assigned the membership values $\mu_{Low}(u) = 0.8$ and $\mu_{High}(u) = 0.2$ (as can be seen $\mu_{Low}(u) + \mu_{High}(u) = 1$).

lineate below the fuzzy-system setup:

- Logical parameters: singleton-type fuzzy systems; min-max fuzzy operators; orthogonal, trapezoidal input membership functions (see Figure 5); weighted-average defuzzification.
- Structural parameters: two input membership functions (*Low* and *High*; see Figure 5); two output singletons (*benign* and *malignant*); a user-configurable number of rules. The relevant variables are one of the evolutionary objectives.
- Connective parameters: the antecedents and the consequent of the rules are searched by the evolutionary algorithm. The algorithm also searches for the consequent of the default rule which plays the role of an `else` condition (note that for the fuzzy-genetic approach presented in Section 4, the consequents are predefined instead of evolved, thus reducing the search space). All rules have unitary weight.
- Operational parameters: the input membership-function values are to be found by the evolutionary algorithm. For the output singletons we used the values 2 and 4, for *benign* and *malignant*, respectively.

4 A Fuzzy-Genetic Approach

The problem, at this stage, consists of searching for three fuzzy-system parameters: input membership functions, antecedents of rules, and relevant variables (consequents of rules are predefined; see Section 3.2). We applied a Pittsburgh-like approach, using a simple genetic algorithm [40] to search for individuals whose genomes encode these three parameters. The next subsection describes the setup of the genetic algorithm, after which subsection 4.2 presents the results obtained using this approach.

4.1 The Evolutionary Setup

The genome encodes three sets of parameters: input membership functions, antecedents of rules, and relevant variables. It is defined as follows:

- Membership-function parameters. There are nine variables ($v_1 - v_9$), each with two parameters *P* and *d*, defining the start point and the length of the membership-function edges, respectively (Figure 5).
- Antecedents. The i-th rule has the form:

  if (v_1 **is** A_1^i) **and** ... **and** (v_9 **is** A_9^i) **then** (*output* **is** *benign*),

  where A_j^i represents the membership function applicable to variable v_j. A_j^i can take on the values: 1 (*Low*), 2 (*High*), or 0 or 3 (*Other*).
- Relevant variables are searched for implicitly by letting the algorithm choose non-existent membership functions as valid antecedents; in such a case the respective variable is considered irrelevant. For example, the rule

  if (v_1 **is** *High*) **and** (v_2 **is** *Other*) **and** (v_3 **is** *Other*) **and** (v_4 **is** *Low*) **and** (v_5 **is** *Other*) **and** (v_6 **is** *Other*) **and** (v_7 **is** *Other*) **and** (v_8 **is** *Low*) **and** (v_9 **is** *Other*) **then** (*output* **is** *benign*),

  is interpreted as:

  if (v_1 **is** *High*) **and** (v_4 **is** *Low*) **and** (v_8 **is** *Low*) **then** (*output* **is** *benign*).

Table 2 delineates the parameter encoding, which together form a single individual's genome.

Table 2. Parameter encoding of an individual's genome. Total genome length is $54 + 18N_r$, where N_r denotes the number of rules (N_r is set *a priori* to a value between 1–5, and is fixed during the genetic-algorithm run).

Parameter	Values	Bits	Qty	Total bits
P	{1,2,...,8}	3	9	27
d	{1,2,...,8}	3	9	27
A	{0,1,2,3}	2	$9 \times N_r$	$18 \times N_r$

To evolve the fuzzy inference system, we used a genetic algorithm with a fixed population size of 200 individuals, and fitness-proportionate selection (Subsection 2.2). The algorithm terminates when the maximum number of generations, G_{max}, is reached (we set $G_{max} = 2000 + 500 \times N_r$, i.e., dependent on the number of rules used in the run), or when the increase in fitness of the best individual over five successive generations falls below a certain threshold (in our experiments we used threshold values between 2×10^{-7} and 4×10^{-6}).

Our fitness function combines three criteria: (1) F_c: classification performance, computed as the percentage of cases correctly classified; (2) F_e: the quadratic difference between the continuous appraisal value (in the range $[2, 4]$) and the correct discrete diagnosis given by the WBCD database (either 2 or 4); and (3) F_v: the average number of variables per active rule. The fitness function is given by $F = F_c - \alpha F_v - \beta F_e$, where $\alpha = 0.05$ and $\beta = 0.01$ (these latter values were derived empirically). F_c, the ratio of correctly diagnosed cases, is the most important measure of performance. F_v measures the linguistic integrity (interpretability), penalizing systems with a large number of variables per rule (on average). F_e adds selection pressure towards systems with low quadratic error.

4.2 Results

This section describes the results obtained when applying the methodology described in Section 4.1. We first delineate the success statistics relating to the evolutionary algorithm. Then, we describe in full a three-rule evolved fuzzy system that exemplifies our approach.

A total of 120 evolutionary runs were performed, all of which found systems whose classification performance exceeds 94.5%. In particular,

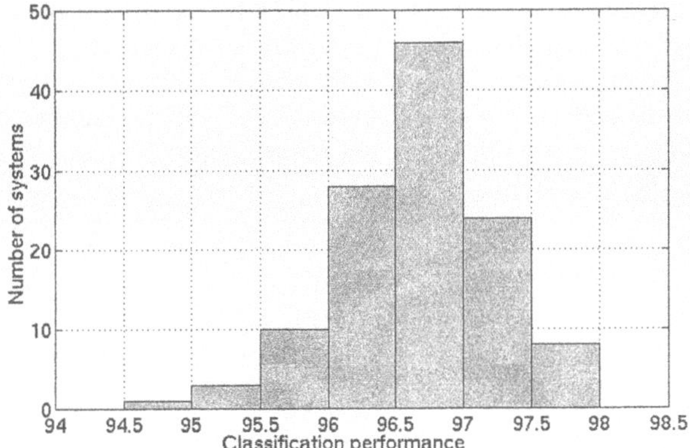

Figure 6. Summary of results of 120 evolutionary runs. The histogram depicts the number of systems exhibiting a given performance level at the end of the evolutionary run. The performance considered is that of the best individual of the run, measured as the overall percentage of correctly classified cases over the entire database.

considering the best individual per run (i.e., the evolved system with the highest classification success rate), 78 runs led to a fuzzy system whose performance exceeds 96.5%, and of these, 8 runs found systems whose performance exceeds 97.5%; these results are summarized in Figure 6.

Table 3 shows the results of the best systems obtained with the fuzzy-genetic approach. The number of rules per system was fixed at the outset to be between one and five, i.e., evolution seeks a system with an *a priori* given number of rules. A comparison of these systems with other approaches is presented in Section 5.4 (see also [26]).

We next describe our top-performance system, which serves to exemplify the solutions found by our evolutionary approach. The system, delineated in Figure 7, consists of three rules (note that the `else` condition is not counted as an active rule). Taking into account all three criteria of performance—classification rate, number of rules per system, and average number of variables per rule— this system can be considered the top one over all 120 evolutionary runs. It obtains an overall classification rate (i.e., over the entire database) of 97.8%.

A thorough test of this three-rule system revealed that the second rule

Table 3. Results of the best systems evolved by the fuzzy-genetic approach. Shown below are the classification performance values of the top systems obtained by these approaches, along with the average number of variables-per-rule. Results are divided into five classes, in accordance with the number of rules-per-system, going from one-rule systems to five-rule ones.

Rules-per-system	Performance	variables-per-rule
1	97.07%	4
2	97.36%	3
3	97.80%	4.7
4	97.80%	4.8
5	97.51%	3.4

Database

	v_1	v_2	v_3	v_4	v_5	v_6	v_7	v_8	v_9
P	3	5	2	2	8	1	4	5	4
d	5	2	1	2	4	7	3	5	2

Rule base

Rule 1 **if** (v_3 **is** *Low*) **and** (v_7 **is** *Low*) **and** (v_8 **is** *Low*) **and** (v_9 **is** *Low*) **then** (*output* **is** *benign*)

Rule 2 **if** (v_1 **is** *Low*) **and** (v_2 **is** *Low*) **and** (v_3 **is** *High*) **and** (v_4 **is** *Low*) **and** (v_5 **is** *High*) **and** (v_9 **is** *Low*) **then** (*output* **is** *benign*)

Rule 3 **if** (v_1 **is** *Low*) **and** (v_4 **is** *Low*) **and** (v_6 **is** *Low*) **and** (v_8 **is** *Low*) **then** (*output* **is** *benign*)

Default **else** (*output* **is** *malignant*)

Figure 7. The best evolved, fuzzy diagnostic system with three rules. It exhibits an overall classification rate of 97.8%, and an average of 4.7 variables per rule. Thorough testing revealed that Rule 2 can be dropped.

(Figure 7) is never actually used; in the fuzzy literature this is known as a rule that never *fires*, i.e., is triggered by none of the input cases. Thus, it can be eliminated altogether from the rule base, resulting in a two-rule system (also reducing the average number of variables-per-rule from 4.7 to 4).

5 A Fuzzy Coevolutionary Approach: Fuzzy CoCo

The fuzzy-genetic approach, even though it obtained good diagnostic systems, plateaued at a certain performance level. In this section we present Fuzzy CoCo, a cooperative coevolutionary approach to fuzzy modeling, capable of obtaining higher-performance systems while requiring less computation than the fuzzy-genetic approach. The next subsection briefly explains cooperative coevolution; after which Section 5.2 presents Fuzzy CoCo; Section 5.3 then describes the setup of Fuzzy CoCo when applied to the WBCD problem, and, finally, Section 5.4 presents the results obtained.

5.1 Cooperative Coevolution

Coevolution refers to the simultaneous evolution of two or more species with coupled fitness. Such coupled evolution favors the discovery of complex solutions whenever complex solutions are required [23]. Simplistically speaking, one can say that coevolving species can either compete (e.g., to obtain exclusivity on a limited resource) or cooperate (e.g., to gain access to some hard-to-attain resource). Cooperative (also called symbiotic) coevolutionary algorithms involve a number of independently evolving species which together form complex structures, well-suited to solve a problem. The fitness of an individual depends on its ability to collaborate with individuals from other species. In this way, the evolutionary pressure stemming from the difficulty of the problem favors the development of cooperative strategies and individuals. Single-population evolutionary algorithms often perform poorly—manifesting stagnation, convergence to local optima, and computational costliness—when confronted with problems presenting one or more of the following features: (1) the sought-after solution is complex, (2) the problem or its solution is clearly decomposable, (3) the genome encodes different types of values, (4) strong interdependencies among the components of the solution, (5) component-ordering drastically affects fitness. Cooperative coevolution effectively addresses these issues, consequently widening the range of applications of evolutionary computation. Potter [32], [33] developed a model in which a number of populations explore different decompo-

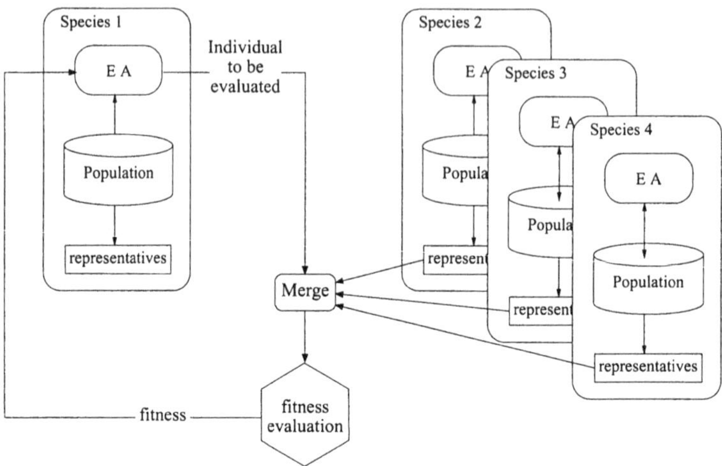

Figure 8. Potter's cooperative coevolutionary system. The figure shows the evolutionary process from the perspective of Species 1. The individual being evaluated is combined with one or more *representatives* of the other species so as to construct several solutions which are tested on the problem. The individual's fitness depends on the quality of these solutions.

sitions of the problem. Below we detail this framework as it forms the basis of our own approach.

In Potter's system, each species represents a subcomponent of a potential solution. Complete solutions are obtained by assembling *representative* members of each of the species (populations). The fitness of each individual depends on the quality of (some of) the complete solutions it participated in, thus measuring how well it cooperates to solve the problem. The evolution of each species is controlled by a separate, independent evolutionary algorithm. Figure 8 shows the general architecture of Potter's cooperative coevolutionary framework, and the way each evolutionary algorithm computes the fitness of its individuals by combining them with selected representatives from the other species. A greedy strategy for the choice of representatives of a species is to use one or more of the fittest individuals from the last generation.

5.2 The Coevolutionary Algorithm

Fuzzy CoCo is a cooperative coevolutionary approach to fuzzy modeling wherein two coevolving species are defined: database (membership functions) and rule base [27]. This approach is based primarily on the framework defined by Potter [32], [33].

A fuzzy modeling process usually deals with the simultaneous search for operational and connective parameters (Table 1). These parameters provide an almost complete definition of the linguistic knowledge describing the behavior of a system, and the values mapping this symbolic description into a real-valued world (a complete definition also requires logical and structural parameters whose definition is best suited for human skills). Thus, fuzzy modeling can be thought of as two separate but intertwined search processes: (1) the search for the membership functions (i.e., operational parameters) that define the fuzzy variables, and (2) the search for the rules (i.e., connective parameters) used to perform the inference.

Fuzzy modeling presents several features discussed earlier which justify the application of a cooperative-coevolutionary approach: (1) The required solutions can be very complex, since fuzzy systems with a few dozen variables may call for hundreds of parameters to be defined. (2) The proposed solution—a fuzzy inference system—can be decomposed into two distinct components: rules and membership functions. (3) Membership functions are continuous and real-valued, while rules are discrete and symbolic. (4) These two components are interdependent because the membership functions defined by the first group of values are indexed by the second group (rules).

Consequently, in Fuzzy CoCo, the fuzzy modeling problem is solved by two coevolving, cooperating species. Individuals of the first species encode values which define completely all the membership functions for all the variables of the system. For example, with respect to the variable $Triglyceridelevel$ shown in Figure 1, this problem is equivalent to finding the values of P_1, P_2, and P_3.

Individuals of the second species define a set of rules of the form:

$$\textbf{if } (v_1 \textbf{ is } A_1) \textbf{ and} \ldots \textbf{ and } (v_n \textbf{ is } A_n) \textbf{ then } (output \textbf{ is } C),$$

where the term A_v indicates which of the linguistic labels of fuzzy variable v is used by the rule. For example, a valid rule could contain the expression:

$$\textbf{if} \ldots \textbf{ and } (Triglyceridelevel \textbf{ is } High) \textbf{ and} \ldots \textbf{ then} \ldots$$

which includes the membership function $High$ whose defining parameters are contained in the first species (population).

begin Fuzzy CoCo
  g:=0
  for each species S
    Initialize populations $P_S(0)$
    Evaluate population $P_S(0)$
  end for
  while not done **do**
    for each species S
      g:=g+1
      $E_S(g) =$ elite-select $P_S(g-1)$
      $P'_S(g) =$ select $P_S(g-1)$
      $P''_S(g) =$ crossover $P'_S(g)$
      $P'''_S(g) =$ mutate $P''_S(g)$
      $P_S(g) = P'''_s(g) + E_S(g)$
      Evaluate population $P_S(g)$
    end for
  end while
end Fuzzy CoCo

Figure 9. Pseudo-code of Fuzzy CoCo. Two species coevolve in Fuzzy CoCo: membership functions and rules. The elitism strategy extracts E_S individuals to be reinserted into the population after evolutionary operators have been applied (i.e., selection, crossover, and mutation). Selection results in a reduced population $P'_S(g)$ (usually, the size of $P'_S(g)$ is $||P'_S|| = ||P_S|| - ||E_S||$). The line "Evaluate population $P_S(g)$" is elaborated in Figure 10.

The two evolutionary algorithms used to control the evolution of the two populations are instances of a simple genetic algorithm [40]. Figure 9 presents the Fuzzy CoCo algorithm in pseudo-code format. The genetic algorithms apply fitness-proportionate selection to choose the mating pool, and apply an elitist strategy with an elitism rate Er to allow some of the best individuals to survive into the next generation. Standard crossover and mutation operators are applied with probabilities P_c and P_m, respectively.

We introduced elitism to avoid the divergent behavior of Fuzzy CoCo, observed in preliminary trial runs. Non-elitist versions of Fuzzy CoCo tended to lose the genetic information of good individuals found during evolution, consequently producing populations with mediocre individuals scattered throughout the search space. This is probably due to the relatively small size of the population which renders difficult the preservation of good solutions while exploring the search space. The introduction of simple elitism produces an undesirable effect on Fuzzy CoCo's performance: populations converge prematurely even with reduced values of the elitism rate E_r. To offset this effect without losing the advantages of elitism, it was necessary to increase the mutation probability P_m by an order of magnitude so as to improve the exploration capabilities of the algorithm. (Increased mutation rates were also reported by Potter [32], [33] in his coevolutionary experiments.)

A more detailed view of the fitness evaluation process is depicted in Figure 10. An individual undergoing fitness evaluation establishes cooperations with one or more representatives of the other species, i.e., it is combined with individuals from the other species to construct fuzzy systems. The fitness value assigned to the individual depends on the performance of the fuzzy systems it participated in (specifically, either the average or the maximal value).

Representatives, called here *co-operators*, are selected both fitness-proportionally and randomly from the previous generation since they have already been assigned a fitness value (see Figure 9). In Fuzzy CoCo, N_{cf} co-operators are selected according to their fitness, usually the fittest individuals, thus favoring the exploitation of known good solutions. The other N_{cr} co-operators are selected randomly from the population to represent the diversity of the species, maintaining in this way exploration of the search space.

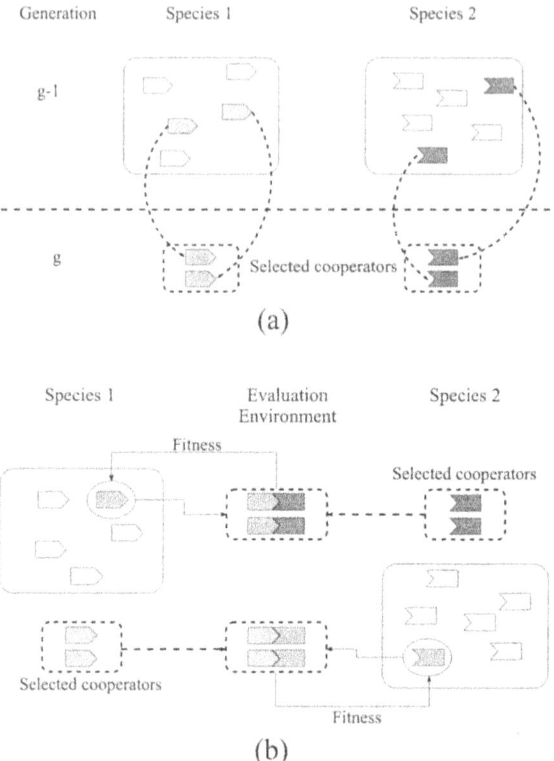

Figure 10. Fitness evaluation in Fuzzy CoCo. (a) Several individuals from generation $g - 1$ of each species are selected according to their fitness to be the representatives of their species during generation g; these representatives are called "co-operators." (b) During the evaluation stage of generation g (after selection, crossover, and mutation—see Figure 9), individuals are combined with the selected co-operators of the other species to construct fuzzy systems. These systems are then evaluated on the problem domain and serve as a basis for assigning the final fitness to the individual being evaluated.

5.3 The Evolutionary Setup

Fuzzy CoCo was set to search for four parameters: input membership-function values, relevant input variables, and antecedents and consequents of rules. These search goals are more ambitious than those defined for the fuzzy-genetic approach (Section 4) as the consequents of rules are added to the search space. The genomes of the two species are constructed as follows:

Table 4. Genome encoding of parameters for both species. Genome length for membership functions is 54 bits. Genome length for rules is $19 \times N_r + 1$, where N_r denotes the number of rules.

Species 1: Membership functions

Parameter	Values	Bits	Qty	Total bits
P	{1,2,...,8}	3	9	27
d	{1,2,...,8}	3	9	27
	Total			54

Species 2: Rules

Parameter	Values	Bits	Qty	Total bits
A	{0,1,2,3}	2	$9 \times N_r$	$18 \times N_r$
C	{1,2}	1	$N_r + 1$	$N_r + 1$
	Total			$19 \times N_r + 1$

- Species 1: Membership functions. There are nine variables ($v_1 - v_9$), each with two parameters, *P* and *d*, defining the start point and the length of the membership-function edges, respectively (Figure 5).
- Species 2: Rules. The i-th rule has the form:

  if (v_1 **is** A_1^i) **and** ... **and** (v_9 **is** A_9^i) **then** (*output* **is** C^i),

  A_j^i can take on the values: 1 (*Low*), 2 (*High*), or 0 or 3 (*Other*). C^i bit can take on the values: 0 (*Benign*) or 1 (*Malignant*). Relevant variables are searched for implicitly by letting the algorithm choose non-existent membership functions (0 or 3) as valid antecedents; in such a case the respective variable is considered irrelevant.

Table 4 delineates the parameter encoding for both species' genomes, which together describe an entire fuzzy system. Note that in the fuzzy-genetic approach (Section 4) both membership functions and rules were encoded in the same genome, i.e., there was only one species.

To evolve the fuzzy inference system, we applied Fuzzy CoCo with the same evolutionary parameters for both species. Values and ranges of values used for these parameters were defined according to preliminary tests performed on benchmark problems (mostly function-optimization problems found in Potter [32]); Table 5 delineates these values. The algorithm terminates when the maximum number of generations, G_{max}, is reached

Table 5. Fuzzy CoCo set-up for the WBCD problem.

Parameter	Values
Population size $\|\|Ps\|\|$	[30-90]
Maximum generations G_{max}	$1000 + 100N_r$
Crossover probability P_c	1
Mutation probability P_m	[0.02-0.3]
Elitism rate E_r	[0.1-0.6]
"Fit" co-operators N_{cf}	1
Random co-operators N_{cr}	{1,2,3,4}

(we set $G_{max} = 1000 + 100 \times N_r$, i.e., dependent on the number of rules used in the run), or when the increase in fitness of the best individual over five successive generations falls below a certain threshold (10^{-4} in our experiments).

Our fitness function combines two criteria: 1) F_c—classification performance, computed as the percentage of cases correctly classified, and 2) F_v—the maximum number of variables in the longest rule. The fitness function is given by $F = F_c - \alpha F_v$, where $\alpha = 0.0015$. F_c, the percentage of correctly diagnosed cases, is the most important measure of performance. F_v measures the linguistic integrity (interpretability), penalizing systems with a large number of variables in their rules. The value α was calculated to allow F_v to occasion a fitness difference only among systems exhibiting similar classification performance. (We did not apply F_e as it proved of little use.)

We stated earlier that cooperative coevolution reduces the computational cost of the search process. In order to measure this cost we calculated the number of fuzzy-system evaluations performed by a single run of Fuzzy CoCo. Each generation, the $||Ps||$ individuals of each population are evaluated N_c times (where $N_c = N_{cf} + N_{cr}$). The total number of fuzzy-system evaluations per run is thus $2 \times G_{max} \times ||Ps|| \times N_c$. This value ranged from 5.28×10^5 evaluations for a one-rule system search, up to 8.16×10^5 evaluations for a seven-rule system (using typical parameter values: $||Ps|| = 80$, $N_{cf} = 1$, and $N_{cr} = 2$). The number of fuzzy-system evaluations required by our single-population approach was, on the average, 5×10^5 for a one-rule system and 11×10^5 for a seven-rule system [26].

Table 6. Comparison of the best systems evolved by Fuzzy CoCo with the top systems obtained using single-population evolution [26] and with those obtained by Setiono's NeuroRule approach [38]. Shown below are the classification performance values of the top systems obtained by these approaches, along with the number of variables of the longest rule in parentheses. Results are divided into seven classes, in accordance with the number of rules per system, going from one-rule systems to seven-rule ones.

Rules per system	Neuro-Rule [38]	Single population GA [26]	Fuzzy CoCo	
	best	best	average	best
1	97.36% (4)	97.07% (4)	97.36% (4)	97.36% (4)
2	–	97.36% (4)	97.73% (3.9)	98.54% (5)
3	98.10% (4)	97.80% (6)	97.91% (4.4)	98.54% (4)
4	–	97.80% (-)	98.12% (4.2)	98.68% (5)
5	98.24% (5)	97.51% (-)	98.18% (4.6)	98.83% (5)
6	–	–	98.18% (4.3)	98.83% (5)
7	–	–	98.25% (4.7)	98.98% (5)

5.4 Results

A total of 495 evolutionary runs were performed, all of which found systems whose classification performance exceeds 96.7%. In particular, considering the best individual per run (i.e., the evolved system with the highest classification success rate), 241 runs led to a fuzzy system whose performance exceeds 98.0%, and of these, 81 runs found systems whose performance exceeds 98.5%.

Table 6 compares our best systems with the top systems obtained by the fuzzy-genetic approach (Section 4) [26] and with the systems obtained by Setiono's NeuroRule approach [38]. The evolved fuzzy systems described in this paper can be seen to surpass those obtained by other approaches in terms of performance, while still containing simple, interpretable rules. As shown in Table 6, we obtained higher-performance systems for all rule-base sizes but one, i.e., from two-rule systems to seven-rule ones, while all our one-rule systems perform as well as the best system reported by Setiono.

Database

	v_1	v_2	v_3	v_4	v_5	v_6	v_7	v_8	v_9
P	2	1	1	1	6	1	3	5	2
d	7	8	4	8	1	4	8	4	1

Rule base

Rule 1	**if** (v_1 **is** *Low*) **and** (v_3 **is** *Low*) **then** (*output* **is** *benign*)
Rule 2	**if** (v_4 **is** *Low*) **and** (v_6 **is** *Low*) **and** (v_8 **is** *Low*) **and** (v_9 **is** *Low*) **then** (*output* **is** *benign*)
Rule 3	**if** (v_1 **is** *Low*) **and** (v_3 **is** *High*) **and** (v_5 **is** *High*) **and** (v_8 **is** *Low*) **and** (v_9 **is** *Low*) **then** (*output* **is** *benign*)
Rule 4	**if** (v_1 **is** *Low*) **and** (v_2 **is** *High*) **and** (v_4 **is** *Low*) **and** (v_5 **is** *Low*) **and** (v_8 **is** *High*) **then** (*output* **is** *benign*)
Rule 5	**if** (v_2 **is** *High*) **and** (v_4 **is** *High*) **then** (*output* **is** *malignant*)
Rule 6	**if** (v_1 **is** *High*) **and** (v_3 **is** *High*) **and** (v_6 **is** *High*) **and** (v_7 **is** *High*) **then** (*output* **is** *malignant*)
Rule 7	**if** (v_2 **is** *High*) **and** (v_3 **is** *High*) **and** (v_4 **is** *Low*) **and** (v_5 **is** *Low*) **and** (v_7 **is** *High*) **then** (*output* **is** *malignant*)
Default	**else** (*output* **is** *malignant*)

Figure 11. The best evolved, fuzzy diagnostic system with seven rules. It exhibits an overall classification rate of 98.98%, and its longest rule includes 5 variables.

We next describe two of our top-performance systems, which serve to exemplify the solutions found by Fuzzy CoCo. The first system, delineated in Figure 11, presents the highest classification performance evolved to date. It consists of seven rules (note that the `else` condition is not counted as an active rule) with the longest rule including 5 variables. This system obtains an overall classification rate (i.e., over the entire database) of 98.98%.

In addition to the above seven-rule system, evolution found systems with between 2 and 6 rules exhibiting excellent classification performance, i.e., higher than 98.5% (Table 6). Among these systems, we consider as the most interesting the system with the smallest number of conditions (i.e., total number of variables in the rules). Figure 12 presents one such two-rule system, containing a total of 8 conditions, and which obtains an overall classification rate of 98.54%; its longest rule has 5 variables.

Database

	v_1	v_2	v_3	v_4	v_5	v_6	v_7	v_8	v_9
P	3		1	3	4	5		7	2
d	8		3	1	2	2		4	1

Rule base

Rule 1	**if** (v_1 **is** *Low*) **and** (v_3 **is** *Low*) **and** (v_5 **is** *Low*) **then** (*output* **is** *benign*)
Rule 2	**if** (v_1 **is** *Low*) **and** (v_4 **is** *Low*) **and** (v_6 **is** *Low*) **and** (v_8 **is** *Low*) **and** (v_9 **is** *Low*) **then** (*output* **is** *benign*)
Default	**else** (*output* **is** *malignant*)

Figure 12. The best evolved, fuzzy diagnostic system with two rules. It exhibits an overall classification rate of 98.54%, and a maximum of 5 variables in the longest rule.

The improvement attained by Fuzzy CoCo, while seemingly slight (0.5-1%) is in fact quite significant. A 1% improvement implies 7 additional cases which are classified correctly. At the performance rates in question (above 98%) every additional case is hard-won. Indeed, try as we did with the fuzzy-genetic approach—tuning parameters and tweaking the setup—we arrived at a performance impasse. Fuzzy CoCo, however, readily churned out better-performance systems, which were able to classify a significant number of additional cases; moreover, these systems were evolved in less time.

6 Concluding Remarks

We presented our recent work which combines the search power of evolutionary algorithms with the expressive power of fuzzy systems to design high-performance, human-interpretable medical diagnostic systems. In particular, we described two approaches for automatically designing systems for breast-cancer diagnosis: (1) a fuzzy-genetic approach and (2) Fuzzy CoCo, our novel cooperative coevolutionary approach to fuzzy modeling.

We applied the two aforementioned algorithms to the Wisconsin breast cancer diagnosis problem. Our evolved systems exhibit both characteristics outlined in Section 1: first, they attain high classification *perfor-*

mance (the best shown to date); second, the resulting systems involve a few simple rules, and are therefore *interpretable*.

We are currently investigating the expansion of Fuzzy CoCo, with two short-term goals in mind: (1) Study the tuning of the genetic-algorithm parameters according to each species characteristics (e.g., encoding schemes, elitism rates, or mutation probabilities). (2) Explore the application of different evolutionary algorithms for each species (e.g., evolution strategies for the evolution of membership functions). In the long term we plan to test some novel ideas that could improve Fuzzy CoCo: (1) Coevolution of $N_r + 1$ species, one species for each of the N_r rules in addition to the membership-function species. (2) Coexistence of several Fuzzy CoCo instances (each one set to evolve systems with a different number of rules), permitting migration of individuals among them so as to increase the exploration and the diversity of the search process. (3) Apply the strategy of rising and death of species proposed by Potter and DeJong [33] in order to evolve systems with variable numbers of rules and membership functions.

References

[1] Alander, J.T. (1997), "An indexed bibliography of genetic algorithms with fuzzy logic," in [29], pp. 299-318.

[2] Bastian, A. (2000), "Identifying fuzzy models utilizing genetic programming," *Fuzzy Sets and Systems*, vol. 113, no. 3, pp. 333-350, August.

[3] Bennett, K.P. and Mangasarian, O.L. (1992), "Neural network training via linear programming," in Pardalos, P.M. (Ed.), *Advances in Optimization and Parallel Computing*, Elsevier Science, pp. 56-57.

[4] Box, G.E.P. (1957), "Evolutionary operation: a method for increasing industrial productivity," *Applied Statistics*, vol. 6, no. 2, pp. 81-101.

[5] Box, G.E.P. and Hunter, J.S. (1959), "Condensed calculations for evolutionary operation programs," *Technometrics*, vol. 1, pp. 77-95.

[6] Cordón, O., Herrera, F., and Lozano, M. (1997), "On the combination of fuzzy logic and evolutionary computation: a short review and bibliography," in [29], pp. 33-56.

[7] Cordón, O., Herrera, F., and Lozano, M. (1999), "A two-stage evolutionary process for designing TSK fuzzy rule-based systems," *IEEE Transactions on Systems, Man, and Cybernetics, Part B: Cybernetics*, vol. 29, no. 6, pp. 703-714, December.

[8] Fogel, D.B. (Ed.) (1998), *Evolutionary Computation: the Fossil Record*, IEEE Press, Piscataway, NJ.

[9] Fogel, L.J. (1962), "Autonomous automata," *Industrial Research*, vol. 4, pp. 14-19.

[10] Friedberg, R.M. (1958), "A learning machine: I," *IBM Journal of Research and Development*, vol. 2, pp. 2-13.

[11] Friedberg, R.M., Dunham, B., and North, J.H. (1959), "A learning machine. II," *IBM Journal of Research and Development*, vol. 3, pp. 282-287.

[12] Herrera, F., Lozano, M., and Verdegay, J.L. (1995), "Generating fuzzy rules from examples using genetic algorithms," in Bouchon-Meunier, B., Yager, R.R., and Zadeh, L.A. (Eds.), *Fuzzy Logic and Soft Computing*, World Scientific, pp. 11-20.

[13] Holland, J.H. (1962), "Outline for a logical theory of adaptive systems," *Journal of the ACM*, vol. 9, no. 3, pp. 297-314, July.

[14] Jang. J.S.R. and Sun, C.T. (1995), "Neuro-fuzzy modeling and control," *Proceedings of the IEEE*, vol. 83, no. 3, pp. 378-406, March.

[15] Karr, C.L. (1991), "Genetic algorithms for fuzzy controllers," *AI Expert*, vol. 6, no. 2, pp. 26-33, February.

[16] Karr, C.L., Freeman, L.M., and Meredith, D.L. (1990), "Improved fuzzy process control of spacecraft terminal rendezvous using a genetic algorithm," in Rodriguez, G. (Ed.), *Proceedings of Intelligent Control and Adaptive Systems Conference*, SPIE, vol. 1196, pp. 274-288.

[17] Mangasarian, O.L., Setiono, R., and Goldberg, W.H. (1990), "Pattern recognition via linear programming: Theory and application to medical diagnosis," in Coleman, T.F. and Li, Y. (Eds.), *Large-Scale Numerical Optimization*, SIAM, pp. 22-31.

[18] Mangasarian, O.L., Street, W.N., and Wolberg, W.H. (1994), "Breast cancer diagnosis and prognosis via linear programming," Mathematical Programming Technical Report 94-10, University of Wisconsin.

[19] Mendel, J.M. (1995), "Fuzzy logic systems for engineering: a tutorial," *Proceedings of the IEEE*, vol. 83, no. 3, pp. 345-377, March.

[20] Merz, C.J. and Murphy, P.M. (1996), UCI repository of machine learning databases.

[21] Michalewicz, Z. (1996), *Genetic Algorithms + Data Structures = Evolution Programs*, Springer-Verlag, Heidelberg, third edition.

[22] Nauck, D. and Kruse, R. (1999), "Neuro-fuzzy systems for function approximation," *Fuzzy Sets and Systems*, vol. 101, no. 2, pp. 261-271, January.

[23] Paredis, J. (1995), "Coevolutionary computation," *Artificial Life*, vol. 2, pp. 355-375.

[24] Peña-Reyes, C.A. and Sipper, M. (1998), "Evolving fuzzy rules for breast cancer diagnosis," *Proceedings of 1998 International Symposium on Nonlinear Theory and Applications (NOLTA'98)*, vol. 2, pp. 369-372, Presses Polytechniques et Universitaires Romandes, Lausanne.

[25] Peña-Reyes, C.A. and Sipper, M. (1999), "Designing breast cancer diagnostic systems via a hybrid fuzzy-genetic methodology," *1999 IEEE International Fuzzy Systems Conference Proceedings*, vol. 1, pp. 135-139, IEEE Neural Network Council.

[26] Peña-Reyes, C.A. and Sipper, M. (1999), "A fuzzy-genetic approach to breast cancer diagnosis," *Artificial Intelligence in Medicine*, vol. 17, no. 2, pp. 131-155, October.

[27] Peña-Reyes, C.A. and Sipper, M. (2000), "Applying Fuzzy CoCo to breast cancer diagnosis," *Proceedings of the 2000 Congress on Evolutionary Computation (CEC00)*, vol. 2, pp. 1168-1175, IEEE Press, Piscataway, NJ, USA.

[28] Peña-Reyes, C.A. and Sipper, M. (2000), "Evolutionary computation in medicine: an overview," *Artificial Intelligence in Medicine*, vol. 19, no. 1, pp. 1-23, May.

[29] Pedrycz, W. (Ed.) (1997), *Fuzzy Evolutionary Computation*, Kluwer Academic Publishers.

[30] Pedrycz, W. and Valente de Oliveira, J. (1996), "Optimization of fuzzy models," *IEEE Transactions on Systems, Man and Cybernetics, Part B: Cybernetics*, vol. 26, no. 4, pp. 627-636, August.

[31] Poli, R. (1996), "Introduction to evolutionary computation," http:-//www.cs.bham.ac.uk/~rmp/slide_book/, October 1996. (Visited: 16 March 1999.)

[32] Potter, M.A. (1997), *The Design and Analysis of a Computational Model of Cooperative Coevolution*, Ph.D. thesis, George Mason University.

[33] Potter, M.A. and DeJong, K.A. (2000), "Cooperative coevolution: an architecture for evolving coadapted subcomponents," *Evolutionary Computation*, vol. 8, no. 1, pp. 1-29, spring.

[34] Rechenberg, I. (1964), "Cybernetic solution path of an experimental problem," Farborough Hants: Royal Aircraft Establishment. Library Translation 1122, August 1965, English Translation of lecture given at the Annual Conference of the WGLR at Berlin in September, 1964.

[35] Russo, F. (1999), "Evolutionary neural fuzzy systems for noise cancellation in image data," *IEEE Transactions on Instrumentation and Measurement*, vol. 48, no. 5, pp. 915-920, October.

[36] Schwefel, H.P. (1965), "Kybernetische Evolution als Strategie der experimentelen Forschung in der Stromungstechnik," Master's thesis, Technical University of Berlin, March.

[37] Seng, T.L., Bin Khalid, M., and Yusof, R. (1999), "Tuning of a neuro-fuzzy controller by genetic algorithm," *IEEE Transactions on Systems, Man and Cybernetics, Part B: Cybernetics*, vol. 29, no. 2, pp. 226-236, April.

[38] Setiono, R. (2000), "Generating concise and accurate classification rules for breast cancer diagnosis," *Artificial Intelligence in Medicine*, vol. 18, no. 3, pp. 205-219.

[39] Taha, I. and Ghosh, J. (1997), "Evaluation and ordering of rules extracted from feedforward networks," *Proceedings of the IEEE International Conference on Neural Networks*, pp. 221-226.

[40] Vose, M.D. (1999), *The Simple Genetic Algorithm*, MIT Press, Cambridge, MA, August.

[41] Vuorimaa, P. (1994), "Fuzzy self-organizing map," *Fuzzy Sets and Systems*, vol. 66, pp. 223-231.

[42] Yager, R.R. and Filev, D.P. (1994), *Essentials of Fuzzy Modeling and Control*, John Wiley & Sons., New York.

[43] Yager, R.R. and Zadeh, L.A. (1994), *Fuzzy Sets, Neural Networks, and Soft Computing*, Van Nostrand Reinhold, New York.

Chapter 15

Genetic Algorithms for Feature Selection in Computer-Aided Diagnosis

B. Sahiner, H.P. Chan, and N. Petrick

One of the important practical applications of computer vision techniques is computer-aided diagnosis (CAD) in medical imaging. It has been shown that CAD can improve the accuracy of breast cancer detection and characterization by radiologists on mammograms. In this chapter, we discuss an important step – feature selection – in classifier design for CAD algorithms. Feature selection reduces the dimensionality of an available feature space and is therefore often used to prevent over-parameterization of a classifier. Many feature selection techniques have been proposed in the literature. We will illustrate the usefulness of genetic algorithms (GAs) for feature selection by comparing GA with a commonly used sequential selection method. A brief introduction to the GA is given and several examples using GA feature selection for the characterization of mammographic lesions are discussed. The examples illustrate the design of a fitness function for optimizing classification accuracy in terms of the receiver operating characteristics of the classifier, the dependence of GA performance on its evolution parameters, and the design of a fitness function tailored to a specific classification task.

1 Introduction

Breast cancer is a major cause of death in women. It is estimated to be the leading cause of cancer death among women globally [1]. In the United States, it is second to lung cancer in mortality among women, with an estimated 41,200 deaths in 2000, accounting for 15% of all cancer deaths [2]. Early treatment of the cancer before it metastasizes is the most promising way to improve the chances of survival of breast cancer patients [3]. Mammography is the most effective method for detection of early breast cancer, and it has been shown that screening mammography reduces breast cancer mortality [4], [5]. However, 10-30% of the breast cancers that are visible on mammograms in retrospective studies are not detected due to various technical or human factors [6], [7], [8], [9]. The specificity of mammography for differentiating lesions as malignant or benign is also very low. In the United States, only 15 to 30% of the patients who have undergone biopsy due to a suspicious finding on mammograms are found to have breast cancer [10], [11].

In an attempt to reduce health care cost and increase the efficacy of screening, various methods are being developed to improve the detection of breast cancer at an early stage. Computer-aided diagnosis (CAD) is considered to be one of the promising approaches that may achieve both goals [12]. Properly designed CAD algorithms can automatically detect suspicious lesions on a mammogram and alert the radiologist to these regions. As a further aid to the radiologist in making diagnostic decisions, the computer can extract image features from suspicious regions containing lesions, and estimate the likelihood that the lesion is malignant or benign. It has been shown in receiver operating characteristic (ROC) studies that both the detection and classification accuracy of radiologists reading with CAD were improved significantly compared to reading without CAD [13], [14], [15], and [16].

Radiologists use certain visual criteria to detect abnormalities on mammograms or to characterize a lesion as malignant or benign. However, it is often difficult to translate these measures into computer algorithms that exactly match what the radiologist visually perceives. Therefore, a common approach to CAD is to extract a number of features, forming a multidimensional feature space, to describe the classes of malignant and benign lesions (or normal and abnormal breast tissue). A classifier is then designed to predict the membership of a given sample based on the class distributions of the feature vectors in the feature space. The features may or may not match what a radiologist uses for the same task. This approach has the advantage that the computer may extract some useful features that are difficult to perceive and are complementary to the image features perceived by a radiologist. In the classifier design process, a subset of features is selected from the entire feature space based on their individual or joint performance. The selected features are then used by the classifier to perform the classification task for the CAD system.

The inclusion of inappropriate features often adversely affects classifier performance, especially when the training set is not sufficiently large. Feature selection is therefore a very important step in CAD [17], [18], [19], and [20], and it is an active area of research in pattern recognition [21], [22]. A number of different techniques have been developed to address this problem. This chapter will describe feature selection techniques based on the genetic algorithm (GA), with special emphasis on applications in CAD. We will discuss how GAs can be tailored to select effective features for high-sensitivity classifier design, for linear classifiers, and for artificial neural networks in a number of applications in computerized breast lesion detection and characterization.

In the next section, we present an overview of GAs and describe the main GA components that are used in this chapter. Section 3 gives a brief review of feature selection methods. The need for feature selection in classification tasks is highlighted, and some of the potential problems with commonly used feature selection algorithms such as sequential feature selection and branch-and-bound methods are mentioned. Section 4 presents the application of GAs to feature selection in three areas in CAD for mammography. Sections 4.1, 4.2, and 4.3 describe the application of GAs to feature selection in computerized characterization of microcalcifications, computerized detection of masses, and computerized characterization of masses, respectively. The data sets, feature extraction and classification methods, and GA implementation for each application are discussed separately. In each application, classification accuracy with GA-based feature selection is compared to that with stepwise feature selection, and specific issues related to GA implementation and classification accuracy are discussed in detail. In Section 5, general conclusions are drawn from the three applications.

2 Genetic Algorithms

GAs are iterative optimization techniques based on the principles of natural evolution. In natural evolution, the basic problem of each population is to find beneficial adaptations to a complex environment. The genetic characteristics that each individual has gained or inherited, the genotype, are carried in the individual's chromosomes. Each individual reproduces more or less in proportion to its fitness within the environment. Reproduction, mutation, and selection are the fundamental mechanisms through which natural evolution takes place.

In a GA, the optimization problem to be solved plays the role of the environment in which a population of individuals live and evolve. The individuals in a GA represent possible solutions to the problem. The genotypes of the individuals are represented by chromosomes, which are fixed-length binary strings. The fitness of an individual, or the quality of a particular solution, as measured by a pre-defined fitness function, reflects the degree of adaptation of the individual to its environment. The population evolves based on the principle of natural selection towards a better solution to the problem. Problem solving techniques that mimic natural evolution were introduced as early as 1950's [23]. In the 1960s Holland [24] and Fogel [25] started developing the techniques that are known as genetic algorithms and evolutionary programming today. Numerous researchers have contributed to the field of evolutionary computation in the past four decades [26].

To solve an optimization problem, a GA requires five components, which are analogous to components of natural selection. These components are described below.

2.1 Encoding

Encoding is a way of representing the decision variables of the optimization problem as chromosomes. Each chromosome is a possible solution to the optimization problem. If there are v decision variables in an optimization problem and each decision variable is encoded as an n-digit binary number, then a chromosome is a string of $n \times v$ binary digits.

2.2 Initial Population

The initial population is a set of chromosomes offered as initial solutions or as a starting point in the search for better chromosomes. The initial population must be large and diverse enough to allow evolution towards better individuals. In general, the population is initialized at random to bit strings of 0's and 1's. However, more direct methods for finding the initial population can sometimes be used to improve convergence time.

2.3 Fitness Function

The fitness function rates chromosomes (i.e., possible solutions) in terms of how good they are in solving the optimization problem. The fitness function returns a single fitness value for each chromosome, which is then used to determine the probability that this chromosome will be selected as a parent to generate new chromosomes. The fitness function is the primary GA component that is used to tailor a traditional GA to suit a specific problem.

2.4 Genetic Operators

Genetic operators are applied probabilistically to chromosomes of a generation to produce a new generation of chromosomes. Three basic operators are parent selection, crossover and mutation. The parent selection operation mimics the natural selection process by selecting chromosomes to create a new generation, where the fittest chromosomes reproduce most often. The crossover operation refers to the exchange of substrings of two chromosomes to generate two new offspring. Crossover occurs between two selected parents with a predefined probability. Mutation simply complements the binary value of each bit on a chromosome according to a predefined mutation probability. Crossover and mutation introduce new chromosomes and new genes to the population.

2.5 Working Parameters

A set of parameters, which includes the number of chromosomes in each generation, the crossover rate, the mutation rate and the stopping criterion, is predefined to guide the GA. The crossover and mutation rates, assigned as real numbers between 0 and 1, are used as thresholds to determine whether the operators will be applied or not. The stopping criterion is predefined as the number of generations the algorithm is to be run or as a tolerance value for the fitness function.

Two forces, exploration and exploitation, interact in the search for better-fit chromosomes. Exploitation occurs in the form of parent selection. Chromosomes with higher degree of fitness exploit this fitness by reproducing more often. Exploration occurs in the form of mutation and crossover, which allow the offspring to achieve a higher degree of fitness than their parents. Crossover is the key to exploration, whereas mutation provides background variation and occasionally introduces beneficial genes into the chromosomes. For a GA to be successful, exploration and exploitation have to be in good balance. If exploitation dominates, the population may be stuck with the same chromosome after a few generations. On the other hand, if exploration dominates over exploitation, good genes may never be able to accumulate in the genetic pool.

GAs are ideal for sampling large search spaces and locating the regions of enhanced opportunity. Although GAs yield near-optimal solutions rather than optimal ones, obtaining such near-optimal solutions are usually the best that one can do in many complex optimization problems involving a large number of parameters.

3 Feature Selection and GAs

In a classification problem, feature selection is usually a necessary first step if the design sample size is not large in comparison to the number available features [27]. Feature selection is theoretically a difficult problem [28], because, depending on the class distributions, it may not be possible to determine the optimal feature subset without exhaustively evaluating all feature combinations. It is well known, for example, that the two independent features that yield the highest classification accuracy in a feature set may not constitute the best pair of features when they are combined [29]. In practice, when the class distributions are not known in advance, but have to be estimated from available data, the problem becomes even more intricate. In the CAD training process, the classifier can be designed so that the probability of training error will not increase when the number of selected features increases. However, optimizing the test performance, which is the goal in practice, complicates the problem because over-training can cause deterioration in test performance when the number of selected features increases [30]. Over-fitting (i.e., over-training) is of the most concern when the number of training cases is small. It is imperative to select a small subset of features to avoid the so-called peaking phenomenon or the curse of dimensionality [31], [32], and [33] (i.e., a decrease in classification accuracy for test cases with increasing number of features) if the ratio of the number of training cases to the number of available features is not sufficiently large.

The exhaustive search method is guaranteed to find the optimal feature subset. The main problem with exhaustive search is that it requires calculating the classification accuracy for 2^M feature sets, where M is the total number of available features. This quickly becomes impractical as M increases in size. The branch and bound method proposed by Narendra [34] requires considerably fewer computations and is guaranteed to find the optimal feature subset if the feature selection criterion is monotonic, i.e., if adding a new feature never decreases the classification accuracy. However, due to the peaking phenomenon described previously, this condition is usually not satisfied, especially when the data set is not large compared to the number of available features.

Sequential selection methods are the most commonly used techniques for feature selection. These methods start with a feature subset and sequentially include or exclude features based on a feature selection criterion, until a stopping criterion is met. For example, backward feature elimination starts with all available features, and excludes the least useful feature one step at a time until the least useful feature meets a significance criterion. Forward feature selection starts with no features and includes the most useful feature one step at a time. Stepwise feature selection combines both approaches, and evaluates both inclusion and exclusion criteria at each step. It has been shown that sequential feature selection methods are not guaranteed to select the optimal feature subset [27]

GAs are well-suited for feature selection problems in large feature spaces, where the optimal solution is practically impossible to compute, and a near-optimal solution is the best alternative. GAs were initially introduced for feature selection by Siedlecki and Sklansky [35]. Brill et al. used a GA to select features for counter-propagation neural networks [36]. Kuncheva et al. used a GA for editing and feature selection in a nearest neighbor classifier [37]. Kudo and Sklansky showed that for nearest-neighbor classifiers, sequential search was suitable for small and medium-scale problems, and GAs were suitable for large-scale problems [38], [39]. In medical imaging applications, Sahiner et al. used GAs for feature selection in classification of breast masses and normal breast tissue [40], and classification of malignant and benign masses on mammograms [41]. Chan et al. used a GA to select features in classification of malignant and benign microcalcifications on mammograms [42]. Zheng et al. used GAs for feature selection in mammographic mass detection using a Bayesian belief network [43]. Yamany et al. used a GA and a backpropagation neural network for computerized classification of endothelial cells [44]. Handels et al. used the same combination for discriminating between malignant melanomas and moles [45]. In addition to problems related to feature selection, GA's have been used for a variety of problems in computerized medical applications [46], [47], [48], [49], and [50].

4 Applications in CAD

In this Section, we illustrate the application of GA to optimization problems by examples. We will discuss some of our experiences with GAs for feature selection in CAD in three areas, namely, classification of malignant and benign microcalcifications, classification of mass and normal breast tissue, and classification of malignant and benign masses on mammograms.

In all three applications a linear discriminant analysis (LDA) classifier was used for classification with the selected features. This is the optimal classifier if the features are distributed as multivariate Gaussian random variables with equal covariance matrices under each class [51]. In addition, a backpropagation neural network was also used in the application described in Section 4.2 below.

The classification accuracy of the classifiers was evaluated by receiver operating characteristic (ROC) methodology [52], which is a commonly used method in CAD. ROC analysis uses the output values from the classifier as the decision variable. The ROC curve represents the relationship between the true-positive fraction (TPF) and the false-positive fraction (FPF) as the decision threshold varies. A computer program was first developed by Dorfman et al. [53] and later modified by Metz et al. [54] to fit an ROC curve to the TPF and FPF data based on maximum likelihood estimation. The ROC curve fitting assumes binormal distributions of the decision variable for the normal and abnormal cases. However, the assumption is satisfied if the distributions can be transformed to normal distributions by a monotonic function.

In our GA, the number of bits (or genes) in a chromosome was equal to the total number of available features in the feature space, and each bit corresponded to an individual feature extracted from the ROIs. The selection of a feature is indicated by a value of 1 for the bit representing the feature in the chromosome. A bit value of 0 indicates that the feature is not selected. The population was initialized at random, with a small probability P_{init} of having a 1 at each bit location. This allowed the GA to start with a few selected features and grow to a reasonable number of features as the population evolves. Figure 1 shows the block diagram of the GA used in all three applications.

For the purpose of comparison with GA-based feature selection, we also studied the classification accuracy of the same classifiers using stepwise feature selection. We employed the Wilks' lambda as our feature selection criterion, which is defined as the ratio of the within-group sum of squares to the total sum of squares of the two classes [55]. The number of features selected by this method is controlled by two parameters, called F_{in} and F_{out}. At each step, the stepwise feature selection algorithm first determines the significance of the change, based on F-statistics, in Wilks' lambda when each available feature is entered, one at a time, into the selected feature pool. If the significance is above the threshold determined by the F_{in} parameter, then the selected feature pool is augmented by including the most significant new feature into the set. Next, the algorithm computes the significance of the change in Wilks' lambda when each variable is removed, one at a time, from the selected feature pool. If the significance is below the threshold determined by the F_{out} parameter, then the least significant variable is removed from the selected feature pool. Increasing either the F_{in} or the F_{out} value makes it more difficult for a feature to be included or makes it easier for a feature to be excluded, thereby decreasing the number of selected features. Similar to GA-based feature selection, stepwise feature selection is a heuristic procedure. For this reason, the optimal values of F_{in} and F_{out} parameters are not known in advance. One has to experiment with these parameters and increase or decrease the number of selected features to obtain the best test performance. A detailed description of the stepwise feature selection [51], [56] and its application to our problems [18], [20], and [57] can be found in the literature.

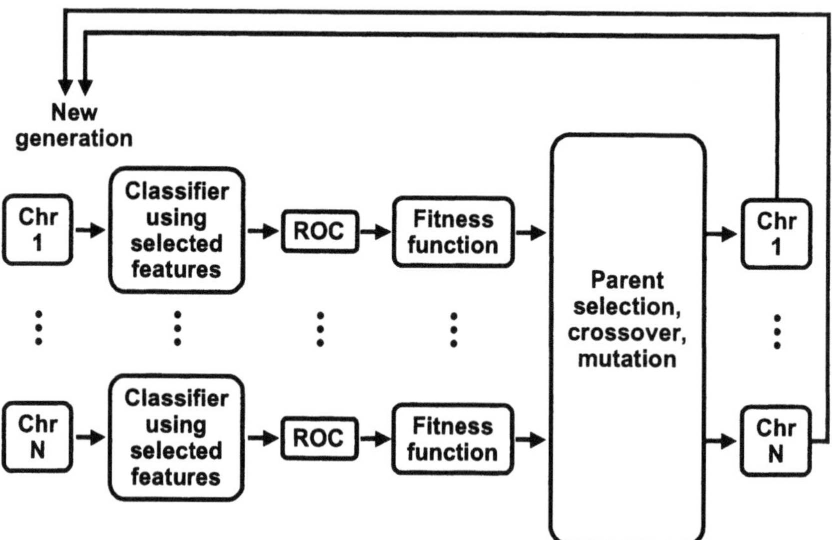

Figure 1. The block diagram of the GA structure used in the applications discussed in this chapter.

4.1 Classification of Malignant and Benign Microcalcifications

Clustered microcalcifications are one of the mammographic signs that indicate the presence of breast cancer. Microcalcifications are calcium deposits in breast tissue that are imaged as small bright spots about 0.1 mm to 0.5 mm in diameter on a mammogram. Many microcalcifications are related to benign breast diseases, but some are caused by malignancy. Malignant microcalcifications often have irregular shapes and sizes, and these individual microcalcifications tend to group together and form a cluster. However, many microcalcifications and microcalcification clusters that do not manifest the "typical" malignant features can still be malignant. Because of the low specificity of microcalcification features, radiologists have to recommend biopsy for most microcalcification clusters in order not to miss breast cancer. We have developed a computer classifier for classifying microcalcification clusters as benign or malignant based on morphological and texture features. GA and stepwise linear discriminant procedures were used for selecting appropriate features and their effectiveness was compared.

4.1.1 Feature Extraction

Two types of features can be extracted from mammographic microcalcifications. Morphological features are used to describe the size, shape, and contrast of the individual microcalcifications and their variation within a cluster. Texture features may be extracted from the breast tissue containing the microcalcification clusters to describe the textural changes of the tissue due to a developing malignancy [58], [59], and [60]. We found that spatial gray level dependence (SGLD) matrices at multiple distances were useful for differentiating malignant and benign masses or microcalcifications on mammograms [61], [62]. In the following, we compared the classification accuracy in the combined morphological and texture feature space with those obtained in the morphological feature space or in the texture feature space alone.

4.1.2 Data Set

The data set for this study consisted of 145 clusters of microcalcifications from mammograms of 78 patients. The only case selection criterion was that it included a biopsy-proven microcalcification cluster. We kept the number of malignant and benign cases reasonably balanced so that 82 benign and 63 malignant clusters were included. All mammograms were acquired with a contact technique using mammography systems accredited by the American College of Radiology. The mammograms were digitized with a laser scanner (Lumisys DIS-1000) at a pixel size of 0.035 mm × 0.035 mm and 12-bit gray levels. The optical density on the film was linearly proportional to the digitized pixel value.

4.1.3 Morphological Feature Space

We have developed an automated signal extraction program to determine the size, contrast, signal-to-noise ratio (SNR), and shape of the microcalcifications from a mammogram based on the coordinate of each individual microcalcification. In a local region centered at a microcalcification, the low frequency structured background is estimated and subtracted from the region. The local root-mean-square (RMS) noise is calculated from the background-corrected region. A gray level threshold is determined as the product of the RMS noise and a pre-selected SNR threshold. With a region growing technique, the signal region is then extracted as the connected pixels above the threshold around the manually identified signal location. An example of a malignant cluster and the microcalcifications extracted at an SNR threshold of 2.0 is shown in Figure 2.

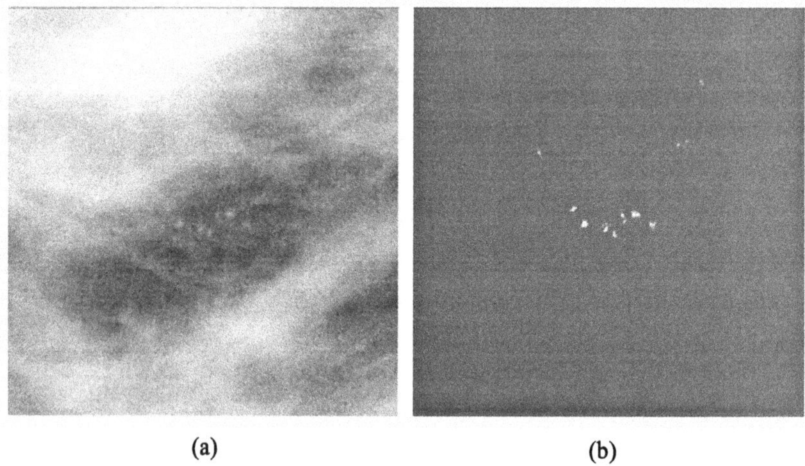

(a) (b)

Figure 2. An example of a cluster of malignant microcalcifications in the data set: (a) the cluster with mammographic background, (b) the cluster after segmentation. Morphological features are extracted from the segmented microcalcifications.

The feature descriptors determined from the extracted microcalcifications are listed in Table 1. The size of a microcalcification (SA) is estimated as the number of pixels above the SNR threshold in the signal region. The mean density (MD) is the average of the pixel values above the background level within the signal region. The second moments are calculated as

$$M_{xx} = \frac{\sum_i g_i (x_i - M_x)^2}{M_0} \tag{1}$$

$$M_{yy} = \frac{\sum_i g_i (y_i - M_y)^2}{M_0} \tag{2}$$

$$M_{xy} = \frac{\sum_i g_i (x_i - M_x)(y_i - M_y)}{M_0} \tag{3}$$

where g_i is the pixel value above the background, and (x_i, y_i) are the coordinates of the ith pixel. The moments M_0, M_x and M_y are defined as follows:

$$M_0 = \sum_i g_i \tag{4}$$

$$M_x = \frac{\sum_i g_i x_i}{M_0} \tag{5}$$

$$M_y = \frac{\sum_i g_i y_i}{M_0} \tag{6}$$

The summations are over all pixels within the signal region. The lengths of the major axis, *2a*, and the minor axis, *2b*, of the effective ellipse that characterizes the second moments are given by

$$2a = \sqrt{2\left[M_{xx} + M_{yy} + \sqrt{(M_{xx} - M_{yy})^2 + 4M_{xy}^2}\right]} \tag{7}$$

$$2b = \sqrt{2\left[M_{xx} + M_{yy} - \sqrt{(M_{xx} - M_{yy})^2 + 4M_{xy}^2}\right]} \tag{8}$$

The eccentricity (EC) of the effective ellipse can be derived from the major and minor axes as

$$\varepsilon = \frac{\sqrt{a^2 - b^2}}{a} \qquad (9)$$

The moment ratio (MR) is defined as the ratio of M_{xx} to M_{yy}, with the larger second moment in the denominator. The axis ratio (AR) is the ratio of the major axis to the minor axis of the effective eclipse.

To quantify the variation of the visibility and shape descriptors in a cluster, the maximum (MX), the average (AV) and the standard deviation (SD) of each feature for the individual microcalcifications in the cluster are calculated. The coefficient of variation (CV), which is the ratio of the SD to AV, is used as a descriptor of the variability of a certain feature within a cluster. Twenty cluster features are therefore derived from the five features (size, mean density, moment ratio, axis ratio, and eccentricity) of the individual microcalcifications. Another feature describing the number of microcalcifications in a cluster (NUMS) is also added, resulting in a 21-dimensional morphological feature space.

Table 1. The twenty-one morphological features extracted from a microcalcification cluster.

	Average	Std Dev	Coef. of Var.	Max.
Area	AVSA	SDSA	CVSA	MXSA
Mean Density	AVMD	SDMD	CVMD	MXMD
Eccentricity	AVEC	SDEC	CVEC	MXEC
Moment Ratio	AVMR	SDMR	CVMR	MXMR
Axis Ratio	AVAR	SDAR	CVAR	MXAR
No. of microcalcifications in cluster	NUMS			

4.1.4 Texture Feature Space

Our texture feature extraction method has been described in detail previously [60]. Briefly, texture features are extracted from a 1024 × 1024 pixel region of interest (ROI) that contains the cluster of microcalcifications. Most of the clusters in this data set can be contained within the ROI. For the few clusters that are substantially larger than a single ROI, additional ROIs containing the remaining parts of the cluster are extracted and processed in the same manner. The texture feature values extracted from the different ROIs of the same cluster are averaged and the average values are used as the feature values for that cluster.

For a given ROI, background correction is first performed to reduce the low frequency gray level variation due to the density of the overlapping breast tissue and the x-ray exposure conditions. The gray level at a given pixel of the low frequency background is estimated as the average of the distance-weighted gray levels of four pixels at the intersections of the normals from the given pixel to the four edges of the ROI [63]. The estimated background image was subtracted from the original ROI to obtain a background-corrected image.

Texture features were derived from the SGLD matrix of the ROI. The SGLD matrix element, $p_{\theta,d}(i,j)$, is the joint probability of the occurrence of gray levels i and j for pixel pairs which are separated by a distance d and at a direction θ [60], [64]. The SGLD matrices were constructed from the pixel pairs in a sub-region of 512×512 pixels centered approximately at the center of the cluster in the background-corrected ROI so that any potential edge effects caused by background correction will not affect the texture extraction. We analyzed the texture features in four directions: $\theta = 0^o$, 45°, 90°, and 135° at each pixel pair distance d. The pixel pair distance was varied from 4 to 40 pixels in increments of 4 pixels. Therefore, a total of 40 SGLD matrices were derived from each ROI. The SGLD matrix depends on the bin width (or gray level interval) used in accumulating the histogram. Based on our previous study, a bin width of 4 gray levels was chosen for constructing the SGLD matrices. This is equivalent to reducing the gray level resolution (or bit depth) of the 12-bit image to 10 bits by eliminating the two least significant bits.

From each of the SGLD matrices, we derived thirteen texture measures including correlation, entropy, energy (angular second moment), inertia, inverse difference moment, sum average, sum entropy, sum variance, difference average, difference entropy, difference variance, information measure of correlation 1, and information measure of correlation 2. The formulation of these texture measures can be found in the literature [60], [64]. As found in our previous study [18], we did not observe a significant dependence of the discriminatory power of the texture features on the direction of the pixel pairs for mammographic textures. However, since the actual distance between the pixel pairs in the diagonal direction was a factor of $\sqrt{2}$ greater than that in the axial direction, we averaged the feature values in the axial directions (0° and 90°) and in the diagonal directions (45° and 135°) separately for each texture feature derived from the SGLD matrix at a given pixel pair distance. The average texture features at the ten pixel pair distances and two directions formed a 260-dimensional texture feature space.

4.1.5 GA Implementation

The available samples in the data set were randomly partitioned into a training set and a test set. The training set was used to formulate a linear discriminant function with each of the selected feature subsets. The effectiveness of each of the linear discriminants for classification was evaluated with the test set. The classification accuracy was determined as the area, A_z, under the ROC curve. To reduce biases in the classifiers due to case selection, training and testing were performed many times, each with a different random partitioning of the data set. In this study, we chose to partition the data set 80 times and the 80 test A_z values were averaged and used for determination of the fitness of the chromosome.

The fitness function for the ith chromosome, $F(i)$, was formulated as

$$F(i) = \left[\frac{f(i) - f_{min}}{f_{max} - f_{min}} \right]^k, \quad i = 1, ..., n \tag{10}$$

where

$$f(i) = \overline{A_z(i)} - \alpha N(i), \tag{11}$$

and $\overline{A_z(i)}$ was the average test A_z for the ith chromosome over the 80 random partitions of the data set, f_{min} and f_{max} were the minimum and maximum $f(i)$ among the n chromosomes, $N(i)$ was the number of features in the ith chromosome, and α was a penalty factor, whose magnitude was less than $1/M$, to suppress chromosomes with a large number of selected features, where M was the number of available features. The value of the fitness function $F(i)$ ranged from 0 to 1. We have explored other forms of the fitness function based on the A_z value in this and other applications [41]. It was found that using the A_z value alone as the fitness function was not effective, because chromosomes with a relatively small A_z value were assigned a relatively high fitness. The smallest and largest values for A_z in our applications were on the order of 0.5 and 1.0, respectively. Therefore, if the A_z value alone was used as the fitness function, the fitness ratio of the best and the worst chromosomes would be at most 2.0. When Eqs. (10)-(11) are used instead for defining the fitness function, the chromosome with the largest A_z value is assigned a fitness of 1, the chromosome with the smallest A_z value is assigned a fitness of 0, and the ratio described above would be infinity. The exponent k in Eq. (10) defines how fast the fitness decreases from 1 to 0 in a population as the A_z changes from the largest to smallest. In this application, we found by experimentation that k=2 is a good choice. In the applications described in Sections 4.2 and 4.3, the same form of fitness function was used with k=2 and k=4, respectively.

The probability of the ith chromosome being selected as a parent, $P_s(i)$, was proportional to its fitness:

$$P_s(i) = \frac{F(i)}{\sum_{i=1}^{n} F(i)}, \quad i = 1, \ldots, n \tag{12}$$

A random sampling based on the probabilities, $P_s(i)$, allowed chromosomes with higher value of fitness to be selected more frequently.

For every pair of selected parent chromosomes, X_i and X_j, a random decision was made to determine if crossover should take place. A uniform random number in (0,1] was generated. If the random number was greater than P_c, the probability of crossover, then no crossover occurred; otherwise, a random crossover site was selected on the pair of chromosomes. Each chromosome was split into two strings at this site and one of the strings was exchanged with the corresponding string from the other chromosome. Crossover results in two new chromosomes of the same length.

After crossover, another chance of introducing new features was obtained by mutation. Mutation was applied to each gene on every chromosome. For each bit, a uniform random number in (0,1] was generated. If the random number was greater than P_m, the probability of mutation, then no mutation occurred; otherwise, the bit was complemented. The processes of parent selection, crossover, and mutation resulted in a new generation of n chromosomes, $X_1^{'}$, ..., $X_n^{'}$, which was evaluated with the 80 training and test set partitions as described above. The chromosomes were allowed to evolve over a pre-selected number of generations. The best subset of features was chosen to be the chromosome that provides the highest average A_z during the evolution process.

In this study, 500 chromosomes were used in the population. Each chromosome had 281 gene locations (bits). P_{init} was chosen to be 0.01 so that each chromosome started with 2 to 3 features on the average. We varied P_c from 0.7 to 0.9, P_m from 0.001 to 0.005, and α from 0 to 0.001.

4.1.6 Classification

The training and testing procedure described above was used for the purpose of feature selection only. After the best subset of features as determined by either the GA or the stepwise feature selection procedure was found, we performed the classification as follows.

LDA was used to classify the malignant and benign microcalcification clusters. We used a cross-validation resampling scheme for training and testing the classifier. The data set of 145 samples was randomly partitioned into a training set and a test set by an approximately 3:1 ratio. The partitioning was constrained so that ROIs from the same patient were always grouped into the same set. The training set was used to determine the coefficients (or weights) of the feature variables in the linear discriminant function. The performance of the trained classifier was evaluated with the test set. In order to reduce the effect of case selection, the random partitioning was performed 50 times. The results were then averaged over the 50 partitions. The average performance of the classifier was estimated as the average of the 50 test A_z values from the 50 random partitions.

To obtain a single distribution of the discriminant scores for the test samples, we performed a leave-one-case-out resampling scheme for training and testing the classifier. In this scheme, one of the 78 cases was left out and the clusters from the other 77 cases were used for formulation of the linear discriminant function. The resulting LDA classifier was used to classify the clusters from the left-out case. The procedure was performed 78 times so that every case was left out of training once to be the test case. The test discriminant scores were accumulated in a distribution, which was then analyzed by the LABROC program [54]. Using the distributions of discriminant scores for the test samples from the leave-one-case-out resampling scheme, the CLABROC program could be used to test the statistical significance of the differences between ROC curves [65] obtained from different conditions. The two-tailed p value for the difference in the areas under the ROC curves was estimated.

4.1.7 Results

The variations of best feature set size and classifier performance in terms of A_z with the GA parameters were tabulated in Table 2(a)-(c) for the morphological, the texture, and the combined feature spaces, respectively. The number of generations that the chromosomes evolved was fixed at 75 in these tables. The training and test A_z values were obtained from averaging results of the 50 partitions of the data sets using the selected feature sets.

The results of feature selection using the stepwise LDA procedure with a range of F_{in} and F_{out} thresholds were tabulated in Table 3(a)-(c). The thresholds were varied so that the number of selected features varied over a wide range. The average A_z values obtained from the 50 partitions of the data set using the selected feature sets were listed.

Table 4 compares the training and test A_z values from the best feature set in each feature space for the two feature selection methods. The GA parameters that selected the feature set with best classification performance in each feature space after 75 generations (Table 2) were used to run the GA again for 500 generations. The A_z values obtained with the best GA selected feature sets after 75 generations are listed together with those obtained after 500 generations. The A_z values obtained with the leave-one-case-out scheme are also shown in Table 4. The differences between the corresponding A_z values from the two resampling schemes are within 0.01. The two feature selection methods provided feature sets that had similar test A_z values in the morphological and texture feature spaces. In the combined feature space, there was a slight improvement in the test A_z value obtained with the GA selected features. Although the difference in the A_z values from the leave-one-case-out scheme between the two feature selection methods did not achieve statistical significance (p = 0.2), as estimated by CLABROC, the differences in the paired A_z values from the 50 partitions demonstrated a consistent trend (40 out of 50 partitions) that the A_z from the GA selected features were higher than those obtained by the stepwise LDA.

Table 2. Dependence of feature selection and classifier performance on GA parameters: (a) morphological feature space, (b) texture feature space, and (c) combined feature space. The number of generations that the GA evolved was fixed at 75. The best result for each feature space is identified with an asterisk.

(a)

P_c	P_m	α	No. of features	A_z (Training)	A_z (Test)
0.7	0.001	0	6	0.84	0.79
0.8			3	0.77	0.76
0.9			4	0.80	0.77
0.7	0.003		7	0.82	0.78
0.8			6	0.82	0.79
0.9			6	0.84	0.79
0.7	0.001	0.0005	3	0.77	0.76
0.8			4	0.80	0.77
0.9			3	0.77	0.76
0.7	0.003		6	0.84	0.79*
0.8			6	0.84	0.79
0.9			6	0.82	0.79
0.7	0.001	0.0010	3	0.77	0.76
0.8			4	0.80	0.77
0.9			3	0.77	0.76
0.7	0.003		6	0.84	0.79
0.8			7	0.84	0.79
0.9			4	0.80	0.77

Table 2 (b)

P_c	P_m	α	No. of features	A_z (Training)	A_z (Test)
0.7	0.001	0	7	0.87	0.82
0.8			8	0.88	0.84
0.9			8	0.88	0.84
0.7	0.003		17	0.91	0.82
0.8			9	0.88	0.79
0.9			10	0.88	0.79
0.7	0.001	0.0005	9	0.88	0.85*
0.8			7	0.86	0.82
0.9			8	0.87	0.84
0.7	0.003		13	0.90	0.81
0.8			10	0.87	0.81
0.9			12	0.88	0.81
0.7	0.001	0.0010	7	0.87	0.83
0.8			9	0.88	0.83
0.9			8	0.88	0.83
0.7	0.003		10	0.88	0.83
0.8			21	0.94	0.82
0.9			12	0.88	0.80

(c)

P_c	P_m	α	No. of features	A_z (Training)	A_z (Test)
0.7	0.001	0	13	0.93	0.88
0.8			12	0.92	0.88
0.9			12	0.92	0.89
0.7	0.003		12	0.91	0.86
0.8			16	0.94	0.88
0.9			17	0.95	0.88
0.7	0.001	0.0003	12	0.92	0.87
0.8			12	0.92	0.86
0.9			12	0.93	0.88
0.7	0.003		13	0.93	0.87
0.8			13	0.93	0.88
0.9			12	0.94	0.89*
0.7	0.005		12	0.89	0.80
0.7	0.001	0.0010	11	0.92	0.87
0.8			10	0.91	0.87
0.9			11	0.91	0.86
0.7	0.003		10	0.91	0.86
0.8			14	0.93	0.87
0.9			13	0.92	0.87
0.7	0.005		11	0.89	0.81
0.8			12	0.88	0.82
0.9			12	0.89	0.81

Table 3. Dependence of feature selection and classifier performance on F_{out} and F_{in} thresholds using stepwise LDA: (a) morphological feature space, (b) texture feature space, and (c) combined feature space. The best result for each feature space is identified with an asterisk. When the test A_z is comparable, the feature set with the smallest number of features was considered to be the best.

(a)

F_{out}	F_{in}	No. of features	A_z (Training)	A_z (Test)
2.7	3.8	2	0.76	0.76
1.7	2.8	4	0.79	0.76
1.7	1.8	6	0.83	0.79*
1.0	1.2	7	0.84	0.79
0.8	1.0	9	0.85	0.79
0.4	0.6	10	0.85	0.79
0.2	0.4	12	0.86	0.78

(b)

F_{out}	F_{in}	No. of features	A_z (Training)	A_z (Test)
2.7	3.8	4	0.82	0.80
1.0	1.4	8	0.88	0.83
1.0	1.2	10	0.89	0.82
0.8	1.0	11	0.89	0.83
0.6	0.8	14	0.91	0.85*
0.4	0.6	17	0.92	0.84
0.2	0.4	18	0.92	0.81
0.1	0.2	16	0.90	0.80

(c)

F_{out}	F_{in}	No. of features	A_z (Training)	A_z (Test)
3.0	3.2	6	0.84	0.80
2.0	3.1	6	0.84	0.80
3.0	3.1	10	0.88	0.83
2.9	3.0	10	0.88	0.83
2.7	2.8	10	0.88	0.83
2.0	2.3	11	0.90	0.86
1.7	1.8	11	0.90	0.86
1.3	1.5	14	0.92	0.86
1.0	1.2	19	0.95	0.86
1.0	1.1	23	0.96	0.87*
0.8	1.2	28	0.97	0.86

Table 4. Classification accuracy of a linear discriminant classifier in the different feature spaces using feature sets selected by the GA and the stepwise LDA procedure.

	Training A_z			Test A_z		
	Morph	Texture	Combined	Morph	Texture	Combined
Cross-validation						
GA (75 generations)	0.84 ± 0.04	0.88 ± 0.03	0.94 ± 0.02	0.79 ± 0.07	0.85 ± 0.07	0.89 ± 0.05
GA (500 generations)	0.84 ± 0.04	0.88 ± 0.03	0.96 ± 0.02	0.79 ± 0.07	0.85 ± 0.07	0.90 ± 0.05
Stepwise LDA	0.83 ± 0.04	0.91 ± 0.03	0.96 ± 0.02	0.79 ± 0.07	0.85 ± 0.06	0.87± 0.06
Leave-one-case-out						
GA (75 generations)	0.83 ± 0.03	0.88 ± 0.03	0.94 ± 0.02	0.79 ± 0.04	0.84 ± 0.03	0.89 ± 0.03
GA (500 generations)	0.83 ± 0.03	0.88 ± 0.03	0.95 ± 0.02	0.79 ± 0.04	0.84 ± 0.03	0.89 ± 0.03
Stepwise LDA	0.83 ± 0.03	0.91 ± 0.02	0.96 ± 0.02	0.79 ± 0.04	0.85 ± 0.03	0.87 ± 0.03

The ROC curves for the test samples using the feature sets selected by the GA are plotted in Figure 3. The classification accuracy in the combined feature space was significantly higher than those in the morphological ($p = 0.002$) or the texture feature space ($p = 0.04$) alone. The ROC curve using the feature set selected by the stepwise procedure in the combined feature space was also plotted for comparison. The distribution of the discriminant scores for the test samples using the feature set selected by the GA in the combined feature space is shown in Figure 4(a). If a decision threshold is chosen at 0.3, 29 of the 82 (35%) benign samples can be correctly classified without missing any malignant clusters.

Some of the 145 samples are different views of the same microcalcification clusters. In clinical practice, the decision regarding a cluster is based on information from all views. If it is desirable to provide the radiologist a single relative malignancy rating for each cluster, two possible strategies may be used to merge the scores from all views: the average score or the minimum score. The latter strategy corresponds to the use of the highest likelihood of malignancy score (minimum discriminant score) for the cluster. There were a total of 81 different clusters (44 benign and 37 malignant) from the 78 cases because three of the cases contained both a benign and a malignant cluster. Using the average scores, ROC analysis provided test A_z values of 0.93 ± 0.03 and 0.89 ± 0.04, respectively, for the GA selected and stepwise LDA selected feature sets. Using the minimum scores, the test A_z values were 0.90 ± 0.03 and 0.85 ± 0.04, respectively. The difference between the A_z values from the two feature selection methods did not achieve statistical significance in either case (p = 0.07 and p=0.09, respectively). If a decision threshold is chosen at an average score of 0.2, 22 of the 44 (50%) benign clusters can be correctly identified with 100% correct classification of the malignant clusters. If a decision threshold is set at a minimum score of 0.2, 14 of the 44 (32%) benign clusters can be identified at 100% sensitivity. The distribution of the average discriminant scores of the 81 clusters in the combined feature space is plotted in Figure 4(b).

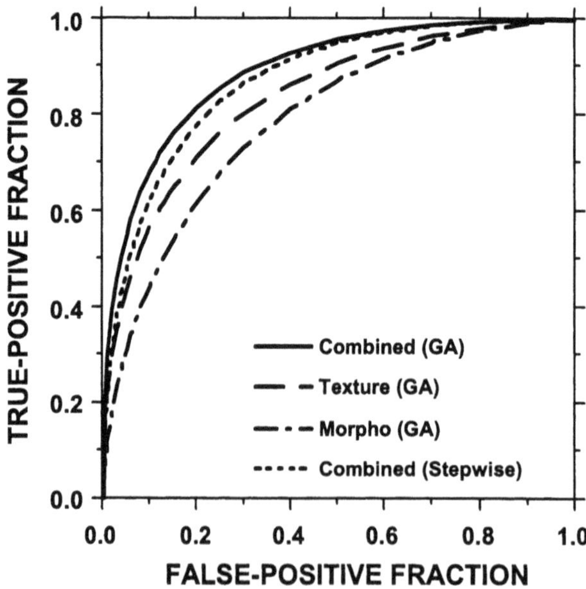

Figure 3. Comparison of ROC curves of the LDA classifier performance using the best GA selected feature sets in the three feature spaces. In addition, the ROC curve obtained from the best feature set selected by the stepwise LDA procedure in the combined feature space is shown. The classification was performed with a leave-one-case-out resampling scheme.

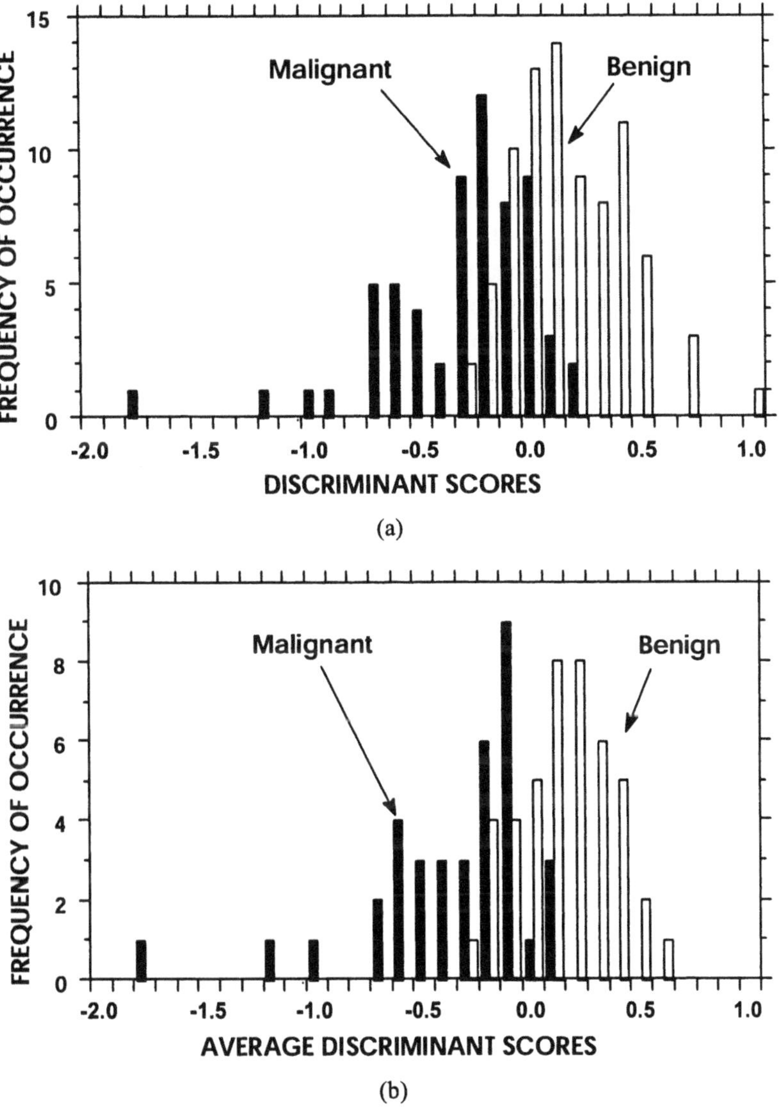

Figure 4. Distribution of the discriminant scores for the test samples using the best GA selected feature set in the combined texture and morphological feature space. (a) Classification by samples from each film, and (b) classification by cluster using the average scores.

4.1.8 Discussion

Classification in multi-dimensional feature spaces is useful because some features that are not useful by themselves can become effective features when they are combined with other features. However, it is impossible to visualize the separation of the classes in high dimensional feature space. An effective feature selection method is therefore essential for identifying the most effective subset of features from a large set of available features. The results of this classification task indicate that both the morphological and texture features have some discriminatory power to distinguish malignant and benign microcalcifications. However, when the morphological feature space is combined with the texture feature space, the resulting feature set selected from the combined feature space can significantly improve the classification accuracy, in comparison with those from the individual feature spaces.

Ideally, the values for the P_m, P_c, and α parameters chosen in the GA only affect the convergence rate; the GA will eventually evolve to the same global maximum regardless of the parameters used. However, when the dimensionality of the feature space is very large and the design samples are sparse, the GA often reaches local maxima corresponding to different feature subsets, as can be seen in Table 2. Similarly, the stepwise feature selection may reach a different local maximum and choose a feature set different from those chosen by the GA. In this application, we systematically varied the parameters over some selected ranges for both the GA and stepwise methods in an effort to search for the feature subset that provided the highest classification accuracy. However, manual search usually can only cover a limited parameter space so that there is no guarantee that the global maximum was found. The proper choice of these parameters when the feature selection methods are applied to an unknown data set is an important area of research.

For the linear discriminant classifier, the stepwise LDA procedure can select near-optimal features for the classification task. We have shown that the GA could select a feature set comparable to or slightly better than that selected by the stepwise LDA. The number of generations that the GA had to evolve to reach the best selection increased with the dimensionality of the feature space as expected. However, even in a

281-dimensional feature space, it only took 169 generations to find a better feature set than that selected by stepwise LDA. Further search up to 500 generations did not find other feature combinations with better performance. Although the difference in A_z between the two methods did not achieve statistical significance, probably due to the large standard deviation in A_z when the number of case samples in the ROC analysis was small, the improvements in A_z in this and our other studies [40], [41] indicate that the GA is a useful feature selection method for classifier design.

4.2 Classification of Mass and Normal Breast Tissue

Masses are important indicators of malignancy on mammograms. The detection of breast masses in mammograms is difficult because masses can be mimicked or obscured by normal breast parenchyma [66]. Our mass detection algorithm uses a detection and classification approach. The advantage of this general approach is that it has the ability to identify masses not having a typical mammographic appearance because the segmentation is not based on any specific mass properties. Our segmentation method utilizes the density-weighted contrast enhancement (DWCE) filter as a preprocessing step. Object-based region growing is then applied to each of the identified structures. Each object is subsequently classified as a breast mass or a normal structure based on extracted morphological and texture features.

4.2.1 Data Set

We conducted a study to evaluate our methods for the classification of masses and normal breast tissues as a step in our mass detection algorithm. The criteria for inclusion of a mammogram in the data set were that the mammogram contained a biopsy-proven mass and that the numbers of malignant and benign cases were approximately equal. All mammograms were acquired as described in Section 4.1.2 except that a pixel size of 0.1 mm × 0.1 mm was used for digitization. Four different ROIs, each with 256 x 256 pixels, were selected from each mammogram. One of the selected ROIs contained the true mass as identified by an experienced radiologist and verified by biopsy. In addition to the ROI that contained the true mass location, the radiologist in the study was asked to select three presumably normal ROIs from the mammogram. The first of these three ROIs contained primarily dense tissue, which could mimic a mass lesion, the second ROI contained mixed dense/fatty tissue, and the third contained mainly fatty tissue. An example of each of these ROIs is shown in Figure 5. Therefore, a total of 168 ROIs containing masses and 504 ROIs containing normal tissue were used in the study. The low frequency gray level variation in the ROIs due to the density of the overlapping breast tissue and the x-ray exposure conditions was reduced using the background correction technique described in Section 4.1.4.

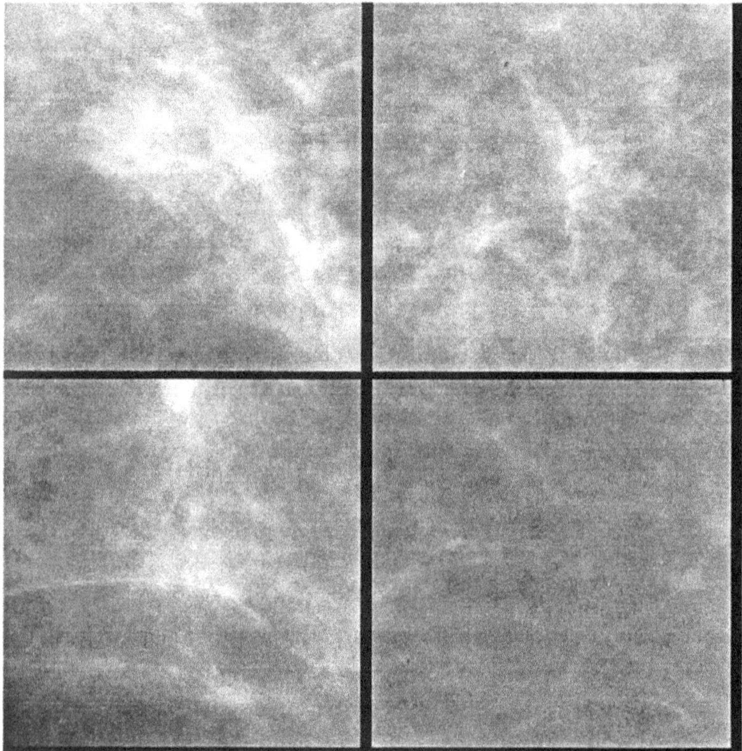

Figure 5. An example of the mass and normal ROIs selected from one of the mammograms used in this study. The four ROIs are upper left—mass; upper right—mixed dense/fatty tissue; lower left—dense tissue; lower right—fatty tissue.

4.2.2 Morphological Features

We have developed an automated algorithm, based on k-means clustering, for segmentation of an ROI into an object region and background tissue [40]. Eleven morphological features and four margin features were automatically extracted from the object region after the segmentation was performed. The morphological features included the number of edge pixels, area, circularity, rectangularity, contrast, the ratio of the number of edge pixels to the area, and 5 normalized radial length features. A detailed discussion of the shape features used in this study can be found in the literature [67]. The margin features were computed as follows. First, the mean and the standard deviation of the pixel values inside the object were computed. Next, pixels in a boundary region outside the object but within a distance of 15 pixels from the object border were thresholded. The values of the thresholds were chosen to be the mean minus 0.5, 1, 1.5, and 2 times the standard deviation. The number of pixels in the boundary region, which were above the thresholds, was defined as the margin features. Thus a total of 15 morphological features were extracted from each ROI.

4.2.3 Texture Features

Similar to the discussion in Section 4.1.4, texture features were extracted from SGLD matrices constructed from the pixel gray levels in the ROI. We computed global texture features, which represented the average texture measures throughout the entire ROI, and local texture features, which represent (i) the texture measure of a denser sub-region inside the ROI that was likely to contain the mass, and (ii) the texture difference between this sub-region and other peripheral regions in the ROI which contain normal breast tissue. The method used for the computation of SGLD matrices and multiresolution texture analysis are explained in detail elsewhere [68]. A total of 364 global texture features and 208 local texture features were extracted.

4.2.4 Classification

In this study, we investigated GA-based feature selection for two kinds of classifiers, namely (i) LDA; and (ii) a multilayer backpropagation neural network (BPN) [69]. The BPN used in this study consisted of an input layer, an output layer, and a single hidden layer. Each layer in the BPN contained a number of nodes, which were connected to previous and subsequent layers by trainable weights. A single feature was applied to each node in the input layer. The net input to each node in the hidden layer and the output layer was a weighted sum of the node outputs from the previous layer. The output of a node was related to its net input by a sigmoidal function. The output layer contained a single node, whose output indicated the degree of suspicion that the ROI contained a mass. The BPN was trained using batch processing and the delta-bar-delta rule for improved rate of convergence and stability [63].

At each run of the GA, the image data set of 672 ROIs was divided into a training and a test set, with ROIs belonging to the same film grouped into the same set. The training set was used in the GA for feature selection. After feature selection, a classifier was trained using the GA-selected features from the training set. The classification accuracy of the procedure was evaluated by applying the classifier to the test group. For studying the effect of GA parameters on the classification accuracy with the LDA, ten random partitionings of training and test sets were obtained for each set of different GA parameters, and the results were averaged in order to reduce the effect of case selection. For experiments with the BPN, fifty random partitionings were used. For both experiments, the number of mass and non-mass ROIs in each training set was 126 and 378 (3/4 of the total), respectively, while the number of mass and non-mass ROIs in each test set was 42 and 126 (1/4 of the total), respectively.

4.2.5 GA Implementation

Inside the GA, the training set was equally divided into two groups, *S1* and *S2*. For each chromosome, two classifiers were trained, with *S1* and *S2* as the training groups, respectively. Only the features present in the chromosome were used as input features for classifier training. The classifier trained with group *S1* was applied to the group *S2*, and vice versa, for calculation of two sets of *pseudo-test* classifier outputs. The accuracy of the pseudo-test classifier outputs, and the number of selected features were then used to define the fitness of the individual chromosome. This process was repeated for each of the n chromosomes in each generation. The fitness function, parent selection, crossover, and mutation operations were the same as defined in Section 4.1.5.

4.2.6 Results

4.2.6.1 Effect of Penalty Term and Number of Generations

To determine a reasonable number of generations for the GA to evolve, we selected several combinations of crossover probability (P_c) and mutation probability (P_m), and monitored the growth of the number of selected features. The initial probability of a feature's presence was fixed at P_{init}=0.002. The GA was allowed to evolve with two different α values of the penalty term in the fitness function. We observed that the crossover probability P_c did not have a major effect on the number of selected features. However, both α in the penalty term and the mutation probability P_m affected the number of selected features. Figures 6 and 7 plot the average number of selected features over 10 training sets versus the generation number for α=0 and α=1/2000, respectively. The average number of selected features is plotted for P_m=0.001 and P_m=0.003 in each figure. The crossover probability is kept constant at P_c=0.7. The test A_z value obtained up to a given generation is plotted against the generation number in Figures 8 and 9 for the same conditions (α=0 and α=1/2000), respectively. The average A_z value over 10 test sets is shown.

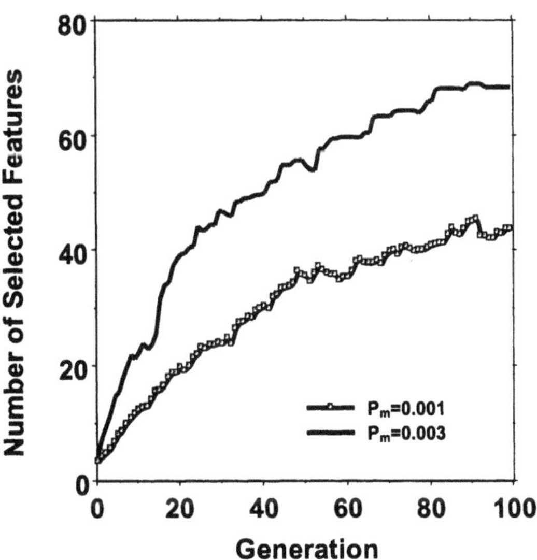

Figure 6. Evolution of the number of selected features for α=0.

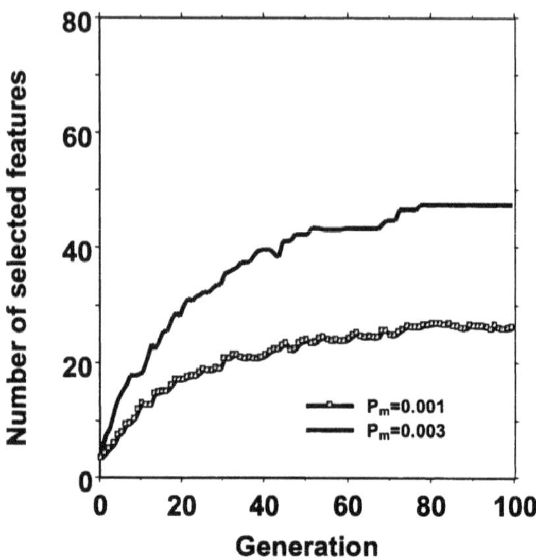

Figure 7. Evolution of the number of selected features for α=1/2000.

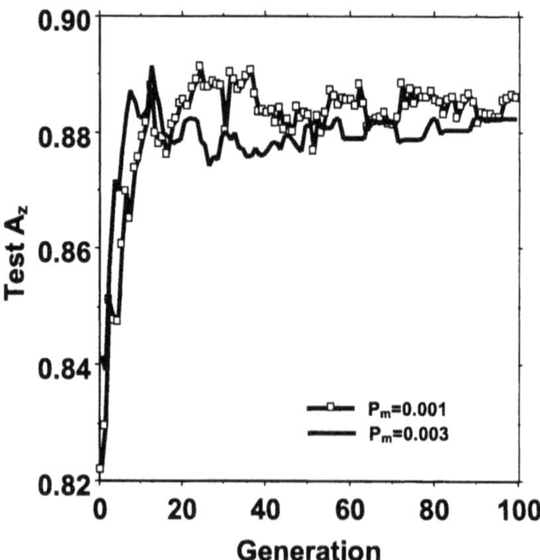

Figure 8. Evolution of the average test A_z for α=0.

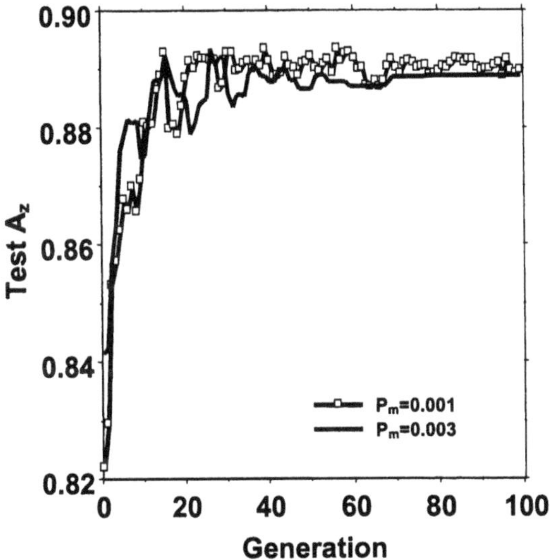

Figure 9. Evolution of the average test A_z for α=1/2000.

It is observed that while the average test A_z value does not increase after the 25th generation, the number of selected features continues to increase beyond the 60th generation for all combinations of GA parameters studied. Since the main component of the fitness function in the GA is the A_z value rather than the number of features, more features may be added into the selected feature pool as long as the area under the ROC curve does not deteriorate. Comparing Figures 6 and 7, it is observed that the penalty term suppressed the number of selected features. The number of selected features eventually leveled off at about the 80th generation when the penalty term was non-zero (Figure 7).

The average test A_z values at the end of 100 generations were 0.89 for the combinations studied in Figure 9, and 0.88 for the combinations studied in Figure 8. The standard deviation of the individual A_z values, as determined by the LABROC program, varied between 0.02 and 0.04.

4.2.6.2 Comparison with Stepwise Feature Selection for LDA

After additional experimentation with P_m, P_c, and P_{init}, we found that the best combination of these parameters for the current classification problem was P_m=0.001, P_c=0.9, and P_{init}=0.002. Using these parameters, the average test A_z value for the 10 partitions was 0.90. To compare the best performance of GA to the best performance of stepwise feature selection, we varied the two threshold values in stepwise feature selection (F_{in} and F_{out}) so that the average test A_z value over the ten partitionings was maximized. The number of selected features and the test results for both methods are compared in Table 5.

Table 5. Test A_z values of an LDA classifier using stepwise feature selection and GA-based feature selection.

	Stepwise		GA	
Test group	A_z	No. of features	A_z	No. of features
1	0.87	19	0.90	20
2	0.91	15	0.89	24
3	0.92	25	0.93	24
4	0.88	22	0.88	20
5	0.86	23	0.84	23
6	0.92	29	0.93	20
7	0.92	15	0.91	17
8	0.84	21	0.88	19
9	0.86	14	0.88	18
10	0.88	20	0.92	16
Average	0.89	19.3	0.90	20.1

4.2.6.3 Comparison with Stepwise Feature Selection for BPN

Since training a BPN is considerably slower than training a linear discriminant classifier, we modified our training strategy for this classifier. There were two basic differences between the experiments with BPN and linear discriminant classifier. (1) In order to handle a smaller feature pool with BPN, we used a single distance, d=20, for texture features. The global texture features computed at this pixel distance, plus the morphological features previously described in Section 4.2.2, constituted the feature pool in this evaluation. Therefore, there were a total of 41 features (26 texture and 15 morphological) from which the feature selection algorithms choose the best subset. (2) In order not to repeat the feature selection process several times with several different training sets, the entire data set was used in the feature selection step of the classification procedure. After feature selection was completed, the classifier was trained and tested with 50 different partitionings of the data set into training and test groups.

The parameters of the BPN and the GA used in this subsection were as follows. The number of input nodes of the BPN was equal to the number of features selected by the GA. Based on our previous experience, the number of hidden layer nodes was chosen as four. The BPN had a single output node. It was trained for 400 iterations for each chromosome in each generation. The GA was allowed to evolve for a total number of 75 generations. The parameters of the GA were selected as P_m=0.02, P_{init}=0.02, P_c=0.9, and α=1/2000.

The final GA-selected pool of variables contained 16 features. The average training and test A_z values over 50 partitionings were 0.92 and 0.90, respectively.

To compare our GA-based feature selection method for a BPN, we also applied stepwise feature selection to the same data set and the 41 features described above. In order to have a fair comparison with the GA-based method, the entire data set was used for feature selection. The final selected pool of variables contained 19 features. The same 50 partitionings used for the GA experiments were used to train and test a BPN with the stepwise-selected features. The average training and test A_z values over 50 partitionings were 0.92 and 0.89, respectively. The distribution of the pairwise difference of the test A_z of the BPN classifiers in combination with stepwise and GA-based feature selection methods is shown in Figure 10. The figure shows that the GA feature selection outperformed the stepwise feature selection in 42 of the 50 partitions while the opposite was true in only 5 partitions.

4.2.7 Discussion

Our results indicate that the classification accuracy with GA-based feature selection is better than that with stepwise feature selection. This is most easily seen from Figure 10, which compares the distribution of the difference in the A_z values for a BPN classifier with GA-based feature selection relative to that with stepwise feature selection. It can be observed that the GA-based feature selection provided higher A_z values. However, we could not perform a paired t-test to evaluate the statistical significance of the differences for the results listed in Table 5 or those shown in Figure 10. The paired *t*-test requires independence among the samples whereas our test (or training) sets in the different partitionings overlapped with each other. We have used the CLABROC program [65] to test the statistical significance of the difference between the corresponding pair of ROC curves for each partitioning. The difference did not achieve statistical significance for the individual pairs because the number of cases in each partitioned data set is small and thus the standard deviation of A_z is large (0.02 to 0.04). However, it should be noted that the improvement in A_z with GA-based feature selection, although small, is consistently observed over the majority of the different partitionings of the data set.

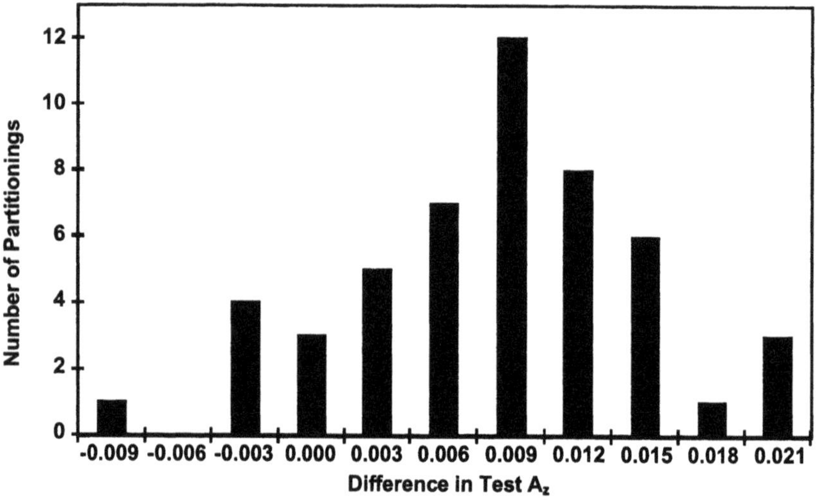

Figure 10. The distribution of the pairwise difference of the test A_z values of the BPN classifiers with GA-based and stepwise feature selection.

As in Section 4.1, the GA provided a small improvement over stepwise feature selection in LDA. As described in the previous section, one possible explanation why GA did not provide a greater improvement is that, for the LDA, the stepwise feature selection procedure is already near optimal. It is actually somewhat unexpected that the GA-based feature selection can still provide an observable improvement in A_z.

Stepwise feature selection is computationally faster than GA-based feature selection. In the present application, stepwise feature selection required 64-sec CPU time for each partition in Table 5 on a 90-MHz

Pentium-based personal computer. The GA-based feature selection required 519 sec for each partition in Table 5 on a 133-MHz Alpha-based workstation. When the GA was used for feature selection for a BPN, the vast majority of the CPU time was spent on training the BPN with a given chromosome. Once BPN training was completed for all chromosomes in a generation, the computation time required for the GA to produce a new generation of chromosomes was minimal.

4.3 Classification of Malignant and Benign Masses

Although mammography is the most sensitive method for detection of breast cancer, the positive predictive value (PPV) (ratio of the number of malignancies found to the total number of biopsy performed) of visual mammographic criteria for breast cancer characterization is relatively low. The PPV of biopsies performed for mammographically suspicious nonpalpable breast masses ranges from 20 to 30% [70], [71], and [72]. To reduce health care costs and patient morbidity, it is desirable to increase the PPV of mammographic diagnosis while maintaining its sensitivity of cancer detection. Computerized mammographic analysis methods can potentially aid the radiologists in achieving this goal.

In the task of lesion characterization, the cost of missing a malignancy is very high because it will cause a delay in treatment. Therefore, the performance of a classifier in the high-sensitivity (high true-positive fraction) region of the ROC curve is more important than the overall area A_z under the ROC curve. In other words, if a classifier is to be designed for breast lesion characterization, the specificity at high levels of sensitivity is much more important than the specificity at low levels of sensitivity. Recently, Jiang et. al. derived an expression for an ROC partial area index above a desired sensitivity level. This index will be useful as a performance measure for lesion characterization problems [73]. Since a feature (or feature combination) that can provide a large overall A_z (or a large Wilks' lambda and Mahalanobis distance) may not provide a large partial ROC area, it is necessary to develop a feature selection method tailored to maximize the partial area in the high sensitivity region. The flexibility of a GA in the design of its fitness function allows the partial area index to be incorporated for feature selection.

In this study, we developed a methodology to design high-sensitivity classifiers. The design process was illustrated by the task of classifying masses on digitized mammograms as malignant or benign. A GA-based algorithm with the ROC partial area index as the feature selection criterion, in combination with LDA, was used for the design of this classifier. Texture features extracted from RBST images [74] were used for classification. The performance of the high-sensitivity classifier was compared to the performance achieved by LDA with stepwise feature selection (LDA_{sfs}) using the Wilks' lambda as the feature selection criterion.

4.3.1 Data Set

The data set used in this study consisted of 255 mammograms from 104 patients. The criteria for inclusion of a mammogram in the data set was that the mammogram contained a biopsy-proven mass and that the numbers of malignant and benign masses were approximately equal. Image acquisition and digitization were the same as Sections 4.1 and 4.2. There were 128 mammograms with benign masses, and 127 mammograms with malignant masses. Examples of two malignant and two benign masses are shown in Figure 11. Of the 104 patients evaluated in this study, 48 had malignant masses. The probability of malignancy of the mass that underwent biopsy was ranked for each mammogram by a Mammography Quality Standards Act (MQSA) approved radiologist on a scale of 1 to 10. A ranking of 1 corresponded to the masses with the most benign mammographic appearance, and a ranking of 10 corresponded to the masses with the most malignant mammographic appearance. The distribution of the malignancy ranking of the masses is shown in Figure 12. The pathology of the masses was determined by biopsy and histologic analysis.

4.3.2 Image Transformation

In this study, the classification of malignant and benign masses was based on the textural differences of their mammographic images. We have designed a rubber band straightening transform (RBST) which was found to facilitate the extraction of texture features from the region surrounding a mammographic mass [74]. The RBST technique transforms a band of pixels surrounding a mass onto the Cartesian plane. The four basic steps in the RBST are mass segmentation, edge

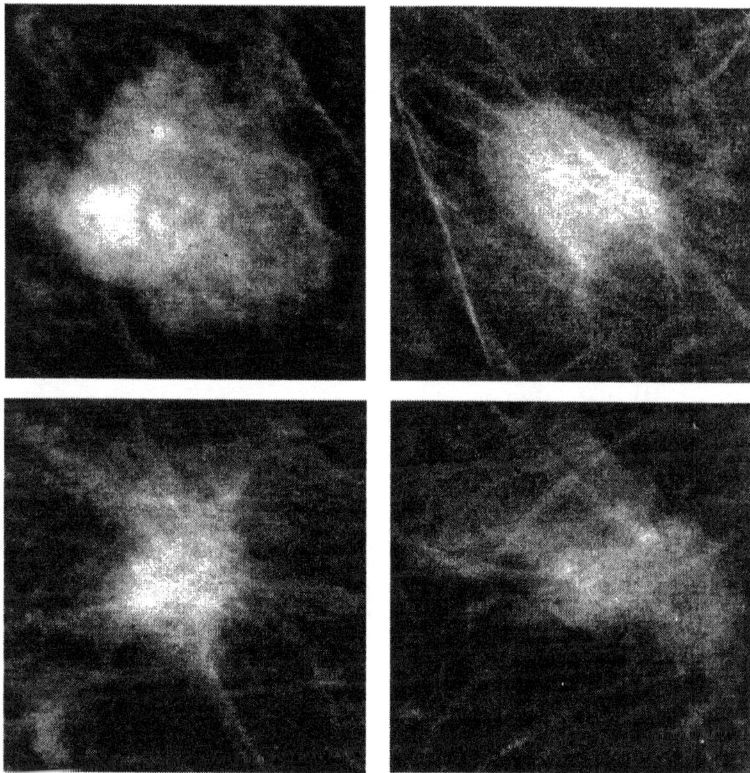

Figure11. Examples of benign and malignant masses from our data set: upper left—lobulated cyst; upper right—adenofibroma; lower left—invasive ductal and intraductal carcinoma; lower right—invasive and intraductal carcinoma.

enumeration, computation of normals, and interpolation. Details of these steps can be found in the literature [74]. A modified k-means clustering algorithm [40] was used for segmentation. After the outline of the mass was obtained, an edge enumeration algorithm assigned a pixel number to each border pixel of the mass, such that neighboring pixels were assigned consecutive numbers. The normal direction to the mass at each boundary pixel was computed based on the border connectivity and the pixel numbers obtained from the edge enumeration algorithm. In the last step, the pixels at a constant distance j from the mass outline were stored as the row j of the RBST image. A point $p(i,j)$ was defined using the original image coordinates, which is at a distance j from boundary pixel i along the normal $L(i)$. Then the (i,j)th element of the RBST image was defined as the distance-weighted average of the two closest pixels to $p(i,j)$ in the original image. The width of the band transformed by the RBST was chosen as 40 pixels in this study, which corresponded to 4 mm on the mammogram.

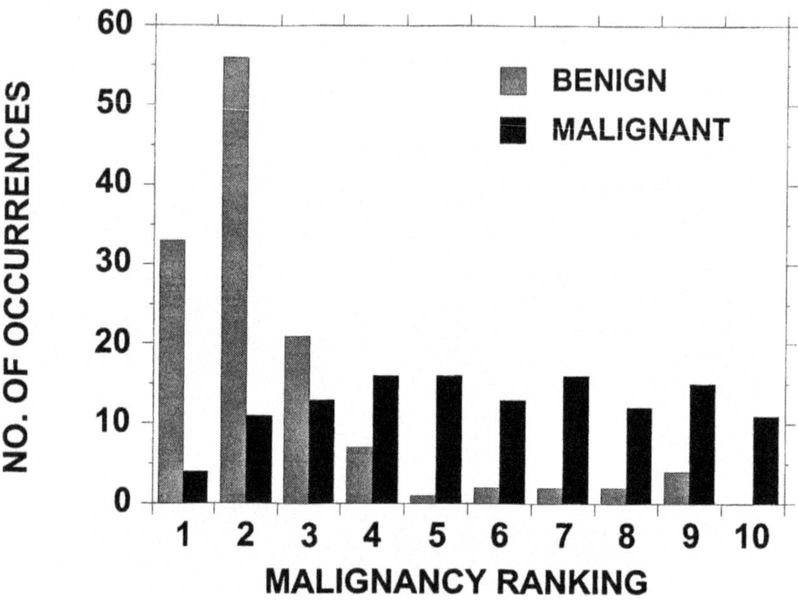

Figure 12. The distribution of the malignancy ranking of the masses in our data set, as visually estimated by a radiologist experienced in mammographic interpretation. 1=definitely benign, 10=definitely malignant.

4.3.3 Texture Features

The texture features used for the classification of the malignant and benign masses were SGLD features, described in Section 4.1, and run length statistics (RLS) features, described next.

A gray level run is defined as a set of consecutive, collinear pixels in a given direction which have the same gray level value. A run length is the number of pixels in a gray level run. The RLS matrix for a given image describes the run length statistics in a given direction for each gray level value in the image. The (i,j)th element of the RLS matrix $r_\theta(i,j)$ represents the number of times that runs of length j in the direction θ consisting of pixels with a gray level i exist in the image [75]. The RLS matrices in this study were extracted from the vertical and horizontal gradient magnitudes of the RBST images. The RLS matrices were obtained from each gradient magnitude image in two directions, $\theta=0^{\circ}$ and $\theta=90^{\circ}$. Therefore, a total of 4 RLS matrices were obtained for each RBST image.

Based on our previous study, a bit depth of 5 was used for the computation of RLS matrices [74]. Five RLS features, namely, short runs emphasis, long runs emphasis, gray level nonuniformity, run length nonuniformity, and run percentage were extracted from each RLS matrix. The definitions of these features can be found in the literature [76]. This resulted in the computation of 20 RLS features per RBST image.

For the construction of SGLD matrices from RBST images, four different directions ($\theta=0^{\circ}$, 45°, 90°, and 135°) and ten different pixel pair distances (d=1, 2, 3, 4, 6, 8, 10, 12, 16, and 20) were used for. The total number of SGLD matrices was therefore 40. Based on our previous studies [18], a bit depth of eight bits was used in the SGLD matrix construction. Eight SGLS features, namely, correlation, difference entropy, energy, entropy, inertia, inverse difference moment, sum average, and sum entropy were extracted from each SGLD matrix. This resulted in the computation of 320 SGLD features per RBST image.

4.3.4 Classification

LDA was used for classification. Unlike Section 4.2, the data set was not partitioned into training and test sets before feature selection. Feature selection was implemented using both GA and stepwise feature selection using the entire data set of 255 ROIs. After the features were selected, the accuracy of the LDA was tested using a leave-one-case-out paradigm, as described in Section 4.1.6

4.3.5 GA Implementation

The fitness function used in this section was different from that used in Sections 4.1 and 4.2. The main component of the fitness function in Sections 4.1 and 4.2 was the area A_z under the ROC curve. However, for applications where the performance at high sensitivity (or high true-positive fraction) is important, such as breast lesion characterization in CAD, the partial area index of the ROC curve [73] is better suited for the design of the fitness function. Therefore, the ROC partial area index, denoted as A_{TPF_0}, was used as the main component of the fitness function in this study.

The partial area index A_{TPF_0} summarizes the average specificity above a sensitivity of TPF_0 (Figure 13), and can be expressed as

$$A_{TPF_0} = 1 - \frac{1}{1 - TPF_0} \int_{TPF_0}^{1} FPF(TPF) d(TPF) \qquad (13)$$

which is the ratio of the partial area under the actual ROC curve to the partial area of the perfect ROC curve. The maximum value for A_{TPF_0} is thus 1. The A_{TPF_0} value for a classifier that operates purely on random guessing is $(1-TPF_0)/2$, which is the area under the chance diagonal normalized to $(1-TPF_0)$.

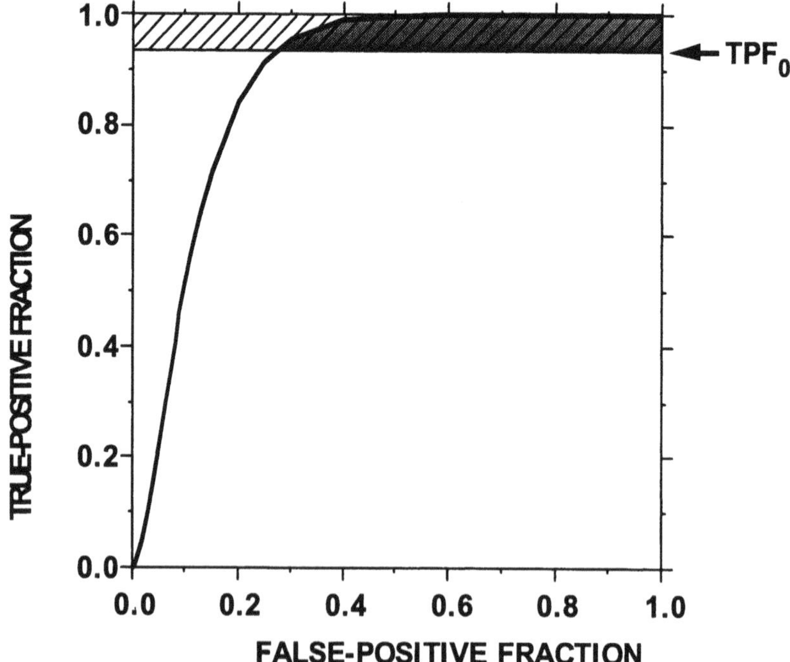

Figure 13. The partial area index A_{TPF_0} is defined as the ratio of the partial area under the ROC curve above a given sensitivity (gray area) to the partial area of the perfect ROC curve (hatched region) above the same sensitivity.

Our goal in this study was to train a GA to select features which would yield high specificity in the high-sensitivity region of the ROC curve. Therefore, the fitness of a chromosome was defined as a monotonic function of A_{TPF_0}, such that the maximization of the fitness function would maximize A_{TPF_0}:

$$\text{Fitness} = \left(\frac{A_{TPF_0} - A_{\min}}{A_{\max} - A_{\min}} \right)^4 \tag{14}$$

where A_{max} and A_{min} were the maximum and minimum values of A_{TPF_0} among all chromosomes in a generation. The classifiers thus designed will be referred to as GA-based high-sensitivity classifiers in the following discussions. It may be noted that we used an exponent of 4 in the definition of the fitness function, different from an exponent of 2 used in the studies described in Sections 4.1 and 4.2. Our experiments with the value of the exponent showed that there is a very small difference between using an exponent of 4 or 2 in this application. However, there is a large difference between using 2 or 1 [41].

The significance of the difference in A_{TPF_0} of different classifiers was determined using a recently developed statistical test [73]. The test is analogous to statistical tests involving the area A_z under the entire ROC curve.

The parameters of the GA were selected as P_c=0.9, P_m=0.0025, P_{init}=0.002 chromosome length=340, number of chromosomes=200, and the GA was stopped at 200 iterations.

4.3.6 Results

To demonstrate the training of high-sensitivity classifiers using GA, we chose two levels of sensitivity thresholds, TPF_0=0.50 and TPF_0=0.95 in Eq. (12). The classification results of these classifiers were compared with those of LDA_{sfs}. GA-based feature selection was also performed without emphasis on high sensitivity (TPF_0=0). The classifier designed with the features thus selected will be referred to as an ordinary GA-based classifier. Its performance was compared to those of the GA-based high-sensitivity classifiers and LDA_{sfs}. Figure 14 shows the evolution of the number of selected features, and Figure 15 shows the total area under the ROC curve (A_z) and the partial area above the true positive fraction of 0.95 (TPF_0=0.95), $A_{0.95}$, for a typical GA training.

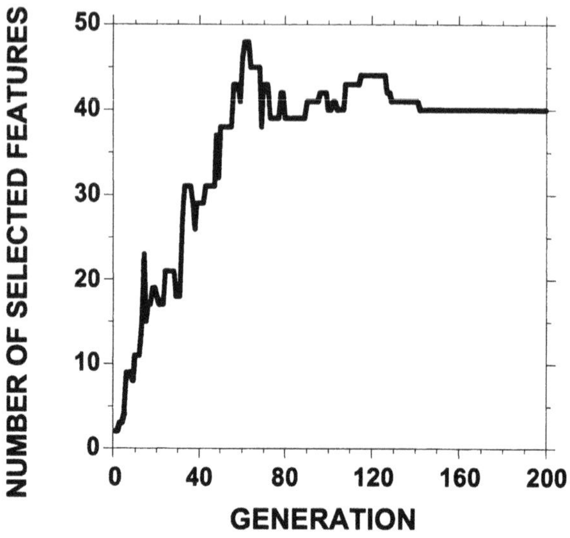

Figure 14. The evolution of the number of selected features for a GA training session (TPF_0=0.95).

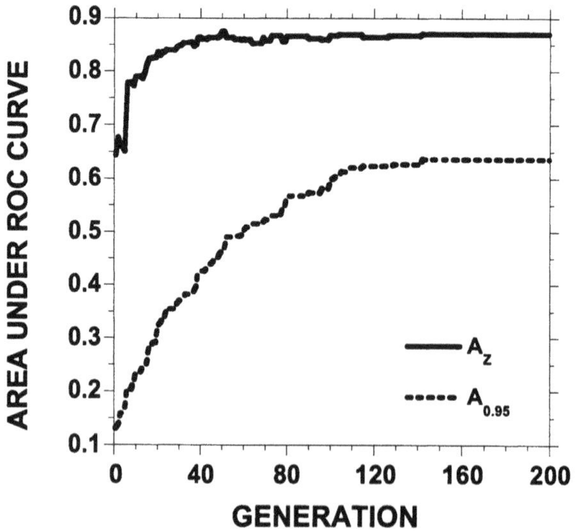

Figure 15. The evolution of the area A_z and the partial area $A_{0.95}$ under the ROC curve for the GA training session of Figure 14 (TPF_0=0.95).

The performance of GA feature selection was compared with that of stepwise feature selection. In LDA_{sfs}, the optimal values of the F_{in} and F_{out} thresholds are not known *a priori*. We therefore varied these thresholds to obtain the feature subset with the best test performance. Table 6 shows the number of selected features, the area A_z under the ROC curve, the partial area above the true positive fraction of 0.5 ($A_{0.50}$), and that above 0.95 ($A_{0.95}$) as these F thresholds are varied. By comparing the A_z values and the performance at the high-sensitivity portion of the ROC curve, the combination F_{in}=1.4, F_{out} =1.2 was found to provide the best feature subset.

Table 6. The number of features, the area A_z under the ROC curve, the partial area above the true positive fraction of 0.5 ($A_{0.50}$), and that above 0.95 ($A_{0.95}$) for various values of F_{in} and F_{out} in the stepwise feature selection method.

F_{in}	F_{out}	No. of selected features	A_z	$A_{0.50}$	$A_{0.95}$
3.8	2.7	9	0.84	0.71	0.22
2.6	2.4	13	0.85	0.72	0.27
2.2	2.0	14	0.86	0.73	0.25
1.8	1.6	26	0.89	0.80	0.38
1.4	1.2	41	0.92	0.83	0.47
1.0	1.0	49	0.92	0.83	0.46

The ROC curve of the best LDA_{sfs} classifier and those of GA-based classifiers (TPF_0=0.50 and TPF_0=0.95) are compared in Figure 16.

To quantify the improvement obtained by the GA-based high-sensitivity classifier, we performed statistical significance tests on the partial area above a true-positive threshold of 0.95 as described in [73]. The difference between the partial areas of the GA-based high-sensitivity classifiers and LDA_{sfs} above a true-positive threshold of 0.95 was statistically significant with two-tailed p-levels of 0.006 and 0.02 for the classifiers trained with TPF_0=0.95 and TPF_0=0.5, respectively.

The performance of the high-sensitivity classifiers and the ordinary GA-based classifiers (TPF_0=0) are also compared in Figure 16. It is observed that the difference between the high-sensitivity and the ordinary GA-based classifiers is less than the difference between the high-sensitivity classifiers and the LDA_{sfs}. Table 7 summarizes the A_z, $A_{0.50}$ and $A_{0.95}$ values, as well as the number of features selected by each classifier.

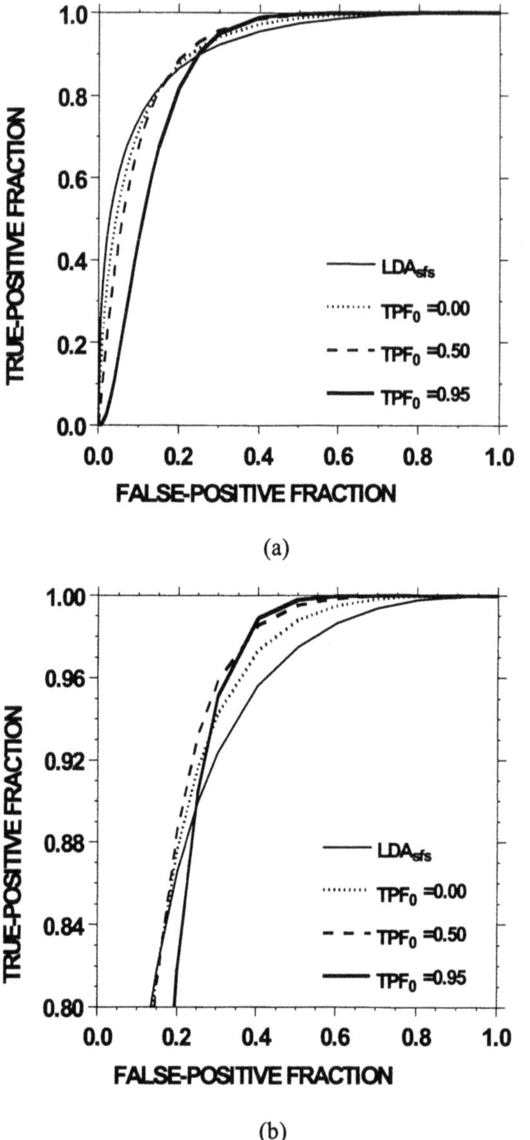

Figure 16. The ROC curves of the LDA_{sfs}, the ordinary GA-based classifier (TPF_0=0), and the GA-based high-sensitivity classifiers trained with TPF_0=0.50 and TPF_0=0.95 (a) the entire ROC curves, (b) enlargement of the curves for TPF>0.8.

Table 7. The number of features, the area A_z under the ROC curve, the partial area above the true positive fraction of 0.5 ($A_{0.50}$), and that above 0.95 ($A_{0.95}$) for the GA-based and stepwise feature selection methods.

TPF_0 value for GA training	No. of selected features	A_z	$A_{0.50}$	$A_{0.95}$
0	40	0.92	0.85	0.56
0.5	39	0.91	0.85	0.62
0.95	40	0.87	0.81	0.64
Stepwise	41	0.92	0.83	0.47

Figures 17 and 18 show the distributions of the classifier outputs for the high-sensitivity classifier (TPF_0=0.95) and the LDA_{sfs}, respectively.

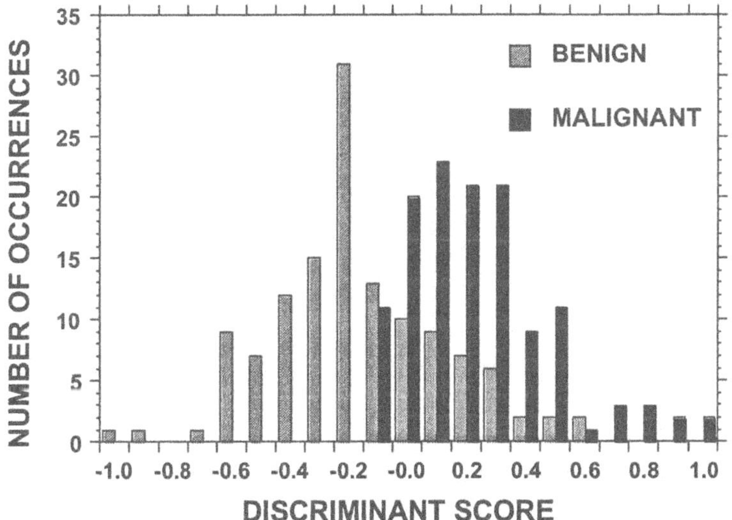

Figure 17. The distribution of the classifier output for the high-sensitivity classifier with TPF_0=0.95. By setting an appropriate threshold on these classifier scores, 61% of masses could correctly be classified as benign without missing any malignancies in this study.

Using the LDA_{sfs}, the distribution of the malignant masses has a relatively long tail that overlaps with the distribution of the benign masses. With the high-sensitivity classifier, this tail seems to be shortened, so that more benign masses may be correctly excluded without missing malignancies. At 100% sensitivity, the specificity with the appropriate choice of the decision threshold was 61% and 34% for the high-sensitivity classifier and the LDA_{sfs}, respectively.

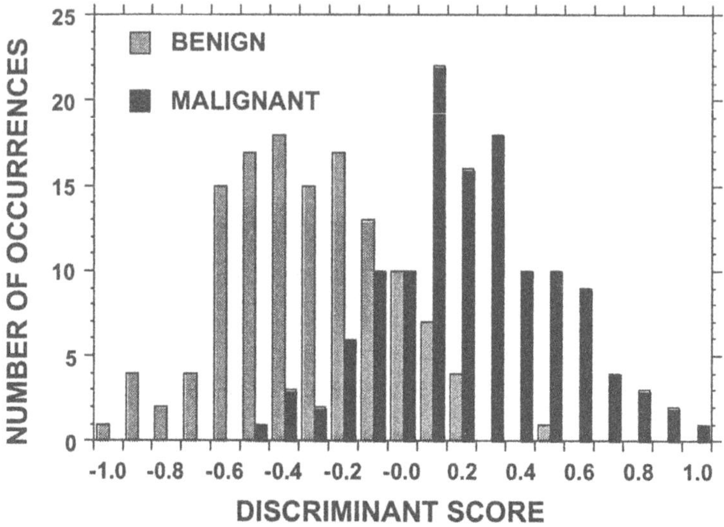

Figure 18. The distribution of the classifier output for LDA_{sfs}. By setting an appropriate threshold on these classifier scores, 34% of masses could correctly be classified as benign without missing any malignancies in this study.

4.3.7 Discussion

Figure 16 demonstrates that when feature selection is performed with a properly designed fitness function in the GA, the designed classifier can be more effective than LDA_{sfs} in the high-sensitivity region of the ROC curve. From Table 7, it is observed that although the A_z value for the properly trained high-sensitivity classifier (e.g., TPF_0=0.5 or 0.95) may be smaller than that of the LDA_{sfs}, the partial area index $A_{0.95}$ is larger. The statistical analysis in this study showed that a significant improvement in the partial area of the high-sensitivity region of the ROC curve can be achieved with properly designed high-sensitivity classifiers.

From Figures 14 and 15, it is observed that the best fitness and the number of features did not change between iterations 140 and 200 for the high-sensitivity classifier with TPF_0=0.95. A similar trend was observed with other values of TPF_0 investigated in this study.

Therefore, 200 generations seems to be sufficient for the GA to complete its evolution in this application. In Figure 15, the best A_z value was attained around the 50th generation, and the A_z value remained largely constant afterwards. However, the $A_{0.95}$ value increased until around 140 generations. This meant that the classification accuracy at high sensitivity continued to increase although the A_z value did not change. The GA thus guided the selection of features that could improve the specificity in the high-sensitivity region of the ROC curve, even at the cost of sacrificing the specificity in the low-sensitivity region of the ROC curve.

Figures 17 and 18 demonstrate the importance of designing a high-sensitivity classifier tailored to the requirements of an algorithm that classifies lesions as malignant or benign. With the high-sensitivity classifier, 61% of masses could correctly be classified as benign without missing any malignancies for this data set. Without this tailoring, an LDA with stepwise feature selection was able to classify only 34% of masses as benign without missing any malignancies.

5 Conclusion

In this chapter, we have addressed the problem of feature selection for classifier design when a large number of features are available for a classification task. We illustrated the use of feature selection techniques to reduce the dimensionality of the feature space by discussing some practical applications in computer-aided diagnosis (CAD) of breast cancer on mammograms. In particular, we investigated and compared genetic algorithm (GA) based feature selection with a more established sequential feature selection approach for the design of either linear classifiers or neural network classifiers. We found that the GA-based methods can be more effective in identifying good features compared to the sequential approach. The GA was found to outperform the sequential technique in our applications. GA-based methods do have limitations because the parameter values selected for the GA can have a significant impact on the features selected and subsequently on the overall performance of the classification task. The population size of chromosomes, the initial number of genes, the crossover and mutation probabilities, and the maximum number of iterations should be selected to match the size and complexity of the design sample set. However, there are currently no rules of thumb to guide the selection of these parameters. Care must be taken to avoid parameters that allow a large number of features to be selected; which will result in over-fitting of the training set with poor generalization to the population at large. Numerous variations of feature selection and the GA method can be found in the literature. We did not attempt to compare the various methods or performances of these different approaches. Our goal was to underscore the importance of feature selection and demonstrate the feasibility of using GA methods to select effective features for a classification task. We also demonstrated the advantage of a GA-based method that allows tailoring its fitness function to select features of desired properties. Further investigation will be needed to determine the usefulness of the GA for general classification problems in high dimensional feature spaces and to determine if guidelines can be found to efficiently select proper GA parameters for a certain evolution process.

References

[1] Parkin, D.M., Pisani, P., and Ferlay, J. (1999), "Global cancer statistics," *CA Cancer J. Clin.,* vol. 49, pp. 33-64.

[2] Greenlee, R.T., Murray, T., Bolden, S., and Wingo, P.A. (2000), "Cancer statistics, 2000," *CA Cancer J. Clin.,* vol. 50, pp. 7-33.

[3] Feig, S.A. and Hendrick, R.E. (1993), *Risk, Benefit, and Controversies in Mammographic Screening*, in *Syllabus: A categorical Course in Physics Technical Aspects of Breast Imaging,* A.G. Haus and M.J. Yaffe, Editors. Radiological Society of North America, Inc, Oak Brook, IL. pp. 119-135.

[4] Duffy, S.W. and Tabar, L. (2000), "Screening mammograpgy re-evaluated," *Lancet,* vol. 355, pp. 747-748.

[5] Smart, C.R., Hendrick, R.E., Rutledge, J.H., and Smith, R.A. (1995), "Benefit of mammography screening in women ages 40 to 49 years: current evidence from randomized controlled trials," *Cancer,* vol. 75, pp. 1619-1626.

[6] Martin, J.E., Moskowitz, M., and Milbrath, J.R. (1979), "Breast cancer missed by mammography," *AJR*, vol. 132, pp. 737-739.

[7] Wallis, M.G., Walsh, M.T., and Lee, J.R. (1991), "A review of false negative mammography in a symptomatic population," *Clin. Radiol.*, vol. 44, pp. 13-15.

[8] Bird, R.E., Wallace, T.W., and Yankaskas, B.C. (1992), "Analysis of cancers missed at screening mammography," *Radiol.*, vol. 184, pp. 613-617.

[9] Harvey, J.A., Fajardo, L.L., and Innis, C.A. (1993), "Previous mammograms in patients with impalpable breast carcinomas: retrospective vs blinded interpretation," *AJR*, vol. 161, pp. 1167-1172.

[10] Adler, D.D. and Helvie, M.A. (1992), "Mammographic biopsy recommendations," *Curr. Op. Radiol.*, vol. 4, pp. 123-129.

[11] Kopans, D.B. (1991), "The positive predictive value of mammography," *AJR*, vol. 158, pp. 521-526.

[12] Shtern, F., Stelling, C., Goldberg, B., and Hawkins, R. (1995), "Novel technologies in breast imaging: national Cancer Institute perspective," *Society of Breast Imaging Conference*, pp. 153-156.

[13] Chan, H.P., Doi, K., Vyborny, C.J., Schmidt, R.A., Metz, C.E., Lam, K.L., Ogura, T., Wu, Y., and MacMahon, H. (1990), "Improvement in radiologists' detection of clustered microcalcifications on mammograms. The potential of computer-aided diagnosis," *Invest. Radiol.*, vol. 25, pp. 1102-1110.

[14] Kegelmeyer, W.P., Pruneda, J.M., Bourland, P.D., Hillis, A., Riggs, M.W., and Nipper, M.L. (1994), "Computer-aided mammographic screening for spiculated lesions," *Radiol.*, vol. 191, pp. 331-337.

[15] Chan, H.-P., Sahiner, B., Helvie, M.A., Petrick, N., Roubidoux, M.A., Wilson, T.E., Adler, D.D., Paramagul, C., Newman, J.S., and Gopal, S.S. (1999), "Improvement of radiologists'

characterization of mammographic masses by computer-aided diagnosis: an ROC study," *Radiol.*, vol. 212, pp. 817-827.

[16] Jiang, Y., Nishikawa, R.M., Schmidt, R.A., Metz, C.E., Giger, M.L., and Doi, K. (1999), "Improving breast cancer diagnosis with computer-aided diagnosis," *Acad. Rad.*, vol. 6, pp. 22-33.

[17] Kilday, J., Palmieri, F., and Fox, M.D. (1993), "Classifying mammographic lesions using computer-aided image analysis," *IEEE Trans. Med. Img.*, vol. 12, pp. 664-669.

[18] Chan, H.P., Wei, D., Helvie, M.A., Sahiner, B., Adler, D.D., Goodsitt, M.M., and Petrick, N. (1995), "Computer-aided classification of mammographic masses and normal tissue: linear discriminant analysis in texture feature space," *Phys. Med. Biol.*, vol. 40, pp. 857-876.

[19] McNitt-Gray, M.F., Huang, H.K., and Sayre, J.W. (1995), "Feature selection in the pattern classification problem of digital chest radiograph segmentation," *IEEE Trans. Med. Img.*, vol. 14, pp. 537-547.

[20] Sahiner, B., Chan, H.P., Petrick, N., Wagner, R.F., and Hadjiiski, L. (2000), "Feature selection and classifier performance in computer-aided diagnosis: the effect of finite sample size.," *Med. Phys.*, vol. 27, pp. 1509-1522.

[21] Jain, A. and Zongker, D. (1997), "Feature selection: Evaluation, application, and small sample size performance," *IEEE Trans. Pat. Anal. Mach. Intell.*, vol. 19, pp. 153-158.

[22] Ferri, F.J., Pudil, P., Hatef, M., and Kittler, J. (1994), "Comparative study of techniques for large-scale feature selection," *Pattern Recognition in Practice,* vol. IV, pp. 403-413.

[23] Box, G.E.P. (1957), "Evolutionary operation: a method for increasing industrial productivity," *Appl. Stat.*, vol. 6, pp. 81-101.

[24] Holland, J.H. (1962), "Outline for a logical theory of adaptive systems," *J. Assoc. Comput. Mach.*, vol. 3, pp. 297-314.

[25] Fogel, L.J., Owens, A.J., and Walsh, M.J. (1966), *Artificial Intelligence Through Simulated Evolution,* Wiley, New York.

[26] Forrest, S. (1993), "Genetic algorithms: principles of natural selection applied to computation," *Science,* vol. 261, pp. 872-878.

[27] Jain, A.K., Duin, R.P.W., and Mao, J. (2000), "Statistical pattern recognition: a review," *IEEE Trans. Pat. Anal. Mach. Intell.,* vol. 22, pp. 4-37.

[28] Cover, T.M. and Campenhpout, J.M.V. (1977), "On the possible orderings in the measurement selection problem," *IEEE Trans. Sys. Man. and Cybern.,* vol. 7, pp. 657-661.

[29] Cover, T.M. (1974), "The best two independent measurements are not the two best," *IEEE Trans. Sys. Man. and Cybern.,* vol. 6, pp. 116-117.

[30] Wu, Y., Giger, M.L., Doi, K., Vyborny, C.J., Schmidt, R.A., and Metz, C.E. (1993), "Artificial neural networks in mammography: application to decision making in the diagnosis of breast cancer," *Radiol.,* vol. 187, pp. 81-87.

[31] Meisel, W.S. (1972), *Computer-oriented approaches to pattern recognition,* Academic Press, New York.

[32] Duda, R.O. and Hart, P.E. (1973), *Pattern Classification and Scene Analysis,* Wiley, New York.

[33] Raudys, S.J. and Jain, A.K. (1991), "Small sample size effects in statistical pattern recognition: recommendations for practitioners," *IEEE Trans. Pat. Anal. Mach. Intell.,* vol. 13, pp. 252-264.

[34] Narendra, P.M. and Fukunaga, K. (1977), "A branch and bound algorithm for feature subset selection," *IEEE Trans. Comput.,* vol. 26, pp. 917-922.

[35] Siedlecki, W. and Sklansky, J. (1989), "A note on genetic algorithm for large-scale feature selection," *Patt. Recog. Let.,* vol. 10, pp. 335-347.

[36] Brill, F., Brown, D., and Martin, W. (1992), "Fast genetic selection of features for neural network classifiers," *IEEE Trans. Neural Net.,* vol. 3, pp. 324-328.

[37] Kuncheva, L.I. and Jain, L.C. (1999), "Nearest neighbor classifier: simulataneous editing and feature selection," *Patt. Recog. Let.,* vol. 20, pp. 1149-1156.

[38] Kudo, M. and Sklansky, J. (2000), "Comparison of algorithms that select features for pattern classifiers," *Patt. Recog.,* vol. 33, pp. 25-41.

[39] Kudo, M. and Sklansky, J. (1998), "A comparative evaluation of medium- and large-scale feature selectors for pattern classifiers," *Kybernetika,* vol. 34, pp. 429-434.

[40] Sahiner, B., Chan, H.P., Petrick, N., Wei, D., Helvie, M.A., Adler, D.D., and Goodsitt, M.M. (1996), "Image feature selection by a genetic algorithm: Application to classification of mass and normal breast tissue on mammograms," *Med. Phys.,* vol. 23, pp. 1671-1684.

[41] Sahiner, B., Chan, H., Petrick, N., Helvie, M., and Goodsitt, M. (1998), "Design of a high-sensitivity classifier based on a genetic algorithm: application to computer-aided diagnosis," *Phys. Med. Biol.,* vol. 43, pp. 2853-2871.

[42] Chan, H.P., Sahiner, B., Lam, K.L., Petrick, N., Helvie, M.A., Goodsitt, M.M., and Adler, D.D. (1998), "Computerized analysis of mammographic microcalcifications in morphological and texture feature space," *Med. Phys.,* vol. 25, pp. 2007-2019.

[43] Zheng, B., Chang, Y.-H., Wang, X.-H., Good, W.F., and Gur, D. (1999), "Feature selection for computerized mass detection in digitized mammograms by using a genetic algorithm," *Acad. Rad.,* vol. 6, pp. 327-332.

[44] Yamany, S.M., Khiani, K.J., and Faraq, A.A. (1997), "Application of neural networks and genetic algorithms in the classification of endothelial cells," *Patt. Recog. Let.,* vol. 18, pp. 1205-1210.

[45] Handels, H., Ross, T., Kreusch, J., Wolff, H.H., and Poppl, S.J. (1999), "Feature selection for optimized skin tumor recognition using genetic algorithms," *Art. Intell. Med.*, vol. 16, pp. 283-297.

[46] Kupinski, M.A. and Anastasio, M.A. (1999), "Multiobjective genetic optimization of diagnostic classifiers with implications for generating receiver operating characteristic curves," *IEEE Trans. Med. Img.*, vol. 18, pp. 675-685.

[47] Anastasio, M.A., Yoshida, H., Nagel, R., Nishikawa, R.M., and Doi, K. (1998), "A genetic algorithm-based method for optimizing the performance of a computer-aided diagnosis scheme for detection of clustered microcalcifications in mammograms," *Med. Phys.*, vol. 25, pp. 1613-1620.

[48] Gudmundsson, M., El-Kwae, E.A., and Kabuka, M.R. (1998), "Edge detection in medical images using a genetic algorithm," *IEEE Trans. Med. Img.*, vol. 17, pp. 469-474.

[49] Fujita, H., Hara, T., Jing, X., Matsumoto, T., Yoshimura, H., and Seki, K. (1995), "Automated detection of lung nodules by using a genetic algorithm technique in chest radiography," *Radiol.*, vol. 197(P), pp. 426-426.

[50] Pena-Reyes, C.A. and Sipper, M. (2000), "Evolutionary computation in medicine: an overview," *Art. Intell. Med.*, vol. 19, pp. 1-23.

[51] Lachenbruch, P.A. (1975), *Discriminant Analysis,* Hafner Press, New York.

[52] Metz, C.E. (1986), "ROC methodology in radiologic imaging," *Invest. Radiol.*, vol. 21, pp. 720-733.

[53] Dorfman, D. and Alf Jr, E. (1969), "Maximum likelihood estimation of parameters of signal detection theory and determination of confidence intervals-rating method data.," *J. Math. Psych.*, vol. 6, pp. 487-496.

[54] Metz, C.E., Herman, B.A., and Shen, J.H. (1998), "Maximum-likelihood estimation of receiver operating characteristic (ROC)

curves from continuously-distributed data," *Stat. Med.,* vol. 17, pp. 1033-1053.

[55] Tatsuoka, M.M. (1988), *Multivariate Analysis, Techniques for Educational and Psychological Research,* 2nd ed. Macmillan, New York.

[56] Norusis, M.J. (1993), *SPSS for Windows Release 6 Professional Statistics,* SPSS Inc., Chicago, IL.

[57] Wei, D., Chan, H.P., Helvie, M.A., Sahiner, B., Petrick, N., Adler, D.D., and Goodsitt, M.M. (1995), "Classification of mass and normal breast tissue on digital mammograpms: multiresolution texture analysis," *Med. Phys.,* vol. 22, pp. 1501-1513.

[58] Chan, H.P., Wei, D., Lam, K.L., Sahiner, B., Helvie, M.A., Adler, D.D., and Goodsitt, M.M. (1995), "Classification of malignant and benign microcalcifications by texture analysis," *Med. Phys.,* vol. 22, pp. 938.

[59] Chan, H.P., Sahiner, B., Wei, D., Helvie, M.A., Adler, D.D., and Lam, K.L. (1995), "Computer-aided diagnosis in mammography: Effect of feature classifier on characterization of microcalcifications," *Radiol.,* vol. 197(P), pp. 425.

[60] Chan, H.P., Sahiner, B., Petrick, N., Helvie, M.A., Leung, K.L., Adler, D.D., and Goodsitt, M.M. (1997), "Computerized classification of malignant and benign microcalcifications on mammograms: texture analysis using an artificial neural network," *Phys. Med. Biol.,* vol. 42, pp. 549-567.

[61] Chan, H.P., Niklason, L.T., Ikeda, D.M., and Adler, D.D. (1992), "Computer-aided diagnosis in mammography: detection and characterization of microcalcifications," *Med. Phys.,* vol. 19, pp. 831.

[62] Chan, H.P., Wei, D., Lam, K.L., Lo, S.-C.B., Sahiner, B., Helvie, M.A., and Adler, D.D. (1995), "Computerized detection and classification of microcalcifications on mammograms," *Proc. SPIE Med. Img.,* vol. 2434, pp. 612-620.

[63] Sahiner, B., Chan, H.P., Petrick, N., Wei, D., Helvie, M.A., Adler, D.D., and Goodsitt, M.M. (1996), "Classification of mass and normal breast tissue: a convolution neural network classifier with spatial domain and texture images," *IEEE Trans. Med. Img.,* vol. 15, pp. 598-610.

[64] Haralick, R.M., Shanmugam, K., and Dinstein, I. (1973), "Texture features for image classification," *IEEE Trans. Sys. Man. and Cybern.,* vol. SMC-3, pp. 610-621.

[65] Metz, C.E., Wang, P.L., and Kronman, H.B. (1984), "A new approach for testing the significance for differences between ROC curves measured from correlated data," in: Deconinck, F. (Ed.), *Information Processing in Medical Imaging,* The Hague, Martinus Nijhoff, pp. 432-445.

[66] Vyborny, C.J. and Giger, M.L. (1994), "Computer vision and artificial intelligence in mammography," *AJR,* vol. 162, pp. 699-708.

[67] Petrick, N., Chan, H.P., Wei, D., Sahiner, B., Helvie, M.A., and Adler, D.D. (1996), "Automated detection of breast masses on mammograms using adaptive contrast enhancement and texture classification," *Med. Phys.,* vol. 23, pp. 1685-1696.

[68] Wei, D., Chan, H.P., Petrick, N., Sahiner, B., Helvie, M.A., Adler, D.D., and Goodsitt, M.M. (1997), "False-positive reduction technique for detection of masses on digital mammograms: global and local multiresolution texture analysis," *Med. Phys.,* vol. 24, pp. 903-914.

[69] Rumelhart, D.E., Hinton, G.E., and Williams, R.J. (1986), "Learning internal representation by error propagation," in: Rumelhart, D.E. and McClelland, J.L. (Eds.), *Parallel Distributed Processing*, vol. 1, MIT Press, Cambridge, MA.

[70] Hermann, G., Janus, C., Schwartz, I.S., Krivisky, B., Bier, S., and Rabinowitz, J.G. (1987), "Nonpalpable breast lesions: Accuracy of prebiopsy mammographic diagnosis," *Radiol.,* vol. 165, pp. 323-326.

[71] Hall, F.M., Storella, J.M., Silverstond, D.Z., and Wyshak, G. (1988), "Nonpalpable breast lesions: recommendations for biopsy based on suspicion of carcinoma at mammography," *Radiol.,* vol. 167, pp. 353.

[72] Jacobson, H.G. and Edeiken, J. (1990), "Biopsy of occult breast lesions: analysis of 1261 abnormalities," *JAMA,* vol. 263, pp. 2341-2343.

[73] Jiang, Y., Metz, C.E., and Nishikawa, R.M. (1996), "A receiver operating characteristic partial area index for highly sensitive diagnostic tests," *Radiol.,* vol. 201, pp. 745-750.

[74] Sahiner, B., Chan, H.P., Petrick, N., Helvie, M.A., and Goodsitt, M.M. (1998), "Computerized characterization of masses on mammograms: the rubber band straightening transform and texture analysis," *Med. Phys.,* vol. 25, pp. 516-526.

[75] Weszka, J.S., Dyer, C.R., and Rosenfeld, A. (1976), "A comparative study of texture measures for terrain classification," *IEEE Trans. Sys. Man. and Cybern.,* vol. 6, pp. 269-285.

[76] Galloway, M.M. (1975), "Texture classification using gray level run lengths," *Comp. Graph. Img Proc.,* vol. 4, pp. 172-179.

Index

- S -

- T -

- U -

- V -

- W -

List of Contributors

H. Billhardt
School of Computer Science
Polytechnical University of Madrid
Boadilla del Monte. 28660 Madrid
Spain

N.D. Black
Medical Informatics
University of Ulster at Jordanstown
Newtownabbey, Co. Antrim
Northern Ireland BT37 0QB
United Kingdom

R. Brause
FB 15, Institute for Informatics
J.W.G.-University
D-60054 Frankfurt
Germany

P. Brockhausen
Universität Dortmund
LS Informatik VIII
44221 Dortmund
Germany

H.P. Chan
Department of Radiology
University of Michigan
Ann Arbor, MI
USA

J. Crespo
School of Computer Science
Polytechnical University of Madrid
Boadilla del Monte. 28660 Madrid
Spain

J.E. Eichner
Department of Biostatistics and Epidemiology
College of Public Health
University of Oklahoma Health Sciences Center
Oklahoma City, OK 73190
USA

R. Fried
Department of Statistics
University of Dortmund
44221 Dortmund
Germany

Y. Fukuoka
Department of Biomedical Information
Tokyo Medical and Dental University
Chiyoda-ku, Tokyo 101-0062
Japan

U. Gather
Department of Statistics
University of Dortmund
44221 Dortmund
Germany

L. Gierl
Institute for Medical Informatics und Biometry
University Rostock
18055 Rostock
Germany

F. Hamker
Caltech
Division of Biology, 139-74
Pasadena CA 91125
USA

J.C. Healy
Creighton University School of Medicine
Omaha, Nebraska 68131
USA

J.W. Huang
Department of Biomedical Engineering
Rensselaer Polytechnic Institute
Troy, NY
USA

M. Imhoff
Surgical Department
Community Hospital Dortmund
44137 Dortmund
Germany

L.C. Jain
School of Electrical and Information Engineering
University of South Australia
Mawson Lakes SA 5095
Australia

T. Joachims
Universität Dortmund
LS Informatik VIII
44221 Dortmund
Germany

J.K.C. Kingston (corresponding author)
AIAI
University of Edinburgh
Edinburgh EH1 1HN
Scotland, United Kingdom

K. Kyriacou
Electron Microscopy / Molecular Pathology
The Cyprus Institute of Neurology and Genetics
Nicosia
Cyprus

J.A. Lopez
Medical Informatics
University of Ulster at Jordanstown
Newtownabbey, Co. Antrim
Northern Ireland BT37 0QB
United Kingdom

V. Maojo (corresponding author)
School of Computer Science
Polytechnical University of Madrid
Boadilla del Monte. 28660 Madrid
Spain

R.P. Marble
College of Business Administration
Creighton University
Omaha, Nebraska 68178
USA

P.A. McKee
Department of Medicine
College of Medicine
University of Oklahoma Health Sciences Center
Oklahoma City, OK 73190
USA

B.A. Mobley
Department of Physiology
College of Medicine
University of Oklahoma Health Sciences Center
Oklahoma City, OK 73190
USA

N. Molony
Royal Infirmary of Edinburgh
Edinburgh
Scotland, United Kingdom

W.E. Moore
Native American Prevention Research Center
University of Oklahoma Health Sciences Center
Oklahoma City, OK 73190
USA

K. Morik
Universität Dortmund
LS Informatik VIII
44221 Dortmund
Germany

C.D. Nugent
Medical Informatics
University of Ulster at Jordanstown
Newtownabbey, Co. Antrim
Northern Ireland BT37 0QB
United Kingdom

J. Paetz
FB 15, Institute for Informatics
J.W.G.-University
D-60054 Frankfurt
Germany

C.S. Pattichis
Department of Computer Science
University of Cyprus
CY-1678 Nicosia
Cyprus

M.S. Pattichis
Department of Electrical and Computer Engineering
The University of New Mexico
Albuquerque, NM 87131-1356
USA

C.A. Peña-Reyes
Logic Systems Laboratory
Swiss Federal Institute of Technology in Lausanne – EPFL
CH 1015, Lausanne
Switzerland

N. Petrick
Department of Radiology
University of Michigan
Ann Arbor, MI
USA

R.J. Roy (corresponding author)
Department of Biomedical Engineering
Rensselaer Polytechnic Institute
Troy, NY
USA

S. Rüping
Universität Dortmund
LS Informatik VIII
44221 Dortmund
Germany

B. Sahiner
Department of Radiology
University of Michigan
Ann Arbor, MI
USA

J. Sanandres
School of Computer Science
Polytechnical University of Madrid
Boadilla del Monte. 28660 Madrid
Spain

E. Schechter
Cardiology Section, Department of Medicine
College of Medicine
University of Oklahoma Health Sciences Center
Oklahoma City, OK 73190
USA

C.N. Schizas
Department of Computer Science
University of Cyprus
CY-1678 Nicosia
Cyprus

R. Schmidt
Institute for Medical Informatics und Biometry
University Rostock
18055 Rostock
Germany

F. Schnorrenberg
Department of Computer Science
University of Cyprus
CY-1678 Nicosia
Cyprus

J. Simpson
Peat Marwick McLintock
Edinburgh
Scotland, United Kingdom

M. Sipper
Department of Computer Science
Ben-Gurion University
Beer-Sheva 84105
Israel
and
Logic Systems Laboratory
Swiss Federal Institute of Technology in Lausanne – EPFL
CH 1015, Lausanne
Switzerland

H.-N. Teodorescu
Faculty of Electronics and Telecommunications
Technical University of Iasi
Iasi 6600
Romania

J.A.C. Webb
School of Electrical and Mechanical Engineering
University of Ulster at Jordanstown
Newtownabbey, Co. Antrim
Northern Ireland BT37 0QB
United Kingdom

X.-S. Zhang
Department of Biomedical Engineering
Rensselaer Polytechnic Institute
Troy, NY
USA

GPSR Compliance
The European Union's (EU) General Product Safety Regulation (GPSR) is a set of rules that requires consumer products to be safe and our obligations to ensure this.

If you have any concerns about our products, you can contact us on

ProductSafety@springernature.com

In case Publisher is established outside the EU, the EU authorized representative is:

Springer Nature Customer Service Center GmbH
Europaplatz 3
69115 Heidelberg, Germany

www.ingramcontent.com/pod-product-compliance
Ingram Content Group UK Ltd.
Pitfield, Milton Keynes, MK11 3LW, UK
UKHW021900190726
13853UKWH00003B/1360
9783662003206